T0213453

Inside Hazardous Technological Systems

Inside Hazardous Technological Systems

Methodological Foundations, Challenges and Future Directions

Edited by

Kenneth Pettersen Gould

and Carl Macrae

CRC Press
Taylor & Francis Group
Boca Raton London New York

CRC Press is an imprint of the
Taylor & Francis Group, an **informa** business

Contents

Foreword
Qualitative Inquiries into Safety Research

Silvia Gherardi[1]

The present collection of chapters organised around the idea of looking inside "hazardous systems" with the lens of qualitative methodologies, offers a distinctive contribution to safety research because it is positioned at the crossroad of two entwined conversations. On one side the organisational approach to risk and disasters, which changed the conception of risk in the early 1980s, and on the other side the challenge of qualitative methodologies for conducting research on safety in the 1990s. Both conversations are deeply indebted and shaped by the legacy of Barry Turner and his Man-Made Disaster (MMD) model (Turner, 1978, Turner and Pidgeon, 1997) that contributed to the development of a type of scientific inquiry which considers organisational factors in the aetiology of accidentally caused disasters. In reading this book, we can see the materialisation of an outstanding contribution that, after more than 40 years, is still able to inspire new ways of doing empirical research and develop new problematisations of ongoing challenges to safety.

Man-Made Disasters, as a book, a model of analysis and as the symbol of Turner's theorisation on safety culture, is a cultural fact in itself because it represents an inscription in the world of culture. It is also a neologism in the vocabulary of risk and safety, since "Man-Made Disaster" is a term which denotes a class of phenomena which constitutes a distinct social reality in the flux of experience. With this term Turner marked out a type of disaster previously undistinguished from natural disasters. He was the first to conceive of disasters in terms of a process of incubation and not as "bolts from the blue" (as he loved to say). The man-made disasters model, both as a descriptive tool and in its later development as a diagnostic and learning aid (Toft and Turner, 1987, Toft and Reynolds, 1994), is about how technical, social, institutional and administrative arrangements can produce disasters. It is about the relationship between information, error and surprise in organisations, "for we know that responsibility for failure can be just as dispersed and fragmented as responsibility for success" (Turner, 1978, p. xv). In this quote we may trace the shift that took place in the 1980s "from technical risk to safety culture" and the direction of such a shift towards the consideration of the complexity induced by organisational and cultural factors in doing research and theorising risk and safety.

Turner provided the first item of evidence to show that crises – like those induced by MMD – may constitute a threat but they also may, from one point of view, be considered as opportunities for pursuing social change. Turner (1978, p. 22) recalls that the two Chinese characters used to express the word "crisis" mean "danger" and "opportunity" in order to show that crises may be considered as opportunities within an organisation, or even as a means of aiding personal growth for individual managers. In fact, Turner at the end of his life when he and Nick Pidgeon were revising the 1978's version of MMD, moved to see crisis as an opportunity for organisational learning, thus exploring how knowledge maybe de-contextualised from the original situation (crisis) and re-contextualised within organisational procedures as knowledge in use.

The complexity inherent in the cultural and organisational approach to disasters is that any epistemic practice (any organisational culture) is at the same time "a way of seeing [that] is always also a way of not seeing" (Turner, 1978, p.49). Organisational culture – as a way of seeing – can be conceived as embedded in the equipment that organisations use in operating upon the world, as cultural elements for relating to the tasks at hand, to the environments in which they are immersed and to the manner in which those within the organisation interact with each. This is plainly a sociomaterial conception of culture which breeds an orientation to action situated in the operational context and which does not undervalue the material culture inscribed in the use of equipment and technology. At the same time a culture is a way of not seeing; it is "a collective blindness to important issues, the danger that some vital factors may be left outside the bounds of organizational perception" (Turner, 1978, p. 47). This union of opposites – culture as a tool for seeing and not seeing – takes us to a topic that Turner resumed later when he proposed a reading of Derrida applied to the use of software for hazard control (Turner, 1994a). In this light, culture is viewed as spoken language, as a system of marks, whether these are the technology, the computer software or the semantic networks he was writing about.

In fact, the sociological literature on disasters was enriched in those years by a further shift away from organisational factors understood in the mainly structural or sociotechnical sense towards the cultural factors used in interpreting safety problems (Gephart et al., 1990; Wright, 1994; Pidgeon, 1995; Gherardi, 2004). The main assumption is that organisations are concerned with intention and the execution of intention. Disasters always represent a failure of intention, a failure of foresight. Turner and Pidgeon view a sociological definition of disaster as raising a challenge against existing cultural assumptions. A disaster was conceived as an "an event, concentrated in time and space, which threatens a society or a relatively self-sufficient subdivision of a society with major unwanted consequences as a result of the collapse of precautions which had hitherto been culturally accepted as adequate" (Turner and Pidgeon, 1997, p. 70). The advantage of this definition is that it covers instances where the amount of physical damage is not great, but the mishap reveals a gap in defences which were regarded as secure and a need for cultural readjustment.

Thus, by adopting the MMD model of analysis, we may follow Turner and Pidgeon by formulating a definition that does not refer to the physical impact or scale of the event. Instead, in sociological terms, an accident can be defined "as the overturning and disruption of cultural norms and expectations for dealing with risk and safety matters" (Turner and Pidgeon, 1997, p. 74). The MMD model suggests that "technical, human, managerial and cultural dimensions interact in a contingent open-ended process that precludes deterministic analysis" (Horlick-Jones, 1996, p. xx).

In those years the linguistic turn was establishing itself stating that language does not describe reality but creates it and, therefore, the borders between ontology and epistemology were blurring. In fact, through language and through epistemological practices the object of study (be it, safety, risk, disaster) is constructed and consequently we can use the term "onto-epistemology" (Barad, 2007). Thus, a new challenge is facing qualitative research: how to conduct qualitative research once the researchers consider themselves and their epistemic practices as part of the research assemblage?

To the first conceptualisation of disasters in sociotechnical terms we may apply notions drawn from actor-network theory and see in the disappearance of the hyphen from "sociotechnical" systems the demise of the distinction between the social and the technical, the social and the material (as in sociomateriality). In this case, the boundaries among the material, technical, human, managerial or cultural vanish, so that an accident can be defined simply as a breakdown or dis-alignment in what hitherto was a way of ordering heterogeneous materials.

In the duality and ambivalence of language-in-use in organisational context we can trace a theme that years later became central in Science and Technologies Studies and was expressed as "multiplicity of objects" (Mol, 2002) or "empirical ontologies" (Law and Lien, 2013). It is the idea that an object (a definition of risk, or hazard or safety) is not a single "reality" but may be constructed as multiple "reals" in different parts of an organisation and in relation to different apparatuses of detection and/or measurement. This theme was already present in the conception of culture in MMD but continues to be overlooked. In MMD the ambiguity in language and of language is a theme inherent to the description of information difficulties in the incubation of disasters, and the risks that organisations cause through operational management or maintenance deficiencies. Plurality of interpretations is present, but what is not stressed is that risk, hazard or safety are multiple and depend on the epistemic practices that identify them.

I stress the point of view that culture may be formulated as a particular knowledge structure and value system that has received normative status. The reason for linking culture to knowledge and to knowing is that a cultural approach to safety and disasters implies the methodological choice of studying culture in socially situated knowledgeable practices. We find this methodological challenge anytime that safety is defined in relation to knowledge and to practice and thus when safety is researched as a collective knowledgeable doing and an emergent social competence (Gherardi, 2006).

What qualitative approaches have in common is in particular that safety, risk and disasters are studied as social in origin and this epistemological framework induces us to give attention to what is known, what is not known, who does not know, and why it is so (Pettersen Gould, 2020).

The contemporary debates that we may read in this book have their roots in the qualitative social science approaches of the 1980s and 1990s that showed that the interaction between human activities and consequences in organisations is more complex than probabilities can capture (Renn, 1998) and that the organisational structures of managing and controlling risks are prone to failures and deficits that may increase actual risk. The chapters in the book take the debate forward and offer a distinctive contribution in putting ethnographic methods at the front: desktop ethnography, conceptual ethnography, historical ethnography, or just qualitative study of normal or daily operations. The reader will engage with struggles between different definitions of ethnography and with different styles of doing ethnographic research. At the core of all of these chapters resides an attention for situated ways of relating with working and organising. This attention to ethnography signals a sensibility towards phenomena as they happen, in contrast with a retrospective style of inquiring into what has already happened. Moreover, we find the same sensibility also in those chapters that deal with other research designs, such as case study or action research.

The complexity of doing empirical research in hazardous systems is inextricably connected to how organisational temporality is linked in the present to past events and to the foresight of future events. The reflection on time in organising was a theme dear to Barry Turner and it still represents an open and problematic issue in relation to methodologies for the study of risk. The semantic field of risk has been historically changing (Turner, 1994b), and this may induce us to reflect on how little we still know about the sociomaterial construction of safety, about reliability and about social responsibility towards present and future generations. Moreover, in considering how a field of study is socially and materially constructed and always changing, we have to keep in mind the influence of the institutional environment. Industrial risks represented the worries of an industrial society; later the "rediscovery" of risk by the most accredited sociologists (Beck, 1992; Giddens, 1991) expressed post-industrial preoccupations with environmental issues. With the incremental onset of new disasters, a post-Anthropocene frame of mind, where digital technologies and artificial intelligence computers construct our world, is again going to change the formulation of the relevant "problems" and how to study them. We have to consider the influence of the institutional environment in the way that problems are framed, in both society and organisations, by those with greatest authority and power in defining the issues at stake.

Therefore, what I am proposing here is to approach the chapters in this book with a "critical cultural reading," situating them against the background of their cultural production. What I am suggesting is a diffractive reading that has dual roots. On one side I wish to remind to the reader the prophetic words of Rosenthal in writing the forewords for Nick Pidgeon's new edition of MMD (Turner and Pidgeon, 1997): "Man-Made Disasters will be a mediating force between those believing in high-reliability organizations and those stressing the problem of accident-prone systems." This tension – as this book illustrates and strengthens – is still at the core of the field of study. On the other side I would like to propose again a telling metaphor that Weick used in appreciation of Barry Turner's work. Weick (1998, p. 72) argued that the staying power of MMD is due, in part, to the fact that it is not a lens but rather a kaleidoscope. The contrast between a lens and a kaleidoscope is that the lens metaphor presumes a realist position, since a lens sizes up something "out there." On the contrary the image of turning a kaleidoscope suggests the changing of patterns in the subjectivist schema, since the patterns of a kaleidoscope maybe internally generated with minimal dependence on information from outside. It was Turner (1995) who used the image of "shaking a kaleidoscope" in reference to his work and Weick re-used it in appreciating how Turner's work continues to influence our current thinking in a kaleidoscope-like fashion. In fact, it "dislodges old patterns, generates new patterns and foster[s] awareness that numerous configurations are possible in the genesis of disasters" (Weick, 1998, p72).

Both images (the mediating force and the kaleidoscope) may be useful in reading the chapters of this book, and I wish to add another metaphor that not only captures the objectivist/subjectivist epistemological debate but brings the conversation forwards. If we wish to fully acknowledge a relational epistemology as expressed by new materialisms and post-humanist practice theory (Gherardi, 2019), we have to use the metaphor of diffraction and diffractive reading. Like the rolling, pushing and transformation of waves in the sea, "diffraction has to do with the way waves combine when they overlap, and the apparent bending and spreading of waves that occurs when waves encounter an obstruction" (Barad 2007, p. 74). It is this movement of overlapping that signifies diffraction. The metaphor of diffraction contemplates materiality and multiplicity and I would like to suggest a diffractive reading of the chapters of this book for appreciating a colourful and diverse picture of its message. In fact, a diffractive methodology is "a practice of reading insights through one another while paying attention to patterns of difference (including the material effects of constitutive exclusions)" (Barad 2011: 445).

This book invites us to practice a diffractive reading of its chapters to find out "what's new" in its fabric.

NOTE

1. Affiliation: Dipartimento di Sociologia e Ricerca Sociale,
   Via Verdi 26, 38122 Trento (Italy). Email: silvia.gherardi@unitn.it

REFERENCES

Barad, K. (2007). *Meeting the universe halfway: Quantum physics and the entanglement of matter and meaning*, Duke University Press.

Barad, K. (2011). Erasers and erasures: Pinch's unfortunate 'uncertainty principle', *Social Studies of Science*, 41(3), 443–454.

Beck, U. (1992). *Risk society: Towards a new modernity*, SAGE, London.

Gephart, R., Steier, L., & Lawrence, T. (1990). Cultural rationalities in crisis sensemaking: A study of a public inquiry into a major industrial accident, *Industrial Crisis Quarterly*, 4, 27–48.

Gherardi, S. (2004). Translating knowledge while mending organizational safety culture, *Risk Management: An International Journal*, 6(2), 61–80.

Gherardi, S. (2006). *Organizational knowledge: The texture of workplace learning*, Blackwell, Oxford.

Gherardi, S. (2019). *How to conduct a practice-based study: Problems and methods*, (second edition), Edward Elgar, Cheltenham.

Giddens, A. (1991). *Modernity and self-identity: Self and society in the late modern age*, Polity Press, Cambridge.

Horlick-Jones, T. (1996). Is safety a by-product of quality management? In D. Jones and C. Hood (eds), *Accident and design*, University College London Press, London.

Law, J., & Lien, M.E. (2013). Slippery: Field notes in empirical ontology, *Social Studies of Science*, 43(3), 363–78.

Mol, A. (2002). *The body multiple: Ontology in medical practice*, Duke University Press, Durham.

Pettersen Gould, K. (2020). Organizational risk: "Muddling through" 40 years of research, *Risk Analysis*, 1–10. Doi.org 10.1111/risa.13460.

Pidgeon, N. (1995). Safety culture and risk management in organizations, *Journal of Cross-Cultural Psychology*, 22(1), 129–140.

Renn, O. (1998). Three decades of risk research: Accomplishments and new challenges, *Journal of Risk Research*, 1(1), 49–71.

Toft, B., & Reynolds, S. (1994). *Learning from disasters: A management approach*, Butterworth-Heinemann, Oxford.

Toft, B., & Turner, B.A. (1987). The schematic report analysis diagram: A simple aid to learning from large-scale failures, *International CIS Journal*, 1(2), 12–23.

Turner, B.A. (1978). *Man-made disaster. The failure of foresight*. Wykeham Science Press, London.

Turner, B.A. (1994a). Software and contingency: The text and vocabulary of system failure, *Journal of Contingencies and Crisis Management*, 2(1), 31–38.

Turner, B.A. (1994b). The future for risk research, *Journal of Contingencies and Crisis Management*, 2(3), 146–156.

Turner, B. (1995). A personal trajectory through organization studies, *Research in the Sociology of Organizations*, 13, 275–301.

Turner, B., & Pidgeon, N. (1997). *Man-made disaster. The failure of foresight*, Second edition, Butterworths-Heinmann, London.

Weick, K.E. (1998). Foresights of failure: An appreciation of Barry Turner, *Journal of Contingencies and Crisis Management*, 6(2), 72–75.

Wright, C. (1994). A fallible safety system: Institutionalized irrationality in the offshore and gas industry, *Sociological Review*, 1, 79–103.

Preface

This book project emerged from a special session held at the Working on Safety (WOS) 2017 conference in Prague recognising and discussing the methodological legacy of the work of Barry A. Turner. The 40th and 20th anniversaries of the publication of Barry Turner's book, *Man-Made Disasters* (Turner 1978; Turner & Pidgeon 1997), provided a historical inspiration for exploring developments in investigating safety and accidents by means of fieldwork, organisational ethnography and socially situated analysis. The foundational nature of Turner's work on the development of disasters is increasingly acknowledged by many within the safety science community. Not only did Turner's research provide fundamental insights that help explain the preconditions of large-scale accidents and disasters; it was equally significant in pioneering the use of qualitative inquiry within our field – particularly in terms of developing theory and describing and conceptualising the complex realities of safety and accidents in industrial systems. With this book, we aim to continue in the footsteps of Turner and explore the applications, implications, challenges and opportunities of applying qualitative, descriptive and socially situated analytical approaches to the study of hazardous technological systems. Such methodological approaches are used in rich and varied ways across the fields of safety and risk, and they have underpinned some of the more substantial theoretical developments of the past few decades. But these methodological strategies – and the practices and challenges associated with them – have not been a major focus within the safety sciences. Accordingly, our intent is that this book provides a focus on this area of research practice, and presents a diverse (though non-comprehensive) survey of the current state of the art. We have aimed to bring together a range of perspectives from different disciplines and domains of safety, including both new voices, established researchers and contributions from Silvia Gherardi and Nick Pidgeon – methodologists who both collaborated closely with Barry Turner. By bringing together a set of critical views on the current state and future directions of this methodological toolbox, this book aims to act as a "companion" volume to provide academics, researchers, students and professionals with broad-ranging insider accounts of the key issues and methodological debates in the fields of safety, accident analysis and risk analysis. We hope that, by situating these debates in the broader tradition of cultural and organisational analysis, and connecting to broader discourses on technological systems, the contributions in this book can help to strengthen the links between the fields of safety sciences on the one hand, and social and organisational studies on the other.

Editors

Kenneth Pettersen Gould is Associate professor in Risk Management and Societal Safety at the University of Stavanger, specializing in organizational risk and safety management. His research concerns hazardous technologies; how risk, reliability, safety and security is analysed and managed in organizational contexts; and how regulatory and management strategies in these areas can be further developed.

Carl Macrae is Professor of Organisational Behaviour and Psychology in the Centre for Health Innovation, Leadership and Learning at Nottingham University Business School. His research examines how organisations manage safety, reliability and resilience; how people make sense of and learn from unexpected events; and how organisational and regulatory systems can be designed to support learning and improvement. Carl regularly advises organisations, regulators and policymakers on challenging safety and risk issues. His work led to the establishment in England of the first system-wide, learning-focused safety investigation body for healthcare, the Healthcare Safety Investigation Branch (HSIB). His book, *Close Calls*, examines the work of airline flight safety investigators and the practical work that is involved in learning from "near-miss" incidents in complex, safety-critical systems. Carl is also a Professor II in the SHARE Centre for Resilience in Healthcare at the University of Stavanger, Norway.

Contributor Biographies

Petter G. Almklov is Professor of Sociology at the Norwegian University of Science and Technology, specialising in the field of work and organisation. Has a PhD in Social Anthropology and also an MSc in Engineering Geology. He has conducted research in a wide range of organisations and industries and published within several different academic fields. His main research interests are work in Hi-Tech contexts, organisational and societal safety, typically focusing on representation in practice and representations of practice.

Stian Antonsen is an organisational sociologist and works as Research Professor at NTNU Social Research, and Adjunct Professor in Safety Management at the Norwegian University of Science and Technology. His research centres around issues of organisational culture, societal security and various forms of digital vulnerability.

Mathilde Bourrier holds a PhD in Sociology from SciencePo Paris and a Habilitation from the University of Technology in Compiègne (France). She is full Professor of Sociology at the University of Geneva and she is also Prof II at the University of Stavanger. She initially worked on the social construction of safety, focusing on the conditions under which organisational reliability can be achieved and sustained. She is especially interested in organisational design and resource allocation during severe and challenging conditions.

Jeffrey Braithwaite, BA, MIR (Hons), MBA, DipLR, PhD, FIML, FCHSM, FFPHRCP (UK), FAcSS (UK), Hon FRACMA, FAHMS is Founding Director of the Australian Institute of Health Innovation, Director of the Centre for Healthcare Resilience and Implementation Science, and Professor of Health Systems Research, Faculty of Medicine and Health Sciences, Macquarie University, Sydney, Australia. He has appointments at six other universities internationally, and is President of the International Society for Quality in Health Care (ISQua) and consultant to the World Health Organization (WHO). His research examines the changing nature of health systems. He is particularly interested in healthcare as a complex adaptive system and applying complexity science and resilience thinking to health care problems.

Geir Sverre Braut is a medical practitioner, with specialisation in community medicine; Senior Advisor at Stavanger University Hospital and Professor at Western Norway University of Applied Sciences; and formerly Deputy Director General at Norwegian Board of Health Supervision and Chief County Medical Officer in Rogaland County, Norway.

Sidney W.A. Dekker (PhD, Ohio State University, USA, 1996) is Professor and Director of the Safety Science Innovation Lab at Griffith University in Brisbane, Australia, and Professor at the Faculty of Aerospace Engineering at Delft University in the Netherlands. Sidney has lived and worked in seven countries across four continents and won worldwide acclaim for his ground-breaking work in human factors and safety. He coined the term "Safety Differently" in 2012, which has since turned into a global movement for change. It encourages organisations to declutter their bureaucracy and set people free to make things go well, and to offer compassion, restoration and learning when they don't. An avid piano player and pilot, he has been flying the Boeing 737 for an airline on the side.

Silvia Gherardi is Senior Professor of sociology at University of Trento and one of the founders of the research unit RUCOLA (Research Unit on Communication, Organizational Learning, and Aesthetics). She has been present for several years in the principal international networks of organisational studies and she is a board member of various international journals. She has been President of the European Group for Organizational Studies (1997–2000). In September 2005 she received the degree of "Doctor Honoris Causa" from the Department of Social Sciences of the Danish Roskilde University, in June 2010 from East Finland University, in June 2014 from St Andrews University, Scotland.

Torgeir Kolstø Haavik (PhD) has a background in geological engineering, social geography and organisational sociology. He holds positions as Research Professor at NTNU Social Research and Adjunct Associate Professor at NTNU (Norwegian University of Science and Technology) – where he teaches organisational safety. His research interests include addressing the sustainability dimension of societal resilience, to strengthen the political dimensions of societal resilience in research areas where science and technology are deeply intertwined – for example, climate change.

Jan Hayes is Professor in Organisational Accident Prevention at School of Property, Construction and Project Management, RMIT University. She has 30 years' experience in safety and risk management. She is Program Leader for the safety research activities of the Future Fuels Cooperative Research Centre and Board Safety Change Assurance Advisor for Airservices Australia. Her research interests include organisational accident prevention, expert decision-making, engineering professional practice, use of standards and narrative-based learning; email: jan.hayes2@rmit.edu.au.

Andrew Hopkins is Emeritus Professor of Sociology, Australian National University, Canberra. He was an expert witness at the Royal Commission into the 1998 Exxon gas plant explosion near Melbourne. He was a consultant to the US Chemical Safety Board in its investigation of the BP Texas City Refinery disaster of 2005, and also for its investigation into the BP Gulf of Mexico oil spill of 2010. He has written books about these accidents, with over 100,000 copies sold. He has been involved in various government reviews of Work Health and Safety regulation and regulators and has done consultancy work for major companies in the mining, petroleum, chemical and electrical industries, as well as for Defence. He speaks regularly to audiences around the world about the human and organisational causes of major accidents.

Trond Kongsvik (PhD) is Professor in Safety Management at the Department of Industrial Economics and Technology Management at Norwegian University of Science and Technology (NTNU). His background is in sociology and psychology, and his published research during the last 20 years has involved topics such as organisational safety, safety culture, safety indicators, safety audits and human factors. His work has mainly been within petroleum, the maritime industry, transport and health; email: trond.kongsvik@ntnu.no.

Christian Henrik Alexander Kuran is currently Assistant Professor and Researcher in Societal Safety at the Institute for Safety, Economics and Planning at the University of Stavanger. He holds master's degrees in Social and Cultural Anthropology, and in Societal Safety. He is an avid proponent of ethnographic methodology and qualitative methods in safety research. His PhD project is on the bending and breaking of safety related rules and regulation in the road-based transport sector.

Jean-Christophe Le Coze is a safety researcher (PhD, Mines ParisTech) at INERIS, the French national institute for environmental safety. His activities combine ethnographic studies and action research in various safety-critical systems, with an empirical, theoretical, historical and epistemological orientation. He is the editor of the book *Safety Science Research: Evolution, Challenges and New Directions* (CRC Press, 2019) and *Post Normal Accident: Revisiting Perrow's Classic* (CRC Press, 2020). He is Associate Editor of the journal *Safety Science*.

Graham Martin is Director of Research at The Healthcare Improvement Studies Institute (THIS Institute), a new unit funded by the Health Foundation at the University of Cambridge to develop the evidence base for improving healthcare quality and safety. His research focuses on social, organisational and professional issues in healthcare system change.

Sarah Maslen is Associate Professor of Sociology at University of Canberra. Her research addresses epistemic cultures and cognition, principally in the engineering and medical domains. In a hazardous industry context, recent studies have addressed core capabilities in engineering practice, abstract and narrative-based reasoning strategies among engineers, professional liability and the impact of incentive arrangements on major accident risk management.

Jane O'Hara is Professor of Healthcare Quality & Safety, within the School of Healthcare at the University of Leeds. Her research focuses on how to support the monitoring, management, measurement and improvement of quality and patient safety within health services.

Nick Pidgeon is Professor and Director of the Understanding Risk Research Group within the School of Psychology at Cardiff University and Professor of Environmental Risk. His research looks at public attitudes, risk perception and public engagement with environmental risks and energy technologies and infrastructures.

Dr Andrew J. Rae is Senior Lecturer in the Safety Science Innovation Lab at Griffith University, where he teaches courses on research methods and safety engineering, and manages the lab's research programme. Andrew's research uses a mix of ethnography, field experiments and theory-building to investigate organisational safety practices. He is particularly interested in understanding the myths, rituals and bad habits that surround the work of managers and safety practitioners, and how this work influences front-line operations. Andrew co-hosts the Safety of Work podcast and is Associate Editor of the journal *Safety Science*; email: d.rae@griffith.edu.au.

Paul R. Schulman (Ph.D. Johns Hopkins University) is a Senior Research Associate at the Center for Catastrophic Risk Management at the University of California, Berkeley and Professor Emeritus of Government at Mills College, Oakland, California (U.S.A.). He has written extensively and consulted on managing hazardous technical systems to high levels of reliability and safety within organizations and across networks of organizations.

Justin Waring is Professor of Medical Sociology and Healthcare Organisations at the Health Services Management Centre, University of Birmingham. His research focuses on the changing organisation and governance of health and care systems.

David E. Weber is a Human Factors, Health & Safety professional with experience in varied industry segments. David holds a PhD in Aviation Human Factors (Safety Science Innovation Lab & Griffith Aviation), a Graduate Diploma in Occupational Health and Safety (Queensland University of Technology), and a Master and Bachelor of Science in Psychology (Basel University). David works as a coach and consultant with Southpac International, where he applies Human Factors and OHS knowledge and methodologies to health and safety challenges within organisations to advance safety management and promote safety innovation.

Siri Wiig, PhD, MSc, is Centre Director of the SHARE – Centre for Resilience in Healthcare at the University of Stavanger (UiS), and full Professor of Quality and Safety in Healthcare Systems at the Faculty of Health Sciences, UiS, Norway. Key research interests are resilience in healthcare, patient safety, quality improvement, safety investigations, user involvement, human factors, risk regulation, leadership, learning, sociotechnical systems. Wiig is Honorary Professor at Australian Institute of Health Innovation, Faculty of Medicine and Health Sciences, Macquarie University, Australia.

Jane O'Hara is Professor of Healthcare Quality & Safety, within the School of Healthcare at the University of Leeds. Her research focuses on how to support the monitoring, management, measurement and improvement of quality and patient safety within health services.

Alberto Zanutto, PhD, researcher at Fondazione Bruno Kessler is Lecturer of Economic Sociology at the University of Verona. He carries out research projects and training activities in the field of organisation studies, with a peculiar focus on cyber security in the industrial sector, as well as digital health and digital transition in the public health sector.

1 Hazardous Technological Systems from the Inside Out

An Introduction

Kenneth Pettersen Gould and Carl Macrae

CONTENTS

TECHNOLOGICALLY SITUATED STUDIES OF RISK AND SAFETY

Some of the most pressing problems facing modern societies in the twenty-first century involve understanding and controlling the risks posed by the hazardous technological systems that our societies create and depend upon. From the safety of our hospitals and transportation networks to the reliability of our utility services and the security of our financial institutions, complex technological systems underpin every aspect of our lives – and when these systems break down they can have devastating consequences for our social, economic and environmental worlds. Understanding hazardous technological systems is critical both for those who design, regulate and work in those systems and for those who study them. But understanding these systems is difficult. As technological systems have become more complex, widespread and interconnected, they have also become more inscrutable and harder to comprehend – and their failures more surprising. These difficulties are likely to become even more pronounced in the coming decades as opaque cognitive technologies, such as autonomous vehicles and medical artificial intelligence, become ubiquitous. Understanding the risks posed by hazardous technological systems and developing appropriate preventive and precautionary strategies therefore increasingly depends on developing a granular and sensitive understanding of the processes and mechanics that explain how these systems work – and why they fail. This, in turn, requires a sophisticated appreciation of the social dynamics, cultural patterns and organisational practices through which complex technological systems are built and enacted in specific contexts. New forms of cooperation and domains of operations are also developing due to the widespread growth in sophisticated information and communication technologies and an increasing emphasis on security threats and associated security risk reduction measures, leading to an intersection between safety and security management in hazardous industries (Bieder and Gould 2020). In short, the challenge – both in research and in practice – is to "get inside" hazardous technological systems: to understand them from the inside out.

This book is about the methods, practices and strategies that help researchers develop knowledge about hazardous technological systems, and the challenges that are faced in the process. By "technological system" we mean the combined technical, cultural and organisational activities that are brought together to achieve some productive goal, such as the operation of an international airline network (Hughes 1994; Leonardi and Barley 2010). This encompasses the broad array of sociotechnical systems comprising people, material objects, management techniques, cultural norms and shared practices that are engaged in the development, assessment and use of technological objects and processes. In the safety sciences, there is a long tradition of developing close accounts of hazardous technological systems that seek to explain what goes on in those systems by theorising the social and technical interactions through which risk and safety emerge. Some of these theoretical accounts have become touchstone works that have helped to shape entire fields and decades of further research. From Barry Turner's (1978) groundbreaking analyses of the social incubation of organisational disasters to the revisionist history of Diane Vaughan's (1997) definitive account of the *Challenger* shuttle explosion, and from the original Berkeley studies of the social and organisational patterns underpinning high-reliability organisations (La Porte, 1982; La Porte 1996) to the new wave of work that theorises reliability-seeking practices (Roe and Schulman 2008; Schulman and Roe, 2016; Macrae, 2014), our knowledge of hazardous technological systems has been deeply shaped by research that combines rigorous analytical methods with careful empirical exploration of the social worlds that are enmeshed in technological systems. What is more, many of the concepts that have been developed and deployed in these technologically situated studies have become familiar parlance in the hardnosed world of practical management. Ideas like accident incubation, the normalisation of deviance, latent factors, practical drift, sensemaking – and, indeed, the very notion of safety culture itself – have made the journey from the realms of social science to the sharp-end of management practice.

In-depth, up-close research that seeks to explain what happens in hazardous systems has produced a wealth of insights of both theoretical and practical value, but the methodologies associated with these modes of research are not often considered in similar depth. These methodologies are typically qualitative, theory-generative and founded on constructivist perspectives of sociotechnical systems that emphasise social, interactive and contextual contributions to risk and safety. Explorations of context, social dynamics and organisational practice appear in relatively fragmented and isolated ways across the literature (Gould 2020), which can hamper efforts to build cumulative knowledge and form more integrated methodological approaches, in contrast to other areas of the field such as psychometrics or technical risk analysis. This represents an important methodological gap with tangible consequences. There has also been a tendency in the safety sciences to emphasise conceptual novelty and theoretical interest over more foundational concerns of methodological coherence and empirical rigour (Le Coze et al. 2014). This, in turn, can remove opportunities to evaluate the utility and explore the interrelation of different methods, approaches and epistemologies (see Schulman, Chapter 10, this volume), which can limit the power of our methodological tool kits. And it means that research activities, methodological principles and the theoretical perspectives and epistemological assumptions that underlie them – which represent the social identities, situated practices and cultural norms of knowledge production – are not always fully accounted for or properly explained. Our theories of organisations and hazardous systems as well as our risk assessments rest on epistemological and methodological foundations that are indeterminate (Wynne 1992) and sometimes hidden from view.

This volume aims to help rebalance this situation by bringing attention back onto the methodological foundations of research that seeks to explain what goes on in hazardous technological systems. The intent is to provide a compendium and a companion that researchers, students and more methodologically minded practitioners can use to explore some of the broader methodological landscape of the safety sciences and to better understand the history, diversity, challenges and opportunities that are associated with the more organisationally and socially oriented research approaches and methods that underpin this field. And, while this is not a textbook or a handbook, the aim is that the book offers insights into some of the pragmatic tactics and practical wisdom that can help to get access to hazardous technological systems and explain them from the inside out. This book is therefore concerned both with lifting the lid on hazardous technological systems and with the methodological strategies and practical activities that help us to do that. It aims to take us inside the methodological practices that can, in turn, take us inside technological practices in hazardous domains.

INSIDE SYSTEMS AND INSIDE RESEARCH: OBJECTIVES, PERSPECTIVES AND COMMITMENTS

What does it mean to do research "inside" hazardous technological systems? This book explores a meaning that is threefold. First, it represents an objective: a core aim to understand the inner workings and social processes that unfold within and around hazardous technological systems. This, in turn, implies practical aims of getting access to these systems and the people that work within them to gather data and develop insight into the way organisational life unfolds in these technological systems – both as they function and as they fail. Second, it represents a perspective: a foundational view that risk, safety and hazardous systems are constructed through and emerge from the interactions and practices of people and technologies within particular organisational contexts and social structures. Such a perspective emphasises the importance of researching and analysing the situated practices, technical objects, organisational processes and cultural characteristics that shape hazardous systems over time. And third, it represents a set of methodological commitments that enact these objectives and this perspective. These include conducting research, as far as possible, "in close" to examine hazardous systems and practices as they have developed and unfold; seeking to understand systems in terms of those who work in and around them by allowing participants to define the contexts and practices of importance and avoiding the imposition of theories and concepts by researchers; and engaging with rich and textured empirical data that typically focuses on the detailed study of a small group of research sites, rather than abstracted analysis of large numbers of cases (Weick 1995). These methodological commitments are typically – though not exclusively – associated with epistemological assumptions that tend towards one or other flavour of constructionism, broadly premised on an acceptance that "research is concerned neither with the production of fantasies about the world, nor with mere mechanical fact-gathering. In social inquiry, there is an interaction between the researcher and the world" (Turner 1981, p. 227).

This book is concerned with research that gives us a view inside hazardous systems, and our intent with this collection is to bring together chapters that themselves offer an exploration inside those research methods and practices. Such "inside" accounts are intended to provide insight into the actual practical work of risk and safety research, exploring tactics, challenges and solutions based on the researchers' relationships with the research process itself. Collections of inside accounts are not new in social research (Hammond 1964; Habenstein 1970; Bell and Roberts 1984), risk analysis (Krimsky and Golding 1992) or in organisation and work studies (Bryman 1988; Cefkin 2010; Ybema et al. 2009). But, despite the increasing importance of social and organisational research in the safety sciences, methodological reflexivity remains relatively under-developed in this field. Some of the more notable and important contributions include sensitive considerations of the role of grounded theory in relation to risk and safety studies and the importance of analytical expertise (Pidgeon et al. 1991) – the use of which was pioneered by a founding father of the field, Barry Turner (1978). Processes of analogical theorising and other analytical practices have been explored by Vaughan (1999, 2004, 2006), particularly considering the often invisible work of building explanations from data. Action research has been explored as an important orientation for researching and engaging with real-world practices of organisational risk management (Horlick-Jones and Rosenhead 2002). The influence of different methodological choices and research strategies on explanations of organisational culture and safety have begun to be explored (Hopkins 2006), along with the empirical and theoretical consequences of underlying research assumptions that privilege abstract cultural characteristics and consider these in isolation from the practical work that goes on in different organisational contexts (Reiman and Odewald 2006). And, reflecting on his experiences with fellow high-reliability researcher Todd LaPorte, Gene Rochlin (2011) examined the relationship between their fieldwork and other ethnographic means of sociological research, and the particular challenges of being present in hazardous field sites where participation is deeply circumscribed. Despite the rather limited number of published explorations into social and organisational methodologies in the safety sciences, the broader field of risk and safety science nonetheless has a rich and extensive history of social and organisational analysis. This history is both explored by and frames many of the chapters in this collection. In turn, understanding where the field has "come from" helps to explain current challenges and future directions.

WHERE HAVE WE COME FROM? A BRIEF HISTORY OF THE ORGANISATIONAL FIELD

Four decades ago, Barry Turner published *Man-Made Disasters* (1978), an innovative and multi-layered analysis of the social and organisational processes underlying disasters. A posthumous second edition co-authored with Nick Pidgeon appeared 20 years later (Turner and Pidgeon 1997). *Man-Made Disasters*, and its two anniversaries, provided a historical inspiration for this collection and in many ways represented the birth of a new arena within the safety sciences. Many within the risk and safety research communities relate Turner's work to "disaster incubation," his theory of the gradual accumulation and progression of disaster-provoking conditions in complex organisational settings (Turner 1978, Turner and Pidgeon 1997). However, not only did Turner's pioneering research provide fundamentals for understanding the preconditions for large-scale accidents and organisational disasters; it was also a significant forerunner and foundation for socially oriented and organisationally sensitive inquiry within this field. Over 40 years ago, Turner was grappling with the methodological challenges of building theories of risk and safety from data on the complex realities of practice associated with accidents in industrial systems, as well as addressing questions which remain urgent and cutting-edge today, such as the cultural challenges of handling – and mishandling – complex information and technical objects, and the social construction and maintenance of ignorance in industrial organisations. Following Turner, a growing field of researchers have become engaged with the organisational and social emergence of risk and safety in complex systems, as well as with critical examinations of associated social processes such as power and trust. Some of these studies have grown out of the more established fields of organisation and management studies (Jeffcutt 1999), which has helped to bring issues of safety and risk more strongly into touch with more traditional sociological, interpretative and ethnographic methodologies such as those of symbolic anthropologists (Turner 1971, 1983, 1990). However, while a wide variety of research into organisational risk and safety is produced each year, and many students and researchers are tackling many facets of this subject in a growing number of research centres and groups around the world, there is still relatively little systematic examination and discussion of the experiences, practices and knowledge status of social and organisational research methods within the safety sciences. This is particularly surprising given the cultural turn that appears to have been underway in the field more broadly and which is bringing greater attention to bear on contextual, cultural and practical issues.

Accidents and disasters associated with organisational innovations or technological developments typically receive considerable public attention and give rise to various processes of blame, reassurance and change. Major accidents such as the 2017 Grenfell tower fire in London, the Fukushima nuclear disaster in 2011 and the 2013 Savar building collapse in Bangladesh provide focal occasions for the emergence of social distrust and criticism regarding technical safety measures and institutional mechanisms of control. These social responses to disaster, and inquiries into the social causes of disaster, illustrate how social processes are always a fundamental consideration in the control of risk – and in the failures of those controls (Pate-Cornell and Murphy 1996). In technological systems, the management of these social processes and associated problems have traditionally been subsumed within efforts to manage "human error": deviations of human behaviour from expected and intended patterns of activity (Reason, 1990). This focus has developed over several decades into broader methods for identifying and analysing human and organisational factors in risk assessment (Davoudian et al. 1994; Flin et al. 2002). The field of human factors has a long history and emerged as an applied psychological field of scientific inquiry in the 1940s to provide scientific input into aircraft accident investigations and the nature of what was labelled at the time as "pilot error": that is, it focused on analyses of the failures of professional flight crew in relation to the tasks, tools and contexts of piloting aircraft. The science of human factors has been based on this premise ever since, and at its core it is concerned with analysing the systems in which people work to identify where errors or behavioural fluctuations may occur (Dekker 2005). Local human–system interactions remain the primary concern of much human factors research and are an important facet of safety science. However, many of the pressing problems that are faced in hazardous systems are broader in scope, relating to political, economic and institutional developments that are changing the systemic contexts and social structures of operational work (Le Coze 2017; Gould et al. 2020). These changes require a close analysis of the relationships between individual actors, their social and organisational contexts and the broader social order to understand how actions and decisions unfold in technological systems.

Accordingly, over the past few decades, social and organisational science has become an increasingly important contributor to the fields of risk analysis and safety science. The 1980s and 1990s saw heightened public concerns and research attention on industrial disasters, with analyses of the organisational processes and characteristics underlying disaster emerging alongside efforts to more precisely specify the nature of hazards (Turner 1978; Perrow 1984; Short 1984). In particular, the argument that systemic accidents are inevitable in certain technological systems became influential with the publication of Perrow's *Normal Accidents* (1984), spurring explorations of fundamental limits to safety and organisational competence and control (Sagan 1995). Researchers increasingly acknowledged the limitations of scientific approaches rooted solely in engineering, systems analysis, cognitive science and economics and drew on broader social sciences to analyse sociotechnical risk and safety. Short (1984) clarified the social characteristics of sociotechnical systems and emphasised the importance of using social and cultural context as the starting point for risk analysis, highlighting the normative nature of judgements about limits of acceptable safety and tolerable risk. As risk analysis became more central in industries such as nuclear power, aviation and the chemical industries through the 1990s, the need for more sophisticated social and organisational theories of risk production and risk acceptability became clearer (Short and Clarke 1992).

By the early 1990s, the central concerns of both researchers and professionals in risk and safety had begun to heavily focus on "people problems." Initially, analysis of human and organisational factors focused on individual factors (Short 1992), communication issues and the role of human error in technological breakdown (Reason 1990). Some of this work had a considerable impact in modifying rational decision theories (Short and Clarke 1992), shifting attention to understanding how people actually think and behave in different contexts. This work rapidly expanded to encompass more sociological and organisational concerns, particularly the interactions and complexities that emerge between humans, machines and organisations (Short and Clarke 1992; Clarke and Short 1993; Reason 1997; Turner and Pidgeon 1997). In parallel, studies of organisational high reliability began shifting attention from the dark to the light side of organisational life by closely examining the organisational processes and cultures which characterised organisations that appeared able to sustain safe performance in hazardous domains (La Porte 1982; La Porte 1996; LaPorte and Consolini 1991; Roberts 1993; Schulman 1993; Hopkins 1999). The emphasis on organisational contexts and local contingencies in this work afforded a wider perspective on organisational behaviour and rationality than existed in technical and psychological theories of human factors. High-reliability research emphasised the importance of organisational-specific knowledge and ongoing interactions between adaptive organisational practices and formalised technical processes in the face of unpredictable operations (Roe and Schulman 2008). Similar perspectives have been developed in more recent research on organisational resilience in hazardous systems (Dekker et al. 2011; Byrne and Callaghan 2013; Macrae, 2014), on resilience as a strategy for complexity management (Kendra and Wachtendorf 2003; Hollnagel et al. 2006; Comfort et al. 2010) and – significantly for this volume – the associated methodological contributions and implications of such approaches for studying safety in hazardous systems (Provan et al. 2019; Rae and Provan 2019; Rae et al. 2020) (see Rae et al., Chapter 8, this volume). Ideas of reliability, robustness and resilience are now well-established as central phenomena of study in the safety sciences, alongside more traditional categories of accidents, failure and disaster (Woods 2015; Weick and Sutcliffe 2011).

THE SHAPE OF THIS BOOK: PRACTICES, METHODS, THEORY

This collection aims to provide a broad exploration of the practices and methods of doing research, conducting fieldwork and developing theory of hazardous technological systems, drawing on a range of different approaches and traditions. The book purposefully includes contributions from both established leaders and emerging voices to provide a diverse set of perspectives and reflections on the nature of research inside hazardous systems, and on the different methodological principles and practical tactics that can underpin this. These span from critical accounts of applying interpretive and social research methods in organisational settings, to explorations of the opportunities and importance of integrating qualitative and quantitative methods, to practical reflections on the challenges associated with negotiating access to research sites and building theory from data. The book is organised as follows and broadly progresses in three thematic phases, beginning with a set of chapters that explore broader methodological strategies and histories, progressing to a set of chapters that consider particular research methods, and concluding with a set of chapters that reflect on the challenges and opportunities for integrating and combining methods and looking to new empirical domains of safety and risk.

The first three chapters provide broad explorations of social and organisational approaches to understanding safety and risk, situating current methods in their historical context and proposing innovative directions for the future. In Chapter 2, Hopkins explores the development of sociological approaches to disaster analysis, particularly drawing on the pioneering work of Barry Turner. Hopkins argues for the value of what he characterises as "desktop ethnographies": in-depth and ethnographically oriented analyses of disasters that make secondary use of the vast amount of evidence that can be collected through formal inquiries. In Chapter 3, Bourrier explores the role of field methodologies in understanding the operational activities that unfold in organisations that handle, and produce, significant risks. She considers both the history and current application of these methodologies and reflects on the uniqueness – or otherwise – of high-hazard organisations and the practical challenges of accessing and entering field sites. While Hopkins and Bourrier focus on broader methodological strategies, Le Coze in Chapter 4 develops a detailed account of the contributions and context of Barry Turner's methodological approaches to understanding organisations and disaster. Le Coze explores how Turner's work made pioneering contributions to our understanding of the organisational processes of disasters and the qualitative methodologies through which those processes can be theorised. Le Coze concludes with a proposal for the development of "conceptual ethnographies" to take forward theory-oriented empirical studies of organisational safety and address the complex relationship between data and theory in relation to 'historical ethnography' (Vaughan, 2004) and 'desktop ethnography' (Hopkins, Chapter 2, this volume).

The next four chapters consider specific methodological approaches that can be deployed to explore social and organisational processes in specific safety-critical settings, including traditional ethnographic and case study methods to novel ways of exploring interpretive activities and work practices. Antonsen and Haavik in Chapter 5 provide a review of the use of case studies in safety science, considering the role of cases in some of the definitive work in the field as well as the range of methodological strategies that can be applied in case study work. They explore the value of cases in theoretical development as well as the practical and political challenges that can arise when selecting and conducting case study research. Haavik provides a different set of linguistic and conceptual tools in Chapter 6, drawing on the methods of Bruno Latour and Actor Network Theory (ANT) to explore the work of studying how meaning is created and sustained in safety-focused activities. Haavik explores how the situated "sensework" that unfolds in networks of safety can be described and explained in meaningful ways, illustrated with a case from his own research in hospital operating theatres. In Chapter 7 Kuran discusses some of the practical challenges and normative and quality criteria for conducting ethnographies in organisational safety. He describes a personal journey of learning to adapt and apply traditional anthropological techniques in the more distributed settings of complex safety-critical transport organisations, and the similarities and differences that emerge compared to established anthropological practice. In Chapter 8, Rae, Weber and Dekker introduce a novel methodological approach for studying normal work practices in safety-critical settings, as a way of understanding the typical operational activities that underpin safety performance. They build on the concepts of work-as-imagined and work-as-done to define a framework for developing everyday work explanations, which engage with the described, assumed and generalised activities of work of different professional groups.

The final six chapters engage with issues of methodological integration and mixed-methods evaluation, as well as illustrating new horizons for safety research. In Chapter 9, Kongsvik and Almklov explore pragmatic strategies for integrating research methods through the use of Action Research (AR). They document the value of combining different methodological approaches and illustrate how the core premises and practices of Action Research can support more methodologically integrated analysis of organisational safety, allowing what they call interpretive extrapolation through the combination of different methods. In Chapter 10, Schulman takes a broad view across social scientific and engineering research to consider how the lack of connection between these fields can underpin failures of integration and specification in our models and theories of safety – and argues that greater integration of social and engineering methodologies could improve our understanding and management of safety. Schulman illustrates his argument by drawing on his personal research experiences in the field of high-reliability research, and demonstrates that conceptual ambiguity is a characteristic of much descriptive research that impacts specificity and precision in our theories of organisational safety – which in turn constrains how we define, use, measure and compare characteristics of safety across organisations and technologies. In Chapter 11, Martin, O'Hara and Waring trace the history and consider the status of mixed-methods research and evaluation in healthcare safety – a relatively new domain of study but one that nonetheless employs some of the most sophisticated empirical analysis of organisationally-based safety interventions. In this chapter they illustrate the principles and complexities that shape mixed-methods research in safety, reflect on the challenges of integrating different modes of knowledge production in the shadow of long-established epistemological hierarchies, and offer a promising prescription for developing empirically rigorous and context-sensitive evaluation of safety interventions. In Chapter 12, Hayes and Maslen focus on the role of senior managers and organisational leaders in safety, revealing this to be a persistent gap in the extant safety literature despite the emphasis placed on leadership in many theories of accidents and safety. To address this, Hayes and Maslen explore the opportunities and practical challenges associated with elite interviewing and reflect on their own experiences of studying the work of senior executives in the oil and gas industry. In Chapter 13, Wiig, Braithwaite and Braut use the development of a new Norwegian safety investigation body to explore how theories of organisational disasters and public and political pressure can inform professional practice and institutional design, and encourage the practical application of more systemic methods of safety analysis.

In Chapter 14, Zanutto offers an account of a qualitative exploration of the organisational management of cybersecurity, illustrating how qualitative analysis of professional discourse can reveal the stories, knowledge and ambiguities involved in understanding and controlling threats to technical infrastructures.

WHERE ARE WE GOING? SKETCHING AN AGENDA FOR THE FUTURE

Taken together, the chapters in this volume provide a tour through some of the key areas of research practice and methodological thinking associated with efforts to study what goes on inside hazardous systems. A pattern of common organising principles and shared challenges emerges from these discussions. And a range of under-researched gaps are evident, providing the outline for a future agenda. Some of the foundational organising principles and associated challenges that underpin work in this area include the value placed on contextually sensitive explanation, the importance of integrating diverse sources of evidence, an acceptance of epistemological flexibility, a concern with understanding systems at different scales, and prioritising efforts to get empirically close to operational activities. Understanding context and developing explanations that are contextually sensitive is a central priority when seeking to understand safety, security and risk within hazardous systems. The importance of this is indicated by the emphasis in many of these chapters on research approaches that include or are premised on ethnographic principles of close engagement with the cultural patterns and organisational settings in which risks are managed. The inherent tension that is apparent in this approach is the related struggle to produce theory that is both situated in and accounts for the uniqueness of a particular local context while also providing insights that are applicable and useful beyond one particular setting at one point in time. This is a reframing of the tension inherent in evaluation studies described by Martin, O'Hara and Waring (Chapter 11) that is echoed across the entire field: how to avoid developing concepts and theories that on the one hand are so abstract and generic they apply to everything but tell you almost nothing, or on the other hand are so specific and particular that they apply to almost nothing except a single, overspecified moment in one particular place. Negotiating this tension animates many of the struggles and innovations in this methodological space and presents one of the core methodological challenges for the safety sciences in the coming years.

The challenge of understanding context points at the importance of drawing on and integrating diverse sources of knowledge in research. The contributions in this volume are either open to, or propose approaches that actively encourage, the collection and use of a plurality of knowledge to look inside hazardous systems. This is a particular focus of efforts to develop sophisticated mixed-methods approaches that combine the strength of quantitative and qualitative techniques, but it is also implied in other approaches, such as those that seek to maximise the use of secondary data collected during formal inquiries or provide embodied access by the researchers to different organisational settings. Drawing on different sources of data is of course common in a range of qualitative and organisational research. But what seems important here is the underlying implication and broader assumption that singular modes of engaging with organisational reality are inherently limited and unlikely to reveal the complexity of thought and diversity of practice that is assumed to characterise activities in hazardous systems. Aligned with and supporting the use of diverse evidence is a considerable degree of flexibility regarding the epistemological assumptions underpinning research implied in many of the discussions in this volume. There are no calls for epistemological purism and indeed many of the contributions implicitly support a degree of epistemological equivocality, arguing for better integration of qualitative and ethnographically oriented work with analyses from the engineering and scientific disciplines, along with the flexibility and openness that is required to bring together different modes of knowledge production – particularly in the face of established epistemological hierarchies that can devalue the findings of contextually sensitive research, or that imposes methodological and empirical criteria that ethnographically oriented research could not possibly meet. A continuing challenge in the coming years will be building a broader-based acceptance across the safety sciences that there are different ways of engaging with and understanding the world and that each brings with it a set of strengths and limitations. It will be important to build on some of the pioneering examples in healthcare that use longitudinal mixed-methods studies (Martin, O'Hara and Waring, Chapter 11, this volume) to examine processes, contexts and outcomes in relation to organisational safety – while at the same time developing more sophisticated understanding of how to blend or switch between different modes of knowledge production in the study of hazardous systems.

A further common thread that is implicit throughout this volume is the importance of studying, and understanding, what happens in hazardous systems at different scales of activity. Hazardous technological systems are typically complex, operate at a range of different sites, are subject to a variety of national and international regulations and interact with a wide array of other organisations. Understanding these systems therefore involves focused, micro-level analysis of context and practice combined with meso- and macro-level analyses of such things as institutional structure, governmental regulation, technological design and economic dynamics (Macrae 2019). For instance, little research has been done on how the global trends of the risk society bring with them unanticipated and "hidden" effects on organisational safety and security practices (Gould and Fjaeran 2020; Pidgeon 2019; Le Coze 2017). The methodological challenges of working across these different scales are considerable and can pose particular difficulties for ethnographically oriented methods that traditionally focus on localised research settings. Understanding context and practice at scale, and understanding how context and practice interact across different scales, presents some of the more pressing methodological challenges in the coming years. A final and related common principle apparent within these chapters is the emphasis on getting empirically as close as possible to the operational activities and processes of hazardous systems. The contributions in this volume all either emphasise the importance of direct research access – in the form of ethnographic observations, interviews or other primary research practices – or focus on the need to gather other forms of data that allow a rich insight into the practical work and operational details of organisational life inside hazardous systems. Several of the chapters explore the practical challenges and tactics associated with gaining access to field settings and professional communities which, when it comes to hazardous systems, can involve particularly sensitive negotiations and trade-offs. Two of the more insidious challenges of working close to operational activity are hinted at through this book. The first is the need to balance the tensions inherent in working in close proximity to practitioners and being dependent on those practitioners for data, whilst also maintaining an appropriate distance and impartiality in analysing their work. The other is to balance the tensions in generating insights and theories that are recognisable and useful to those working in the systems being studied, while remaining alert to the risks of nuanced theoretical accounts being overly simplified or entirely appropriated for instrumental purposes by policymakers or consultants. One of the benefits of doing research up-close in the field is producing theory that is practical and relevant to those who might use it, but one of the hazards is that much of the explanatory power and value of this theory can be lost when it is simplified for broader consumption or packaging in saleable products. Balancing this tension between utility and nuance is likely to remain a defining challenge.

One of the striking things about these shared methodological principles and associated challenges is the degree to which they echo many of the core principles and challenges that characterise the control and management of hazardous systems themselves. Common topics in the safety literature highlight the efforts that some organisations make to ensure that they understand and carefully design and manage the local contexts in which people work; the value of deploying safety information systems and associated cultural practices that integrate and interpret safety data from a range of different sources; the work that safety professionals do to engage with different modes of knowledge production and remain sensitive to their own epistemic weaknesses; the nested processes that have been developed to oversee and make sense of organisational activities from local practices to global operations; and the emphasis that is placed on closely understanding the operational realities and situated practices that underpin the safety of an organisation. In part, these parallels may be expected, given that researching the activities inside hazardous systems and managing and controlling the activities inside hazardous systems are both inherently representational and interpretive processes that depend upon the development of rich pictures of organisational activities. Nonetheless, these parallels have been under-examined to date, and there may be valuable insights to be gained in both directions from understanding the interconnected challenges, practices and tactics of research methodologies and management activities.

In addition to these common principles and challenges running through the book, it is notable that this collection, and the research approaches they broadly represent, point to a set of important emerging issues that would seem key priorities in shaping a future agenda. The first gap concerns sociomateriality and the role of technical objects in risk, safety and security. The role of materiality in the construction of social and organisational activities is an increasingly important topic of organisational research (Leonardi and Barley 2008, 2010; Orlikowski and Scott 2008) but remains peripheral in the safety sciences. Given the focus of the field is on the safety, or otherwise, of technical systems and the objects these systems are instantiated in, the safety sciences would seem well positioned to engage more fully with the nature and role of objects in the construction of risk – and to lead the development of methodological approaches that engage with and theorise the sociomaterial processes that unfold around risk and safety. The second gap is ignorance and its role in organisational safety. Ignorance in all its many forms has been a foundational issue in disaster research, from the early work of Turner (1978) through to emerging theories of disaster (Downer 2011), resilience (Macrae 2014) and risk (Aven 2019; Macrae 2009). Methodologies for studying and theorising the construction and maintenance of ignorance are being deployed across the social sciences (Smithson 1989; McGoey 2019; Grey and Costas 2016). With their theoretical concern with ignorance and practical concern with the management of uncertainty, the risk and safety sciences would likely benefit from – and provide a prime location for – the development of sophisticated methodologies for analysing and explaining the emergence, persistence and consequences of ignorance in organisational life. The third gap is power. There has been relatively limited engagement with the reality that some organisational actors have much greater ability to impose their views on others, and that the limitations and constraints that exist in organisational life are often imposed by leaders, executives and other elites. There are major opportunities for developing and applying methodologies that better engage with the role of more coercive forms of power and constraint in organisations, and how this effects the enactment of hazardous systems (see also Hayes and Maslen, Chapter 12 this volume). Fourth is the gap that exists around our understandings of social orders and the way these influence, and are influenced by, the organisation of hazardous technological systems. This is, again, a foundational topic in the organisational and social sciences (Barley 1986; Leonardi and Barley 2010), but the implications remain under-explored in safety-critical settings. And, while local vulnerabilities in norms and processes for dealing with hazards are increasingly understood as connected to global events and processes, the challenges and problems that globalisation presents for organisational risk and safety remain largely unaddressed (Le Coze 2017; Pidgeon, this volume). There are opportunities to develop longitudinal and comparative methodologies to understand how occupational arrangements, employment relations and other authority structures are shaped by systemic risks such as financial breakdown, climate change and pandemics (Renn et al. 2019), and how the management of safety in turn impacts on broader social orders with societal consequences. Subcontracting and just-in-time principles are examples of myriad organisational changes that challenge traditional forms of social organisation and associated models of safety. With the increasing commoditisation of work (Almklov and Antonsen 2010), operational work – the activities at the sharp-end of safety – are often among the first roles shifted outside the boundaries of organisational ownership and responsibility. In some cases the traditional organisation (understood as the entity controlling the division of labour and coordination) is now almost disappearing in the case of the "gig" or platform economy (Kalleberg and Dunn 2016), posing a range of new challenges for both managing and researching risk and safety in these highly distributed and fluid social forms. Taken together, the broad range of common principles, challenges and gaps that emerge from this volume help sketch out the shape of a future agenda. There is much to be done, both methodologically and theoretically, and reflecting on how we can do better research inside hazardous systems seems a productive route to travel.

REFERENCES

Almklov, P. G., & Antonsen, S. (2010). The commoditization of societal safety. *Journal of Contingencies and Crisis Management, 18*(3), 132–144.

Aven, T. (2019). *The Science of Risk Analysis: Foundation and Practice*. New York: Routledge.

Barley, S. R. (1986). Technology as an occasion for structuring: Evidence from observations of CT scanners and the social order of radiology departments. *Administrative Science Quarterly, 31*(1), 78–108.

Bell, C., & Roberts, H. (Eds.) (1984). *Social Researching: Politics, Problems, Practice*. London: Routledge Kegan & Paul.

Bieder, C., & Gould, K. P. (2020). Exploring the interrelations between safety and security: Research and management challenges. In: Bieder, C. & Gould, K.P. (Eds.), *The Coupling of Safety and Security*, 105–113. Cham: Springer Open.

Bryman, A. (Ed.) (1988). *Doing Research in Organizations*. New York: Routledge.

Byrne, D., & Callaghan, G. (2013). *Complexity Theory and the Social Sciences: The State of the Art*. London: Routledge.

Cefkin, M. (Ed.) (2010). *Ethnography and the Corporate Encounter: Reflections on Research in and of Corporations*. New York: Berghahn Books.

Clarke, L., & Short, J. F. (1993). Social organization and risk: Some current controversies. *Annual Review of Sociology, 19*(1), 375–399.

Comfort, L. K., Boin, A., & Demchak, C. C. (Eds.) (2010). *Designing Resilience. Preparing for Extreme Events*. Pittsburgh: University of Pittsburgh Press.

Davoudian, K., Wu, J. S., & Apostolakis, G. (1994). Incorporating organizational factors into risk assessment through the analysis of work processes. *Reliability Engineering and System Safety, 45*(1–2), 85–105.

Dekker, S. W. A. (2005). *Ten Questions About Human Error*. New Jersey: Lawrence Erlbaum Associates.

Dekker, S. W. A., Cilliers, P., & Hofmeyr, J. H. (2011). The complexity of failure. Implications of complexity theory for safety investigations. *Safety Science, 49*(6), 939–945.

Downer, J. (2011). "737-Cabriolet": The limits of knowledge and the sociology of inevitable failure 1. *American Journal of Sociology, 117*(3), 725–762.

Flin, R., O'Connor, P., & Mearns, K. (2002). Crew resource management: Improving team work in high reliability industries. *Team Performance Management. Anais an International Journal*.

Gould, K. Pettersen (2020). Organizational risk: "Muddling through" 40 years of research. *Risk Analysis: An Official Publication of the Society for Risk Analysis*, doi:10.1111/risa.13460.

Gould, K. Pettersen & Fjaeran, L. (2020). Drift and the social attenuation of risk. In: Le Coze, J. C. (Ed.), *Safety Science Research: Evolution, Challenges and New Directions*, 119–132. Boca Raton: CRC Press.

Grey, C., & Costas, J. (2016.). *Secrecy at Work: The Hidden Architecture of Organizational Life*. Stanford: Stanford University Press.

Habenstein, R. W. (Ed.) (1970). *Pathways to Data – Craft and Methods for Studying Social Organizations*. New York: Routledge.

Hammond, P. E. (Ed.) (1964). *Sociologists at Work: Essays Om the Craft of Social Research*. New York: Basic Books, Inc.

Hollnagel, E., Woods, D. D., & Leveson, N. (2006). *Resilience Engineering – Concepts and Precepts*. Boca Raton: CRC Press.

Hopkins, A. (1999). The limits of normal accident theory. *Safety Science, 32*(2), 93–102.

Hopkins, A. (2006). Studying organisational cultures and their effects on safety. *Safety Science, 44*(10), 875–889.

Horlick-Jones, T., & Rosenhead, J. (2002). Investigating risk, organisations and decision support through action research. *Risk Management*, 4(4), 45–63.

Hughes, T. P. (1994). Technological momentum. In: Marx, L. & Roe Smith, M. (Eds.), *Does Technology Drive History? The Dilemma of Technological Determinism*. Cambridge, MA: MIT Press.

Jeffcutt, P. (1999). From the industrial to the post-industrial subculture. *Organization Studies*, 20(4), VII.

Kalleberg, A. L., & Dunn, M. (2016). Good jobs, bad jobs in the gig economy. *Perspectives on Work*. 20, (1–2), 10–74.

Kendra, J. M., & Wachtendorf, T. (2003). Elements of resilience after the world trade center disaster. Reconstituting New York City's emergency operations center. *Disasters*, 27(1), 37–53.

Krimsky, S., & Golding, D. (Eds.) (1992). *Social Theories of Risk*. Westport: Praeger Publishers.

La Porte, T. R. (1982). On the design and management of nearly error free organizational control systems. In: Sills, D. L., Wolf, C. P., & Shelanski, V. B. (Eds.), *The Accident and Three Mile Island: The Human Dimensions*, 185–200. Boulder, CO: Westview Press.

La Porte, T. R. (1996). High reliability organizations: Unlikely, demanding and at risk. *Journal of Crisis and Contingency Management*, 4(2), 60–71.

La Porte, T. R., & Consolini, S. M. (1991). Working in practice but not in theory: Theoretical challenges of "High Reliability Organizations". *Journal of Public Administration Research and Theory*, 1(1), 19–47.

Le Coze, J. C. (2017). Globalization and high-risk systems. *Policy and Practice in Health and Safety*, 15(1), 57–81.

Le Coze, J. C., Pettersen, K., & Reiman, T. (2014). The foundations of safety science. *Safety Science*, 67, 1–5.

Leonardi, P. M., & Barley, S. R. (2008). Materiality and change: Challenges to building better theory about technology and organizing. *Information and Organization*, 18(3), 159–176.

Leonardi, P. M., & Barley, S. R. (2010). What's under construction here? Social action, materiality, and power in constructivist studies of technology and organizing. *The Academy of Management Annals*, 4(1), 1–51.

Macrae, C. (2009). Making risks visible: Identifying and interpreting threats to airline flight safety. *Journal of Occupational and Organizational Psychology*, 82(2), 273–293.

Macrae, C. (2014). *Close Calls: Managing Risk and Resilience in Airline Flight Safety*. London: Palgrave Macmillan.

Macrae, C. (2019). Moments of resilience: Time, space and the organisation of safety in complex sociotechnical systems. In: Wiig, S. & Fahlbruck, B. (Eds.), *Exploring Resilience: A Scientific Journey from Practice to Theory*, 15–24. London: Springer.

McGoey, L. (2019). *The Unknowers: How Strategic Ignorance Rules the World*. London: ZED Books.

Orlikowski, W. J., & Scott, S. V. (2008). Sociomateriality: Challenging the separation of technology, work and organization. *The Academy of Management Annals*, 2(1), 433–474.

Pate-Cornell, M. E., & Murphy, D. M. (1996). Human and management factors in probabilistic risk analysis: The SAM approach and observations from recent applications. *Reliability Engineering and System Safety*, 53(2), 115–126.

Perrow, C. (1984). *Normal Accidents – Living with High-Risk Technologies*. Princeton: Princeton University Press.

Pidgeon, N. F. (2019). Observing the English weather: A personal journey from safety I to IV. In: J. C. Le Coze (Ed.), *Safety Science Research: Evolution, Challenges and New Directions*, 269–280. Boca Raton: CRC Press.

Pidgeon, N. F., Turner, B. A., & Blockley, D. I. (1991). The use of grounded theory for conceptual analysis in knowledge elicitation. *International Journal of Man-Machine Studies*, *35*(2), 151–173.

Provan, D. J., Rae, A. J., & Dekker, S. W. (2019). An ethnography of the safety professional's dilemma: Safety work or the safety of work? *Safety Science*, *117*, 276–289.

Rae, A., & Provan, D. (2019). Safety work versus the safety of work. *Safety Science*, *111*, 119–127.

Rae, A., Provan, D., Aboelssaad, H., & Alexander, R. (2020). A manifesto for reality-based safety science. *Safety Science*, *126*, http://www.ncbi.nlm.nih.gov/pubmed/104654.

Reason, J. (1990). *Human Error* (1st ed.). Cambridge: Cambridge University Press.

Reason, J. (1997). *Managing the Risks of Organizational Accidents*. Burlington: Ashgate Publishing Limited.

Reiman, T., & Oedewald, P. (2006). Assessing the maintenance unit of a nuclear power plant – Identifying the cultural conceptions concerning the maintenance work and the maintenance organization. *Safety Science*, *44*(9), 821–850.

Renn, O., Lucas, K., Haas, A., & Jaeger, C. (2019). Things are different today: The challenge of global systemic risks. *Journal of Risk Research*, *22*(4), 401–415.

Roberts, K. H. (Ed.) (1993). *New Challenges to Understanding Organizations*. New York: Macmillan Publishing Company.

Rochlin, G. I. (2011). How to hunt a very reliable organization. *Journal of Contingencies and Crisis Management*, *19*(1), 14–20.

Roe, E., & Schulman, P. (2008). *High Reliability Management: Operating on the Edge*. Stanford: Stanford University Press.

Sagan, S. (1995). *The Limits of Safety: Organisations, Accidents and Nuclear Weapons*. Princeton: Princeton University Press.

Schulman, P. R. (1993). The negotiated order of organizational reliability. *Administration and Society*, *25*(3), 353–372.

Schulman, P. R., & Roe, E. (2016). *Reliability and Risk: The Challenge of Managing Interconnected Infrastructures*. Stanford: Stanford University Press.

Short, J. F. (1984). The social fabric at risk: Toward the social transformation of risk analysis. *American Sociological Review*, *49*(6), 711–725.

Short, J. F. (1992). *Organizations, Uncertainties, and Risk*. Boulder: Westview Press.

Short, J. F., & Clarke, L. (1992). Social organization and risk. In: Short, J. F. & Clarke, L. (Eds.), *Organizations, Uncertainties, and Risk*, 309–321. Oxford: Westview Press.

Smithson, M. (1989.). *Ignorance and Uncertainty: Emerging Paradigms*. London: Springer.

Turner, B. A. (1971). *Exploring the Industrial Subculture*. London: Macmillan.

Turner, B. A. (1978). *Man-Made Disasters*. London: Wykeham Press.

Turner, B. A. (1981). Some practical aspects of qualitative data analysis: One way of organizing the cognitive processes associated with the generation of grounded theory. *Quality and Quantity*, *15*(3), 225–247.

Turner, B. A. (1983). The use of grounded theory for the qualitative analysis of organizational behavior. *Journal of Management Studies*, *20*(3), 333–348.

Turner, B. A. (1990). *Organizational Symbolism*. Berlin: De Gruyter.

Turner, B. A., & Pidgeon, N. (1997). *Man-Made Disasters* (2nd ed.). London: Butterworth-Heinemann.

Vaughan, D. (1997). *The Challenger Launch Decision*. Chicago: The University of Chicago Press.

Vaughan, D. (1999). The role of the organization in the production of techno-scientific knowledge. *Social Studies of Science*, *29*(6), 913–943.

Vaughan, D. (2004). Theorizing disaster: Analogy, historical ethnography, and the challenger accident. *Ethnography*, *5*(3), 315–347.

Vaughan, D. (2006). NASA revisited: Theory, analogy, and public sociology. *American Journal of Sociology, 112*(2), 353–393.

Weick, K. E. (1995). *Sensemaking in Organisations*. London: Sage.

Weick, K. E., & Sutcliffe, K. M. (2011). *Managing the Unexpected. Resilient Performance in an Age of Uncertainty*. San Francisco, CA: John Wiley & Sons.

Woods, D. (2015). Four concepts for resilience and the implications for resilience engineering. *Reliability Engineering and System Safety, 141*, 5–9.

Wynne, B. (1992). Uncertainty and environmental learning: Reconceiving science and policy in the preventive paradigm. *Global Environmental Change, 2*(2), 111–127.

Ybema, S., Yanow, D., Wels, H., & Kamsteeg, F. H. (Eds.) (2009). *Organizational Ethnography: Studying the Complexity of Everyday Life*. London: Sage.

2 Turner and the Sociology of Disasters

Andrew Hopkins

CONTENTS

INTRODUCTION

Barry Turner trained as a chemical engineer. However, his PhD was in sociology and his work on disasters was thoroughly sociological. Indeed, his book *Man-Made Disasters*, first published in 1978, was the beginning of a modern tradition of sociological disaster studies. It cannot be said to have *given rise* to that tradition, since it was not widely read at the time. Turner himself wrote later that it was as if "the book had disappeared without a trace" (1997: xi). Some of the subsequent writers in the tradition wrote with little or no reference to Turner's work, but in retrospect his book was the first of a genre that has proved highly influential in safety science.

My first aim in this chapter is to sketch very briefly the substance of Turner's contribution. Second, I shall identify some of the writers in this tradition. Third, I want to explore why the sociological approach seems so appropriate for disaster studies. Finally, the chapter will contain an account of a particular methodology used by Turner and some of the subsequent writers in this tradition, which I shall label "desktop ethnography." This term is almost an oxymoron, but I believe that, when properly understood, it is illuminating.

THE SUBSTANCE OF TURNER'S CONTRIBUTION

The starting point for Turner's analysis is this:

> *Disaster equals energy plus misinformation.* (1997: 157, italics in original)

19

This very succinct statement captures nicely Turner's training as both an engineer and a sociologist. Engineers often start with the proposition that accidents are the result of misplaced or misdirected energy and then go on to classify the types of energy that can be involved (e.g. Viner, 1991). It is, however, the second part of Turner's definition that he pursues in his book – the idea that sociotechnical disasters involve an information or communication failure of some kind. The central, socio-logical question he poses is this:

> What stops people from acquiring and using appropriate advance warning information so that large-scale accidents and disasters are prevented? (1997: 162)

It is hard to overstate the importance of this question. It has guided, explicitly or implicitly, many subsequent disaster analyses.

Turner suggests a number of reasons why information may not be available to the relevant people (1997: 162–165).

1. The information is completely unknown. This is a very rare situation, par-ticularly in relation to disasters in recent decades. It is nearly always the case that information that might have averted disaster is available some-where within the organisation concerned.
2. Information is noted but not fully appreciated. This may occur for a number of reasons, including a false sense of security that leads people to discount danger signs, pressure of work that diverts attention from warning signs, or difficulty in sifting the information from a mass of other irrelevant facts.
3. Prior information may not be correctly assembled. This may be because it is buried in other material, distributed among several organisations or distrib-uted among different parties within the same organisation. The failure to assemble relevant information is a particularly common cause of disasters. On Piper Alpha, the North Sea oil platform destroyed by explosion and fire in 1988 with the loss of 167 lives, information about maintenance work done by one shift was not passed on to the next shift because of inadequa-cies in the permit-to-work system. This was the start of the chain of events that led to [the] disaster.
4. Failure to pass bad news up the line to senior managers. Turner includes this in the previous category, but it is a sufficiently distinctive situation to war-rant separate treatment. This is exemplified in the *Challenger* space shuttle disaster. Information that earlier shuttle rockets had experienced techni-cal problems was confined to one section of NASA and neutralised by a process of reinterpretation. Had this information been passed upwards to the highest-ranking NASA officials, a different set of decisions would have been made (Vaughan, 1996).
5. There is no place for the information in existing categories. The hazards of slag heaps and tailings dams exemplify this problem. Slag heaps are great piles of reject material from underground coal mines. One of the disasters that Turner analysed was the collapse of a slag heap that engulfed the Welsh

village of Aberfan in 1966, killing nearly 150 people (1997: 42). Tailing dams are modern day equivalents. They consist of embankments holding back huge quantities of semi-liquid waste. If these embankments collapse the waste escapes, engulfing all in its path. Brazil has suffered two such disasters in recent years, killing hundreds of people and doing immense environmental damage. Too often, mine safety regimes do not extend to the safe disposal of mine waste, with the result that warnings of danger may go unrecognised or ignored (Hopkins, 2019: 73–75).

Turner sums all this up in his notion of an incubation period, "which begins when the first of the ambiguous or unnoticed events which will eventually accumulate to provoke the disaster occurs" (1997: 160).

His work provides a rich body of hypotheses that can be applied to the analysis of any disaster of interest. That is precisely what I did in my first disaster book, a study of the Moura coal mine explosion in Queensland that killed 11 men (Hopkins 1999). I concluded:

Turner's theory ... turns out to have been a particularly fruitful approach ... The Moura disaster occurred because vital information was rendered ineffective by an inadequate information processing system and by a culture that neutralised it. (1999: 135–136)

Turner's analysis does not highlight the issue of power. As Pidgeon notes in the preface to the second edition, "the model does not deal explicitly with power relations and vested interests in the run-up to major failures" (1997: xvii). Given the centrality of the concept of power in sociology, this is a surprising omission. Unsurprisingly, therefore, many of the sociologists of disaster who came after Turner did indeed accord power a prominent place, particularly in understanding why relevant information did not end up in the hands of those who could do something about it.

THE MODERN SOCIOLOGICAL TRADITION OF DISASTER ANALYSIS

Here are some of the later works in the sociological tradition of disaster analysis. I will make brief comment on the extent to which they have been influenced by Turner.

In 1984 sociologist Charles Perrow published his high-profile book, *Normal Accidents*. He apparently wrote his book in ignorance of Turner's work, since he makes no reference to it at the time. The book offers an account of the Three Mile Island nuclear power station near disaster in the US in 1979, on the basis of which he developed his Normal Accident Theory. This theory dominated thinking about disasters for decades and gave rise to a prolonged debate with the theorists of HRO (high reliability organisations) about whether major accidents were inevitable. The debate was inconclusive, in part because the protagonists were talking past each other (La Porte and Rochlin, 1994; Perrow, 1994), but one of the results of the controversy was to enhance the visibility of these theories to the point that it would be hard to find anyone working in the field of safety science who had not heard of them (see also Hopkins, 1999a, 2001).

In 1996 Diane Vaughan published her study of the 1986 *Challenger* space shuttle disaster. This is a monumental work, in which she introduces sociological ideas, such as the normalisation of deviance, that have now become commonplace in understanding major accidents. She also highlights the idea of organisational culture. As a result of her work, Vaughan was invited to be part of a Board of Inquiry into the Columbia space shuttle accident, which occurred 17 years after *Challenger* (CAIB, 2003). The Board's analysis was heavily influenced by her earlier work and much of the report was focused on the question of organisational culture. The concept of organisational culture is now widely used in discussions of organisational accidents. Vaughan comments on *Man-Made Disasters* in her own book, but there is no suggestion that it influenced her work.

In 2000, sociologist Scott Snook published *Friendly Fire*, an account of the accidental shoot down of US black hawk helicopters by US jets over Iraq. This was an exhaustive sociological treatment which used as one of its key concepts the idea of "practical drift," meaning "the slow steady uncoupling of local practice from written procedure" (2000: 194, 225). It is now commonplace to identify one of the causes of disaster as drift, that is, the divergence of "work as done," from "work as imagined" by procedure writers. Snook refers to Turner only in a footnote, and then only to a journal article, not the book.

My own work is situated in this tradition. As already mentioned, my Moura book drew explicitly and heavily on Turner's insights. My later case studies, such as the Esso Longford gas plant accident (2000), the BP Texas City refiner disaster (2008), and the BP Gulf of Mexico blowout (2012), barely mention Turner, since by then, his thinking was also my own.

THE BROADER CONTEXT

The preceding works are largely case studies of disaster. However, they sit within a broader sociological context that should be understood.

More or less in parallel with studies of disaster, sociologists have been writing about industrial accidents more generally. I mention two here. In 1981 Kit Carson published an important work, *The Other Price of Britain's Oil: Safety and Control in the North Sea*. The Piper Alpha disaster was still seven years in the future, and Carson's book focuses on the large number of mundane and preventable accidents occurring on North Sea platforms. These were occurring, he argued, because of the over-riding emphasis on speed of production, at the expense of safety. The analysis amounts to a political economy of injury, which highlights the power of capital and the role of the state. In 1997 Theo Nichols published his *Sociology of Industrial Injury*. His particular point was that injuries are not accidents; they are caused by the system of production. Again, this was a work of political economy, a perspective that fits easily within the sociological framework. These two works highlight the concept of power which, as we noted, is missing from Turner's model.

Ron Westrum is a sociologist whose work provides further context. He argues that "the most critical issue for organisational safety is the flow of information" (Westrum, 2004: 22). This statement could easily have been written by Turner himself, but Westrum makes no mention of Turner. The fact that these two writers come independently to the same conclusion must be attributed to their common disciplinary background.

Westrum goes on to divide organisational cultures into three types, on the basis of how they process information: the pathological, the bureaucratic and the generative (2004: 23). Hudson, a psychologist, has extended this into a five-category classification: pathological, calculative, bureaucratic, proactive and generative (Parker et al., 2006). He describes the transition in the model from pathological to generative as a process of becoming "increasingly informed," so in that sense, the model retains Westrum's emphasis on information flow. Interestingly, then, the Hudson model embodies the same principles that animate Turner's work. Nevertheless, in Hudson's hands, there is a subtle shift. Westrum (2014) conceived of his model as a set of ideal types (Max Weber's term), whose use was to describe and explain organisational life. The Hudson model, on the other hand, has become a consultancy tool – a way of identifying where a company sits on an organisational ladder, and what it needs to do to move up the ladder. It is one of the most popular consultancy tools in use today.

Another person who has contributed significantly to this tradition is Jim Reason. Reason is a psychologist, but a dissident one. Here is his view on experimental psychology:

> if you can control the variables, it has to be trivial. … That's my personal prejudice having been an academic psychologist for nigh on forty years. (Macrae, 2014: viii)

Reason's professional trajectory has seen him move away from psychology to what I would describe as more sociological style of analysis.

"I ceased being a kosher cognitive psychologist and started working in the real world of hazardous industries" (2013:118), he says in his penultimate book, an intellectual autobiography entitled, *A Life in Error*. He does not explain this title, but it is surely a pun. Clearly, a first meaning is that he spent much of his life writing about error. As for a second meaning, does he mean his intellectual life was somehow an error? Is he saying that his earlier writings on the psychology of human error amounted to a side-track until he finally found the path for which he is most famous? That would seem to be a plausible reading.

The moment when he transcended his disciplinary origins can be most clearly seen in his book, *Human Error*. Most of the book is cognitive psychology. Then comes chapter 7, a leap into "the real world of hazardous industries." Here he introduces his famous idea of latent errors, soon changed to latent failures, and finally to latent conditions. This subtle linguistic transformation is itself further evidence of his move away from psychology to a broader perspective.

I have chosen to call that broader perspective a sociological style of analysis, although I have no idea whether Reason himself would accept that description. I do so for various reasons. First, the 1980s were marked by a series of very high-profile sociotechnical disasters, among them *Challenger*, Three Mile Island, Bhopal, Zeebrugge, Kings Cross and Chernobyl. Naturally, these claimed Reason's attention. His aforementioned chapter 7 identifies the latent failures that lay behind all these disasters. He then goes on to cite both Turner and Perrow in support of his analysis (Reason, 1990: 197). Indeed he has quite a lot of appreciative things to say about Perrow. Moreover, he likens latent failures to "resident pathogens" that incubate until they finally manifest themselves with disastrous consequences. As Pidgeon notes, this is precisely the incubation period that Turner talks about and which is so central to his analysis (1997: 180). Reason's theorising at this point, then, is strongly reminiscent of Turner's sociological ideas.

A second reason for calling his analysis sociological is the nature of "the general failure types" which he identifies as contributing to accidents. These types are as follows: hardware, design, maintenance management, error enforcing conditions, housekeeping, incompatible goals, communications, organisation and training (1997: 134). Without labouring the point, it is obvious that many of these are standard topics in organisational sociology.

Finally, consider Jens Rasmussen, undoubtedly one of the most influential disaster analysts (Le Coze, 2013, 2015). Rasmussen was a control engineer, but in the 1970s he made a major contribution to the psychology of human error, so much so that Jim Reason sees his own work on error as building on Rasmussen's framework (2013: 14–15). But just as Reason switched his attention to major accidents, so too did Rasmussen, in the 1990s. Rasmussen knew nothing of Turner and seems to have been little influenced by Perrow or Vaughan. It was the major accidents of the 1980s noted above that set his new direction. He argued that a full understanding of these accidents depended on recognising the contribution, not only of errors by operators but also of the organisational layers above them, in particular, management, the company, the regulators, the government and even society. This is precisely the kind of understanding that was being propounded by self-identified sociologists of disaster. Moreover, Rasmussen (1997) went on to depict these layers in his famous Accimaps – causal maps of accidents. These were very effective representations of the holistic thinking about accident causation characteristic of sociological inquiry. I myself immediately recognised the close fit between the Accimap and the sociological work I was doing, and I believe I was the first person, after Rasmussen, to use the method, in my explanation of the Longford Gas plant accident (2000). This connection between sociology and the Accimap method is also demonstrated in Snook's work (2000: 21). He developed a "causal map" for his analysis in *Friendly Fire*, without any knowledge of Rasmussen's work, and without the degree of elaboration that can be seen in in Rasmussen's Accimaps.

This raises the question of why Rasmussen did not himself make the connection between sociology and his new perspective. Le Coze (2015) has raised this very question. His answer is that, at least in part, this is because Rasmussen had no training in the discipline of sociology. He did not have the tools to translate what he was doing into the language of sociology, and perhaps did not realise the value of doing so. That indeed seems to be the most plausible explanation. But whether Rasmussen made the connection or not, we can. It is perfectly reasonable to say that Rasmussen's approach to major accident analysis was implicitly sociological.

To sum up the discussion to this point, sociology has proved to be a valuable disciplinary starting point for understanding major accidents. This work has been done by self-identified sociologists, and also by others using a sociological approach.

THE FIT BETWEEN SOCIOLOGY AND MAJOR ACCIDENT ANALYSIS

Consider now the question of why the discipline of sociology has been so much to the fore in understanding organisational accidents, as Reason calls them.

One way to answer the question is by means of a comparison with the contrasting discipline of psychology. What is it about sociology as opposed to psychology that makes it a fruitful approach to understanding major accidents?

Here is my definition of sociology.

Sociology is about humans in groups. Among other things, it is about the organisational structures and cultures that humans collectively create, and about how these in turn shape the behaviour and the ideas of individuals. (Hopkins, 2016: 36)

Psychology, on the other hand, is the science of the human mind and how the mind affects behaviour. (This definition is a composite drawn from many sources.)

It is immediately apparent that a central difference between sociology and psychology is that psychology focuses on the individual, while sociology is concerned with groups. These may range from very small, face-to face-groups, such as the workgroup, or the family, through large organisations such as the corporation, to society or the nation, at the other end of the spectrum. While psychology is interested in the characteristics of individuals, sociology focuses on the properties of groups. To take just one obvious example, culture is a group property; so we may have the culture of the work group, the culture of a corporation or the culture of a nation – a national culture. Culture therefore is a central concept in sociology in a way that it is not in psychology.

The distinction is not quite as absolute as I have made out. Social psychology is a sub-discipline within both psychology and sociology that is concerned with the behaviour of individuals in small groups. Ideas such as groupthink, ironically a term coined by a political scientist (Janis, 1982), are discussed in both disciplines and have been important in understanding how major accidents occur (Hopkins, 2016: 36).

It is also appropriate to note that sociology has much in common with social anthropology. Both are concerned with group phenomena, with perhaps the main difference being that anthropology has traditionally been concerned with small scale societies, while sociology began as an attempt to understand modern industrial, capitalist society. However, this distinction is now very blurred and self-identified anthropologists often focus on aspects of modern industrial society (e.g. Haukelid, 2008).

A useful way to think about the difference between sociology and psychology is that they differ in their units of analysis. For psychology, the unit of analysis is the individual, and for sociology, it is the group. An individual human being is a well-defined unit of analysis. We can assemble large numbers of such units, control for relevant variables and draw conclusions about variables of interest, using statistical methods. It is another matter when groups are the unit of analysis. Take industrial organisations. These vary enormously in size, purpose, internal organisational structure and ownership, to name just of few of the variables. Their boundaries may be hard to define, and they are in a continual state of flux – growing, contracting, changing their business, and changing their organisational structures, all of which makes it very difficult to categorise them for comparative purposes or to conduct longitudinal analysis. Faced with this reality, sociologists often choose to focus on a single organisation or organisational unit and carry out a descriptive case study. The case study method is certainly not the only sociological method, and some sociological problems lend themselves to quantification and statistical analysis. The point is simply that there is a natural fit between the case study method and much of sociology's subject matter.

To complete the argument, major accidents are truly organisational phenomena. They are, as Reason says, accidents that happen to organisations (1997: 1). The case study method, as practiced by sociology, is thus an obvious way to approach the analysis of major accidents (see also Antonsen and Haavik, this volume).

ETHNOGRAPHY

I want to argue now that the particular method Turner uses can be described as *desktop ethnography*. To develop this argument, I first provide a brief account of ethnography.

"Ethno" refers to a group of people accustomed to living together, while "graphy" denotes description. An ethnography is thus a descriptive study of a group of people who live (or work) together. Ethnographic research originates in the discipline of anthropology, where researchers immerse themselves for long periods in the culture of interest and provide detailed qualitative descriptions of what they observe (Geertz, 1973). Sociologists, too, have used the method to study organisations and their cultures, among other things (e.g. Gouldner, 1954; Kriegler, 1980; Bourrier, 1998; Westrum, 1999). The ethnographic study of an organisation will not normally require the researcher to live within the organisation (except perhaps for what sociologists call "total institutions," such as the military, prisons and mental hospitals, Goffman, 1968). But it does require the researcher to spend a great deal of time in the organisation, either as a participant observer, for instance, an employee, or as a non-participant observer of some sort (see also Kuran, Chapter 7, this volume). A famous Australian example was an ethnographic study of BHP Steel by a sociologist who got a job as a steel worker especially for the purpose doing the research. The company did not know of the worker's ulterior purpose and was furious when he published the book, *Working for the Company* (Kriegler, 1980). I doubt this research would get past university ethics committees these days, although in my view there was nothing unethical about it.

It is not only sociologists who do this kind of research. Some of the most important ethnographic research on organisational culture and its impact on safety has been done by organisational psychologists and political scientists. The initial work on high reliability organisations, for example, was carried out by political scientists who spent time on aircraft carriers, in nuclear power stations and in air traffic control rooms, identifying the elements of the cultures of these organisations that contribute to reliable operations (LaPorte and Consolini, 1991; Weick and Sutcliffe, 2001.)

The ethnographic method raises a number of issues, importantly, the validity of the descriptions provided and conclusions drawn. How can we be sure that the elements of culture identified by the researcher are correct?

One of the foremost theorists of organisational culture, Edgar Schein, has thought long and hard about this. Here is his answer. "The accuracy of the depiction is … judged by the credibility of that description to insiders who live in the culture and, at the same time, to outsiders who are trying to understand it" (Schein, 1992: 187). Most importantly, if the members of the culture recognise the description and agree that that is how it is, then the researcher can feel reasonably confident in the findings. This does not mean, of course, that the ethnographer's description will be the same as that provided by members of the culture. The ethnographer's description is designed to speak to the wider concerns of their discipline. An earlier, anonymous reader of this material provided the following vivid comment:

> The ethnographic description is not a simple mirror image. It is a back and forth effort to take into account how people see things, and how they can actually be interpreted in a larger framework. One needs at the same time to be ready to "go native" and then back! This is the ambivalent posture of the researcher using the ethnographic method.

However, if members of the culture fail to recognise the ethnographer's final description of their culture, that description must be called into question. The anthropologist Margaret Mead is famous for her description of coming of age in Samoa. Yet her informants subsequently denied the reality of her description, and later anthropological research showed that she was indeed wrong (Freeman, 1983).

Nevertheless, cultural descriptions are inevitably subjective to some extent. Leading anthropologist Clifford Geertz takes the view that ethnographic descriptions are "essentially contestable." He comments:

> Anthropology, or at least interpretative anthropology, is a science whose progress is marked less by a perfection of consensus than by a refinement of debate. What gets better is the precision with which we vex each other. (Geertz, 1973: 29)

His point, I think, is that the more one abstracts from the data and uses concepts and categories from one's own discipline, the more contestable is the description.

DISASTER STUDIES AS ETHNOGRAPHIES

Let us now consider to what extent sociological disaster studies are ethnographies. Clearly, the researchers were not present for the events they describe, so in what sense are their books ethnographic?

Major accidents often give rise to multimillion-dollar inquiries. Quite apart from the official reports, the inquiries themselves are a priceless source of information about organisational cultures and the way they impact on safety. Such inquiries sit for many days, taking evidence from a large number of people. Panel members and legal counsel may question individual witnesses for hours. Questioners pursue numerous lines of inquiry, probing, looking for things that might have been overlooked and exploring inconsistencies and conflicts of evidence day after day. Proceedings are recorded, and many thousands of pages of transcript evidence are generated. This is far more material than an individual researcher engaged in an intensive interview process could ever produce. Moreover, the fact that witnesses can often be *required* to give evidence to these inquiries and can be interrogated in quite hostile fashion means that inquiries can gain access to information that no ordinary interviewer could ever hope to uncover.

Inquiry reports make use of a relatively small amount of this material. As a result, the transcript provides the researcher with a largely untapped data source, a ready-made set of interviews which can be mined for insights about the culture of the organisation. The material is often so rich and so diverse that researchers reading it are effectively immersed in the world of that organisation without ever setting foot on its premises. It is for this reason that I call this the method of *desktop ethnography*. It is the way I have written most of my disaster case studies.

As described earlier the validity of ethnographic descriptions is based on a back and forth between the understandings of insiders and outsiders. In my case there is a third relevant category. I speak to people in other companies in the same industry as the company under investigation. They often tell me that many of my descriptions are applicable to their own company; I might as well have been writing about their organisation, they say. This makes them exceedingly uncomfortable since they recognise that their organisation is at risk of suffering the same kind of accident. These people are neither insiders nor outsiders but standing somewhere in between. Nevertheless, theirs is obviously a highly relevant standpoint and the opinions of such people amount to a corroboration of my descriptions and conclusions.

Vaughan's study (1996) of the *Challenger* space shuttle disaster was also ethnographic in this sense. She makes use of inquiry transcripts, but more than this, she draws on numerous documents created before the event, as well as her own interviews. Her very extensive sources meant that she was able to study NASA's culture with an intensity that is normally only possible when the researcher works or even lives within the culture of interest. She describes her work as a "historical ethnography" (1996: 61). She says, "My work was in harmony with the work of many social historians and anthropologists who use documents to examine how cultures shape ways of thinking" (1996, p. 61). Valuable though her method is, it comes at a price: Vaughan's study took her nine years to complete!

GENERALISABLITY

Desktop ethnographies of the type I have described raise a particular problem not raised by more conventional ethnographies. Conventional ethnographies do not normally claim that their findings are generalisable to other contexts. Desktop ethnographies, on the other hand, seek to make recommendations to the company concerned and to other companies that face similar hazards. Such recommendations depend implicitly on the generalisability of the findings. Strictly speaking the findings from a case study cannot be generalised to a wider population in a way that can be done with a statistically representative sample. However, we can argue as follows. The findings from a desktop ethnography highlight things that can go wrong, which companies need to consider carefully to avoid similar disasters. They are signposts to things that may need attention. Acting on these findings cannot guarantee that disasters will be prevented. Nevertheless, if a series of case studies yield similar conclusions, our confidence in those conclusions is correspondingly enhanced.

In the real world, decision-makers must act on the basis of the information available to them. They cannot wait for statistically valid conclusions before acting. The findings from desktop ethnographies may indeed be the best information available.

TURNER'S METHOD

Consider, now, Turner's research methods. He did not set out to explain a single disaster; his purpose was to develop a theory of man-made disasters. Accordingly, he chose to carry out three initial case studies, not one. These three were selected on the basis that "a detailed published account was available in the form of an extensive report" (1997: 39). He did not examine inquiry transcripts or other documents. This meant, as Vaughan puts it (1996: 411), that Turner "did not have access to data that allowed him to examine decision-making within a social context." Yet he was able to immerse himself in the world of the participants sufficiently to identify a range of beliefs and understandings that prevented people from seeing what was there. As he put it so eloquently, employees develop "a way of seeing [that] is also a way of not seeing" (1997: 49). This was an impressive exercise of what we might call the sociological imagination (Mills, 1959). It is what Max Weber called *verstehen* and Karl Weick called *sense-making*. It is this that enables us to talk of Turner's method as ethnographic, albeit, desktop ethnography.

But there is more to Turner's method than this. Because he was engaged in theory building, rather than simply explaining a particular case, the question of generalisability arises. If by chance the cases he selected were highly idiosyncratic, his theory would be correspondingly limited. To deal with this possibility he selected a further ten reports of public inquiries into major accidents and scrutinised them in the same way as he had the first three. "The resulting analyses provided no contradictory cases to suggest that the approach already taken should be revised substantially." (1997: 68). In fact, Turner looked at a total of 84 reports, but it is unclear what if any use he made of them, beyond those mentioned above. It is apparent from his list that many of these additional reports would have been far too brief to yield much in the way of sociological insights.

This aspect of Turner's method can be described as *saturation sampling*. This involves the researcher sampling new data points until no new information is forthcoming. At this point the researcher can be reasonably confident of the broad applicability of the findings. In Turner's case, he needed only three case studies to reach saturation point. Additional cases yielded no new insights.

Pidgeon describes Turner's work is as an exercise in grounded theory (1997: xvii). This is an approach in which a researcher starts with a limited data set and seeks to identify emergent concepts and theories. The process involves the construction of new theory from newly assembled data, rather than the testing of an existing theory with new data. The new theory can then be refined using a larger data set. There is also an elaborate set of procedures that go with this approach. Turner himself does not use the term *grounded theory* in the book and he provides very few clues as to how he constructed his theory from the data. He later wrote about grounded theory in a more explicit way (Turner and Martin, 1986), but my guess is that, in his research for *Man-Made Disasters,* he relied more on the sociological imagination than on the rules and procedures of grounded theory (see also Le Coze, Chapter 4, this volume).

CONCLUSION

Rosenthal, in his foreword to the 1997 edition, argues that *Man-Made Disasters* has been an "influential book." This is somewhat at odds with Vaughan's observations, in *her* foreword to the same edition, that *Man-Made Disasters* was seldom cited, at least in the US, and "failed to become integral to mainstream sociology." Later, perhaps, scholars began to look back and connect their own work with Turner's (e.g. Dekker and Pruchnicki, 2014), but in the years following the original publication, that was hardly the case. Rather than *giving rise to* the series of sociological studies of disaster that followed, Turner was more a forerunner, of whom those who followed had relatively little awareness. They followed the same path but not in his footsteps. That path was the discipline of sociology. The conceptual apparatus of sociology was well suited to studying the sociotechnical disasters that came to prominence in the 1980s, and sociologists applied themselves to this subject, at least initially, without reference to one who had walked this path not long before. However, the very existence of this volume testifies to Turner's increasing influence over time. Perhaps, 40 years on, we can now speak of him as the founding father of sociological studies of disaster.

As to Turner's *methodological* legacy, I suspect that it is only when accident analysts reflect on their work retrospectively that they trace continuities to Turner's work. In so far as that is the case, these continuities are not so much a result of Turner's legacy as they are of the common disciplinary tradition of these analysts. In particular, insofar as ethnography has become a recognised method in the safety science community, this is attributable to the sociologists and anthropologists who are currently part of this community, rather than to the legacy of Barry Turner.

Historians of ideas who trace the development of ideas over time do not automatically assume that this development is causal, in the sense that earlier ideas give rise to later ideas. Rather, they allow for the possibility that ideas might be a product of their times. Similarly, we must be wary of assuming a direct causal connection between the work of earlier scholars and those who come later.

REFERENCES

Bourrier, M., 1998. Elements for designing a self-correcting organisation: Examples from nuclear plants. In: Hale, A., & Baram, M. (Eds.), *Safety Management: The Challenge of Change*. Pergamon, Oxford, pp. 133–146.

CAIB (Columbia Accident Investigation Board), 2003. *Report*, Vol. 1. NASA, Washington.

Carson, W. G., 1982. *The Other Price of Britain's Oil*. Martin Robertson, Oxford.

Dekker, S. & Pruchnicki, S., 2014. Drifting into failure: Theorising the dynamics of disaster incubation. *Theoretical Issues in Ergonomics Science*, 15(6), 534–544.

Freeman, D., 1983. *Margaret Mead and Samoa: The Making and Unmaking of an Anthropological Myth.* Harvard University Press, Cambridge.

Geertz, C., 1973. *The Interpretation of Cultures.* Basic, New York. See especially chapter 1, Thick description: towards an interpretive theory of culture.

Goffman, E., 1968. *Asylums.* Penguin, Harmondsworth.

Gouldner, A. W., 1954. *Patterns of industrial bureaucracy.* Free Press.

Haukelid, K. 2008. Theories of (safety) culture revisited – an anthropological approach. *Safety Science*, 46, 413–426.

Hopkins, A., 1999. *Managing Major Hazards: The Moura Mine Disaster.* Allen & Unwin, Sydney.

Hopkins, A., 1999a. The limits of normal accident theory. *Safety Science*, 32, 93–102.

Hopkins, A., 2000. *Lessons from Longford: The Esso Gas Plant Explosion.* CCH, Sydney.

Hopkins, A., 2001. Was three mile island a normal accident? *Journal of Contingencies and Crisis Management*, 9(2), 65–72.

Hopkins, A., 2008. *Failure to Learn: The BP Texas City Refinery Disaster.* CCH, Sydney.

Hopkins, A., 2012. *Disastrous Decisions: The Human and Organisational Causes of the Gulf of Mexico Blowout.* CCH, Sydney.

Hopkins, A., 2016. *Quiet Outrage: The Way of a Sociologist.* CCH, Sydney.

Hopkins, A., 2019. *Organising for Safety: How Structure Creates Culture.* CCH, Sydney.

Janis, I., 1982. *Groupthink: Psychological Studies of Policy Decisions and Fiascos.* Houghton, Boston.

Kriegler, R., 1980. *Working for the Company.* Oxford UP, Melbourne.

LaPorte, T., & Consolini, P., 1991. Working in practice but not in theory: The theoretical challenges of "high-reliability organisations". *Journal of Public Administration Research and Theory*, 1(1), 19–47.

LaPorte, T., & Rochlin, G., 1994. A rejoinder to Perrow. *Journal of Contingencies and Crisis Management*, 2(4), 221–227.

Le Coze, J. C., 2013. Reflecting on Jens Rasmussen's legacy. A strong program for a hard problem. *Safety Science*, 71, 123–141.

Le Coze, J. C., 2015. Reflecting on Jens Rasmussen's legacy (2) behind and beyond," a "constructivist turn". *Applied Ergonomics*, 59(Pt B), 558–569.

Macrae, C., 2014. *Close Calls: Managing Risk and Resilience in Airline Flight Safety.* Palgrave, New York.

Mills, C. W., 1959. *The Sociological Imagination.* Oxford UP, Oxford.

Nichols, T., 1997. *The Sociology of Industrial Injury.* Mansell, London.

Parker, D., Lawrie, M., & Hudson, P., 2006. A framework for understanding the development of organisational safety culture. *Safety Science*, 44(6), 551–562.

Perrow, C., 1984. *Normal Accidents: Living with High-Risk Technologies.* Basic, New York.

Perrow, C., 1994. The limits of safety: The enhancement of a theory of accidents. *Journal of Contingencies and Crisis Management*, 4(2), 212–220.

Rasmussen, J., 1997. Risk management in a dynamic society: A modelling problem. *Safety Science*, 27(2/3), 183–213.

Reason, J., 1990. *Human Error.* Cambridge University Press, Cambridge.

Reason, J., 1997. *Managing the Risks of Organisational Accidents.* Ashgate, Aldershot.

Reason, J., 2013. *A Life in Error: From Little Slips to Big Disasters.* Ashgate, Farnham.

Schein, E., 1992. *Organisational Culture and Leadership*, 2nd ed. Jossey-Bass, San Francisco.

Snook, S., 2000. *Friendly Fire.* Princeton UP, Princeton.

Turner, B., & Martin, P., 1986. Grounded theory and organisational research. *The Journal of Applied Behavioural Science*, 22(2), 141–157.

Turner, B., & Pidgeon, N., 1997. *Man-Made Disasters*, 2nd ed. Wykeham, Oxford.

Vaughan, D., 1996. *The Challenger Launch Decision: Risky Technology, Culture and Deviance at NASA*. University of Chicago Press, Chicago.

Viner, D., 1991. *Accident Analysis and Risk Control*. Viner, East Ivanhoe.

Weick, K., & Sutcliffe, K., 2001. *Managing the Unexpected: Assuring High Performance in an Age of Complexity*. Jossey-Bass, San Francisco.

Westrum, R., 1999. *Sidewinder: Creative Missile Design at China Lake*. Naval Institute Press, Annapolis.

Westrum, R., 2004. A typology of organisational cultures. *Quality and Safety in Health Care*, 13(Suppl II), ii22–ii27.

Westrum, R., 2014. The study of information flow: a personal journey. *Safety Science*, 67(2014), 58–63.

3 Access, Methods and Knowledge When Exploring High-Hazard Systems

Lessons from the Past for the Future

Mathilde Bourrier[1]

CONTENTS

INTRODUCTION

There has been considerable renewed interest in questioning how we know the things we know (Becker, 2017). Over the past three decades social scientists who use field methods have been at the same time more willing and required to report carefully on how they enter the field and collect data, and what is their position in the field they study. Organisation theorists have especially encountered challenges in a world of so-called institutional transparency, where access to organisations has ironically proven to be more and more time-consuming and controlled, if not at all impossible.

Politics of inquiry and knowledge lie at the core of ethnographic methodologies, which require immersion in the field, to be able to conduct contextualised interviews and observations at the workplace.

High-hazard organisations, where safety is paramount, are one subset of high-profile organisations. A range of researchers and theorists have been influential, reaching back to the early work initiated by Turner, followed by Perrow, La Porte, Rochlin,[2] Schulman, Roberts, Weick and Vaughan, to name a few. However, it should be noted upfront that many of their works contributed greatly to the development of knowledge on organisations in general. Safety and risk scholars have been producing some iconic concepts, such as *normal accident* (Perrow, 1984), *high reliability organisation* (La Porte et al., 1987; La Porte, 1996), *necessary violations* (Reason, 1987), *normalisation of deviance* (Vaughan, 1997), *resilience engineering* (Hollnagel et al., 2007) and *sensemaking* (Weick, 1995). These contributed greatly to documenting both agency failure and organisational factors leading to fatal accidents and disasters.

The methodologies that the studies of accidents, risk and safety have promoted are worth looking at in greater detail. While they are all qualitative in nature, they display different properties. Some scholars have extensively used so-called field methods while others relied on documents, public reports, archives and secondary sources. Some, like Turner, Perrow and Lagadec, have used reports of incidents or public inquiries to build their conceptual narratives. Others have used secondary sources, fieldwork done by others and non-fiction essays to forge their theories. This is the case of Weick's influential work. A third category has conducted fieldwork extensively: for example, La Porte and colleagues conducted "normal operations" studies while Vaughan studied the *Challenger* accident in its aftermath, using oral history methodologies as well as archives. Under the label of qualitative methodologies, different types of data collection exist. Few are "true" ethnographies.

Ethnographic methods have gained a growing attraction at the end of the 1990s throughout the 2000s when researchers embarked on studying high-hazard industries from the inside (Bourrier, 1996; Perin, 2005; Pettersen, 2008). However, these methodologies require specific access and trust that are not easy to obtain. Calling for the development of proper qualitative methods, and especially field methods, needs to be paired with a lucid reflection on the conditions under which social scientists are able to access high-hazard organisations today. This is the purpose of this chapter, to reflect on the conditions under which such development can be fruitfully sustained and developed further.

THE LONG MARCH OF FIELD METHODOLOGIES

This chapter aims at reflecting on the nature of empirical evidence that has been produced over the past four decades on high-hazard organisations. What constitutes a solid piece of evidence? Along with this interrogation comes the context within which evidence is gathered. The data collection process, using quantitative as well as qualitative or mixed methodologies, is paramount to any attempt at producing valid and innovative knowledge (see also Kongsvik and Almklov, Chapter 9; Shulman,

Chapter 10 and Martin et al., Chapter 11, all this volume). The subject of risks, crises and disasters makes it all too important to reflect on the evidence harvested to build adequate contextual knowledge in these matters. It is especially important because this field is often plagued with controversies, pros and cons, vocal advocates and equally vocal opponents, a true *Latourian* battlefield of choice (Latour, 1987). This is the case, for example, for the chemical, nuclear and off-shore drilling industries. Hospitals fall also under the category of high-hazard organisations, where human errors are tracked down and subject to detailed scrutiny. The controversy landscape requires social scientists to document, to inquire, to balance, to double check, to triangulate and to contextualise the situations they are witnessing with the utmost rigor. For example, what could be considered as deviations, short-cuts and small infringements of the rules have to be understood in the proper working context. This is exactly what social sciences are designed for: providing evidence on the complex and imperfect conditions of achieving the most mundane as well as the most critical tasks.

Social science is data driven. Qualitative methodologies are historically rooted in anthropology and ethnography, under the influence of Malinowski (2013). A long battle and co-existence opposed classically "qualitative methodologies" embraced by the Chicago school in the 1940s (Blumer, Hughes, Park, Znaniccki and later Becker) and the *survey research*, when Lazarsfeld created the Bureau of Applied Research at the University of Columbia in 1944. This bureau has been for many years the principal nest of critics against qualitative methodologists. The critics revolved around the positioning in the field of the qualitative researcher, their subjectivity, the difficulty to generalise based on the data collected and the absence of control over the sample observed. Ironically, as many as the monographs published by the Chicago scholars became classics in sociology, up until the 1990s the survey research paradigm remained dominant in the US as well as in Europe, with the exception of ethnomethodologists like Circourel and grounded theorists like Glaser and Strauss, who in the 1970s gave a new impetus to qualitative methodologies. Qualitative methodologies finally regained strength in the 1990s, especially in sociology and notably in the sociology of organisations (Crompton and Jones, 1988).

Social scientists who started using field methods were expected to dutifully report carefully on how they enter the field, on how they collect data and on their positioning in the field they study (Patton, 2002). Yet, for a long time organisation theorists remained silent on their research tactics, with a few notable exceptions (Broadhead and Rist, 1976; Bryman, 1988; Crompton and Jones, 1988; Cifkin, 2010; Feldman et al., 2003). The social proximity with their research subjects (same sociological milieu, same circles) or the contrary (the social distance between social scientists and their informants), the tedious negotiation processes to get access to well-guarded organisations (Monahan and Fisher, 2015), the advisor-to-the prince's position that social scientists seek sometimes to facilitate their access to top managers, are usually not disclosed, or at best mentioned in passing. For a long time (up until the end of the nineties), they were not deemed of any interest and considered as the secret garden of the researcher. They were not even taught to students. Crozier, the French scholar of bureaucratic systems never really disclosed the acquaintances he had in different segments of the French bureaucracy. The creation of a cooperative to help American sociologist Gouldner secure access to fieldwork is not well known. Goffman also had special links to the organisations he was studying, with Saint Elizabeth's Hospital, Washington, DC, where his wife was treated, being the most famous case.

Today, the situation is much to the contrary and things have changed. Scholars do recognise that the kind of access one gets to the field heavily impacts the kind of qualitative analysis that one will be able to produce in the end. The field of organisational ethnography has now produced numerous monographs based on qualitative enquiries, aiming at understanding complex organisations, important for society at large. In 1991, organisational theorist Perrow argued that

> organizations are the key to society because large organizations have absorbed society. They have vacuumed up a good part of what we have always thought of as society, and made organizations, once a part of society, into a surrogate of society. (p. 726)

This is true for Wall Street firms in *Liquidated, an Ethnography of Wall Street* (Ho, 2009), this is true for the British Broadcasting Company, BBC (Born, 2011), this is true for the World Trade Organization (Abelès, 2011) and for the World Bank (Goldman, 2006; Weaver, 2008), to name only a few. Each of these inquiries has required tremendous ingeniousness on the part of the researchers to be able to collect primary data and evidence, capable of shedding light on the working processes of these large and complex organisations. It also included a good dose of risk taking on the part of the researcher. Some of this fieldwork would have very radical consequences: Ho never went back to Wall Street and declared[3] that she destroyed part of her data set; Goldman has been banned and will not be able to set foot again on the World Bank premises; Abelès and his group has not returned to WTO. Part of the explanation of these situations has to do with the critical stand that organisational anthropologists end up adopting. But this does not tell the whole truth. One can also argue that this critical posture grew even stronger because of the challenges (including being stonewalled) that researchers faced while negotiating their access to these organisations.

Field methodologies have also been increasingly applied to study hazardous systems. They clearly have gained their *lettres de noblesse*. At the same time, the conditions under which they can be applied and developed within the organisational landscape we have now are raising numerous difficult questions, not easily solved.

HOW HAVE HIGH-HAZARD ORGANISATIONS BEEN STUDIED?

High-hazard organisations have been studied in a variety of ways, from the reanalysis of official accident reports to ethnographic studies. Before exploring these issues in more depth, it is first important to consider whether high-hazard organisations represent a separate category of organisational phenomena or not.

HIGH-HAZARD ORGANISATIONS: A SPECIAL SUBSET OR NOT?

Scholars in risk and safety have uncovered and described many challenges that these organisations face daily. They are described as "complex" (Perrow, 1983), "unlikely, demanding and at-risk" (La Porte, 1996). Embarking on their study, one can pretend that they do not differ that much from other types, displaying the same traps and intrinsic vulnerabilities as most organisations do. Yet the consequences of organisational failures within high risk organisations make them exceptional. Hence, they do display specific peculiar characteristics that justify a category in itself. Perrow, Lagadec and La Porte would probably agree on this last position.

Based on my own work, I believe that part of what one observes in high-hazard organisations can also be found in many other organisations, providing they are complex enough. A few examples might clarify this point. The relentless drive to more proceduralisation, the importance put on planning and scheduling activities, the burden of social control at the workplace, the heavy hand of regulation, the "risk management of everything" (Power, 2004) and the constant challenge to manage a very diverse workforce are all basic features of many contemporary organisations (Bieder and Bourrier, 2013). If we only examine the heterogeneity element, the workforce is diverse in different ways: (i) diverse in terms of status – in-house employees versus contractors (i.e. "strangers" and "nonmembers" to the workplace in Turner's words, 1976) and (ii) diverse in terms of skills, crafts and professions. The coordination of these groups constitutes a daily challenge. Steffy (2010) argued that in the case of the *Deepwater Horizon*'s blowout, the fragmented workforce was an important contributing factor to the explosion. No less than nine firms had employees on site (BP; Transocean; Halliburton; Cameron; DRIL-QUIP; M-I; Schlumberger; Sperry Drilling; Weatherford). In addition, these specific groups are usually not substitutable.

So, on one hand, some scholars have observed that these high-hazard organisations do not differ that much from any large technical systems that entail less risk. On the other hand, sociologists, political scientists, psychologists and safety scientists produced some iconic concepts, such as *normal accident* (Perrow, 1984), *high reliability organisation* (La Porte et al., 1987), *necessary violations* (Reason, 1987), *normalisation of deviance* (Vaughan, 1997), *sensemaking* (Weick, 1995) and *drift into failure* (Dekker, 2016), to name only a few, tailored to the study of high-hazard organisations. These notions contributed greatly to expand our knowledge on fatal processes leading up to accidents and disasters. They also provided a language and a conceptual understanding of the phenomena at play.

EARLY DATA COLLECTION ON HIGH-HAZARD ORGANISATIONS

These concepts were supported by the collection of specific empirical evidence. Qualitative data analysis software such as *Nudist, Nvivo* or *Atlas Ti* were nonexistent at the time. *Man-Made Disasters* is based upon a detailed analysis of eighty-four official reports into accidents and disasters published by the British Government from 1965 to 1975 (Turner, 1976). The data analysis was carried out by means of the qualitative method of "grounded theory." This qualitative analysis presents itself as a documentary-based analysis of the organisational pre-conditions of large-scale accidents (see also Le Coze, Chapter 4, this volume). Perrow (1984), Lagadec (1982) and Sagan (1993) have used official reports of accidents and public inquiries to build their concepts. They have built a metric to address these different cases across their specificities. Vaughan used historical ethnography to study the *Challenger* accident in its aftermath. She studied the case through the NASA archives and conducted some interviews with participants at the time. Weick has used fieldwork done by others (especially in the flight-deck piece, Roberts and Weick, 1995) or historical account of specific disasters (the Mann Gulch's case, Weick, 1993 and the South Canyon's case, Weick, 1996). These help him forge his own narrative, tailoring medium range concepts capable of addressing broader topics of organisational life. In Weick's work, the accident is a pretext to engage the reader with the allegorical quality of the story.

These well-known studies and famous concepts and notions are not based on the collection of ethnographic materials, strictly speaking. They are mainly based on after-the-fact official reports and archives. The social shaping of such documents has not been overlooked by Vaughan (2006). She signals in her comparison of three different reports (Report to the President by the Presidential Commission on the Space Shuttle *Challenger* Accident; The *Columbia* Accident Investigation Board Report; The 9/11 Commission Final Report of the National Commission on Terrorists Attacks upon the United States) that they are in themselves the social product of choices made inside the commissions in order to tackle the events. They are not raw materials; they present a certain narrative of the events, a certain frame. The Rogers Commission adopted an "accident traditional investigation" frame NASA was familiar with. It focused attention on technical causes and human factors. In the CAIB case, the report adopted an "organizational-system failure causal model," which identified causes at the institutional, organisational and individual levels. Finally, the frame of the 9/11 Commission report was a "historical/war frame [*where*] the causal model is regulatory failure." Consequently, what we learn through accident investigation reports should be put in the perspective of the frames at play.

Obviously, this is not to pretend that ethnographic materials present themselves as "pure" elements, lacking biases. They do not. However, using secondary sources, like official reports, is not exactly the same thing as embarking oneself on ethnographic fieldwork. Twenty years ago, Gherardi commenting on the 20th anniversary of *Man-Made Disasters* in 1998 (p. 83) was openly reflecting "on how little we know about social construction of safety, about reliability and about social responsibility towards present and future generations" (Gherardi, 1998, p. 83). It is probably accurate to acknowledge the fact that HRO literature was hoping to fill in this gap. Methodologies applied to the study of high-hazard organisations have evolved since then; however, the "significant cultural turn" (Pettersen, 2020) within the field has probably not yet delivered its full potential.

FIELD METHODS FOR STUDYING NORMAL OPERATIONS

La Porte and colleagues (Rochlin, Roberts, Schulman) have been among the first to conduct ethnographic fieldwork aboard: (i) The Air Traffic Control System (Federal Aviation Administration); (ii) Electric Operations and Power Generation Departments (Pacific Gas and Electric Company) and (iii) the peacetime flight operations of the

US Navy's Carrier Group 3 and its two nuclear aircraft carriers USS *Enterprise* (CVN 65) and USS *Carl Vinson* (CVN 70). Later, the nuclear production at PGE's Diablo Canyon plant (Pacific Gas and Electric Company) was included. These pioneering fieldworks were carried out from the late 1980s to the early 1990s. They have been followed by subsequent works done by members of the initial research group: Rochlin (with Von Meier, 1994) at Swiss and German nuclear power plants; Roe and Schulman (2008) on California Independent System operator (CAISO), Roberts on fire-fighters (Vidal and Roberts, 2014) and later expanding the territory of the class of high reliability organisations to hospitals (Madsen et al., 2006); and work done by outside members and fellow-travellers (Demchak, 1991; Bourrier, 1999; Perin, 2005; Pettersen, 2008).

Obviously, this kind of ethnographic fieldwork is seldom possible in the midst of a crisis or during an accident. One exception is the work done on CAISO, where Schulman and Roe started their fieldwork in the midst of the California electricity crisis (2000–2001), during which power shortages developed as wholesale markets failed to produce needed capacity. "Normal operations" studies (Bourrier, 2002) gave substantial credibility and further ground to Turner's identification of incubation periods. Since "disasters were not sudden cataclysmic events; they had long gestation periods" (Vaughan, 1997, p. xii), it became urgent to engage in field studies *au long cours*. If agency failures are rooted in organisational daily life, and if equally not all high-hazard organisations encounter severe accidents, then why not carefully document the type of working relationships and agency that lie at the core of reliability seeking organisations (Rochlin, 1993)? Concepts such as *incubation period*, *normalisation of deviance*, *drift into failure*, *latent failure conditions* and *weak signals* urge us to study the functioning of organisations during peacetime operations. Disasters do not happen in a clear blue sky. A complex web of factors interact and produce, once in a while, a dangerous event. Clarke (2001) once explained that if people die the way they live, the same is true about organisations: they fail the way they live. He contends that one should not expect exceptional organisations to save us and automatically come to the rescue. Organisations will fail us. However, we do not have much choice but to accommodate ourselves with fallible organisations. Consequently, disaster mitigation strategies should use the bedrock of more common or mundane strategies, which allow employees to go through the day, resorting to a degraded mode when necessary. In this view, there is a continuum (not a disjunction) between catastrophic *and* reliable outcomes. They share components. Reason (1987) also introduced the field to the idea that good performance and error are two faces of the same coin. The resilience engineering agenda (Hollnagel et al., 2007) is also sensitive to this idea: organisations should be encouraged to report on successes and not only focus on errors.

The HRO scholars have not described in detail what their methodological intentions were, but their willingness to immerse themselves within these organisations at work dramatically renewed the perspective. As Rochlin observed:

The challenge is to gain a better understanding of the interactive dynamics of action and agency in these and similar organizations and the means by which they are created and maintained. These arise as much from interpersonal and intergroup interaction as from more commonly studied interactions with external designers and regulators. The interaction is a social construction anchored in cultural dynamics, therefore there are wide variances in its manifestation even for similar plants in roughly similar settings. (Rochlin, 1999, p. 1558)

Their theoretical attempts, post-TMI, aimed at making sense of the "surprises in the field" (i.e. the complex and largely unpredictable interactions between technology and humans in very demanding systems). The idea was to move away from a stereo-typed view of daily operations in order to create a much more complex and rich view on the social production of safety (Bourrier, 2011; Le Coze et al., 2014).

The focus is now on *work as done* and not *on work as imagined*. HRO's line of research has revealed the potential of studying daily operations as opposed to *a posteriori* analyses of major accidents. Resorting to ethnographic and sociological methodologies gave new vitality to the study of such complex organisations. HRO literature encouraged scholars to venture across industries, public and private and even across countries. However, the conditions under which researchers can inde-pendently access these highly guarded organisations has not been treated in detail by HRO scholars. A current assessment of these conditions follows in the next section.

WHERE ARE WE NOW?

DEALING WITH GATEKEEPERS

Calling for qualitative methods to be used more often brings onto the table the issue of access. This is a challenge where we probably have not made as much progress as we could have. Partly this is because, ethnographic methodology which we think is best positioned to understand the social construction of safety and reliability is still a rarity. Access today is not granted more easily than yesterday. Sociologists or political scientists, anthropologists or safety scientists using ethnographic meth-odologies are not enjoying wide and legitimate access, like engineers or managers do for example. One can recall that Rees (1996) had no authorisation to study the Institute of Nuclear Power Operations from the inside and had to resort to after-work interviews, where some employees accepted to talk to him in cafés. Access is usu-ally more easily granted *after the fact*, *after an accident*. Vaughan recalled how and under what conditions, she has been called to come on board to help out with the official inquiry after the Columbia's accident (Vaughan, 2006). But she had previ-ously reported on the fact that no one at NASA invited her to present her book on the *Challenger* launch decision, when it came out in 1997.

However, some researchers have enjoyed a very privileged access. This is the case of Schulman and Roe (2008) at CAISO operations in California. The researchers stayed at CAISO from 2001 to 2008 and confide that:

During the entire case study period and beyond we have been given extraordinary access, as researchers, to CAISO control room operations on a daily basis, to frequent meetings at many organizational levels, and to many key individuals for interviews (many over multiple periods). At no point were we asked to limit our study or modify our findings. At no point were we asked to leave a meeting or not listen to a conversation. It is unlikely that any large organization with such high reliability mandates has been as intensively and extensively studied. (Roe and Schulman, 2008, p. 18)

La Porte and colleagues used the word "conversations" to talk about their fieldwork, indicating that the level of engagement and trust that they had built was of great quality.

But more often, the conditions under which researchers access these high-hazard organisations is controlled by heavy-handed gatekeepers. This is, by the way, not only the case of high-hazard organisations. Usually, high-profile organisations display the same prevention towards social scientists. As an example, anthropologist Marcus, trying to get access to the World Trade Organization, in Geneva, using his "para-site methodology" confessed that the access his team was granted was far from satisfactory (Deeb and Marcus, 2011). Protective gatekeepers are often in the way, a phenomenon identified long ago (Broadhead and Rist, 1976).

In high-hazard organisations, researchers would generally gain access through three main routes. One is at the invitation of unions or workers' organisations, especially in Europe. This type of access is on the decline. But powerful unions interested in an evaluation of changing working conditions might still turn to social scientists (sociologists of work, psycho-sociologists of work and ergonomists of work) to help them assess a situation in transition. The second route is through management, and is often linked to some counselling activities, or an action-research type of design (see also Kongsvik and Almklov, Chapter 9, this volume; and Hayes and Maslen, Chapter 12, this volume). Issues of change management, corporate culture, norms and values are generally the drivers of such an interest. Sociologists of organisations, anthropologists, political scientists and management specialists are usually consulted by managers and it can end up in fruitful long-term relationship. The third one is generated by collaborations with system safety specialists, including human factors specialists. New thinking, innovation and design around work environments are usually the drivers of this cooperation. All of these groups (union organisations, management and safety specialists) represent powerful gatekeepers. Let us examine in more details the third category (safety specialists), who, we would think, could be the natural allies of social scientists.

Safety and human factors specialists are usually present in high-hazard organisations, centrally located at headquarters and/or decentralised at a plant's site. They are generally interested in the social construction of safety and vulnerability, and have field and case studies close to their interests. They could certainly act as an interesting go-between between qualitative researchers and their colleagues. They could also be natural participants in ethnographic studies, for example. However, most of the time, their position is not that strong inside these organisations. It was the case when Perrow wrote about them in 1983 and it is still the case nowadays.

Their weakened position signals that their worldviews are generally marginalised. Consequently, they cannot always be mobilised to help structure a social science perspective on high-hazard organisations and the independent and legitimate access they might provide to such important organisations seems to be fragile, most of the time. Again, this is not only the case for high-hazard organisations. Let us examine the types of accesses that researchers can enjoy nevertheless.

Types of Accesses

Despite obstacles and difficulties, let us first observe that researchers are nowadays usually better prepared when undertaking fieldwork in these industries and organisations. The externalities as well as opportunities that specific roles entail when studying organisations are better weighed in, accepted and reflected upon more candidly. These issues are now present in teaching curricula, disclosed in research papers and talked about at conferences and other academic forums. And practical orientation to the various hats that researchers might need to wear exists along with sensible pragmatic advice (Bryman, 1988; Deny and Sunderland, 2014; Feldman et al., 2003; Cifkin, 2010).

Second, access is linked to the type of position that the researcher chooses to adopt. This choice is not entirely their own. It can be influenced by requirements that are associated with designing practical interventions of some kind, which may be the only possible way of engaging with the functioning of the organisation of interest. Doors do not open easily to the social scientist who declares *only* a genuine interest in an organisation's operations. It can certainly happen; however it is not often the case.

Anthropologists and sociologists can also learn from other fields. In fact, ergonomists, psychologists and management specialists are used to designing specific interventions tailored to the needs of their interlocutors within an organisation. They usually only intervene at a specific request: they have a mandate. Moreover, this is part of their epistemological stance. But this is less the case in social science disciplines. Science and technology studies and sociological, ethnographical, and anthropological studies in general require and value a maximum freedom of manoeuvring inside an organisation in order to collect data through interviews, observations, attendance at meetings, or on-the-job shadowing of employees. However, such an open-door policy has become a rarity. Nowadays, social scientists have to use cunning and generally compose a different kind of approach. They usually end up designing research protocols closer to "policy advice," "technology assessment" and "action research."

One should add to the process the necessity to include the long ethical review procedures that are required nowadays to satisfy Internal Review Boards, both internally within our own research institutions and externally within the institutions under study (Bosk and De Vries, 2008). It means that there is a contractual nature to the relationship and the expectations on both sides. The researcher is expected to deliver advice, to write recommendations or to help design a programme. This can certainly serve as a Trojan horse, and this type of access, although it is controlled and heavily guided, can later generate evidence to produce a more scholarly work that is free of explicit demands and expectations from the organisation. It also allows some degree of embeddedness to be feasible. The mandate that sustains the researcher's presence can cut both ways: it certainly restrains the free range of data collection; however, it also provides the necessary legitimacy to ask questions and meet employees, with an official purpose. It helps circumvent the too common: "No admittance, except on Business."

We do not advocate for this latter type of accesses and would rather see social scientists welcomed wherever they deem that their perspective might shed light on overlooked aspects. At the same time, hoping for free access does not seem to be a realistic option, particularly for more junior researchers. Hence, finding ways to continue to stay closer to the people working in high-hazard organisations, without compromising one's capacity to remain a benevolent observer, is of crucial importance. One good way to practice this alertness is to stay aware of the conditions under which data are collected and produced. This might seem to be too generic advice. So, to be a bit more specific: the key here is to build a portfolio of possible interventions, roles and hats that the researcher can play with, depending on the type of access that might possibly be negotiated. Another avenue to explore for access to large, technical, high-hazard, high-profile organisations is also to favour comparative studies across industries, countries and regions of the world. The plea for more of this type of studies has been made years ago (Pidgeon, 1997, p. xvi), but it still needs some reinforcement.

A Plea for More International and Cross-Industrial Studies

Field studies are often accused of being time-consuming and the fact that the research methods cannot be entirely described because the work is largely inductive can make this work more difficult. Qualitative social research generates large amounts of non-standard data which make analysis problematic and replication even more problematic. Turner (1976) was advocating for cross-sectoral studies to avoid building peculiar theories based on specific and non-generalisable fieldwork. However, in reality, cross-sectoral studies are still too few. This is still something that has to be supported and further pushed for. The same is true with international studies, across sectors and across continents and countries. This does not materialise often. Comparisons offer great opportunities to get to the bottom of generic issues and help researchers to decentre and see beyond taken for granted notions, artefacts and processes.

What is currently lacking is a strong program to replicate and develop these ethnographic methods across industries, organisations, sectors, cultures and countries with similar questions in mind. Researchers often built their own theories and use local fieldwork to back up their intuitions. This has so far produced many famous case studies of accident investigations. There has, over the last 30 years, been a tendency for the production of monographs driven by major accidents and their detailed reports. In addition, interesting monographs with some elements of comparison – this is true for the HRO cases of study, for example (Rochlin and Von Meier, 1994) – have also been written in the format of case studies, since it is a very elegant way to organise narratively diverse fieldwork elements (Roe and Schulman, 2008; Macrae, 2014; Becker, 2014; see also Antonsen and Haavik, this volume). Other researchers have also succeeded in producing cross-national comparisons (Bourrier, 1999; Wiig et al., 2019; Bochatay, 2018). However, this last category still remains too limited.

We propose that one way to circumvent tedious negotiations with global organisations at the local level would be to academically match their global nature, and present oneself as a member of a larger global consortium. That way, the benchmarking aspects that are possible through applying social science perspective could be more readable to the management of these organisations. At the same time, one should not overlook the fact that sometimes negotiations are easier when conducted at the local level, than at the regional, national or even international level. We also think that for the field to get more traction at the shop-floor level (and gain in practical relevance), there is a need for powerful concepts like normalisation of deviance, drift into failure, necessary violations, normal accidents, high reliability organisations or resilience engineering to be substantiated in many different contexts. The safety science field is not in need of more concepts with respects to the functioning of organisations. It needs more empirical evidence.

CONCLUSION

More than 20 years ago, Turner made a strong statement: "The organizations which pervade society have grown more dominant and more far-reaching over the past 200 years. More aspects of our individual and our social life are influenced by those who run such organizations than ever before" (Turner, 1997, p. 5). Social scientists still have to live up to his statement and make sure they are maintained in the loop. Challenges of access indicate the difficulties that exist for social science expertise to be fully accepted, integrated and welcomed in the circles where organisations and work are debated, proposed and decided upon. As an example, although organisational design lies at the core of the contemporary research agenda concerned with the safety of high-hazard organisations, rarely do we see organisational theorists or scholars of organisations invited to take part in designing new organisations, or reforming older ones. Decades of qualitative research have shown that the conditions under which high-hazard systems are kept safe are far from trivial and can be far removed from strict bureaucratic mechanisms. It entails a great deal of "friction" (Rochlin, 1998), bricolage, social invention and ad-hoc mitigation strategies that need to be documented and studied across industries, countries and methodologies. There is still much work to be done.

NOTES

1. Professor at the Department of Sociology and Institute of Sociological Research, Geneva School of Social Sciences, University of Geneva, mathilde.bourrier@unige.ch.
2. Gene Rochlin, member of the High Reliability Organizations' group at Berkeley, passed away over thanksgiving in 2018. Gene, trained as a nuclear physician, had a passion for anthropology and fieldwork methodologies. He was extremely driven by

their potential to study complex, high risk and demanding organizations. He encouraged and advised so many PhD students, including myself, to continue to stay as close as possible from those who work with truly dangerous stuff. His passion was very contagious. Field methodologies and field researchers in high-hazard industries owe him a great deal.

3. Private conversation with the author before she gave her speech at the University of Geneva, in March 17, 2010.

REFERENCES

Abelès, M. (2011). *Des anthropologues à l'OMC : scènes de la gouvernance mondiale*. Paris: CNRS Éditions.

Becker, H. S. (2014). *What about Mozart? What about Murder? Reasoning from Cases*. Chicago, IL: The University of Chicago Press.

Becker, H. S. (2017). *Evidence*. Chicago, IL: The University of Chicago Press.

Bieder, C., & Bourrier, M. (2013). *Trapping Safety into Rules: How Desirable or Avoidable is Proceduralization?* Farnham: Ashgate & CRC Press.

Bochatay, N. (2018). *Continuities in Health Care Work: Group Processes and Training at Two Academic Medical Centers in Switzerland and California* (Doctoral dissertation, University of Geneva).

Born, G. (2011). *Uncertain Vision: Birt, Dyke and the Reinvention of the BBC*. London: Random House.

Bosk, C., & De Vries, R. (2008). Bureaucracies of mass deception: Institutional review boards and the ethics of ethnographic research. In: C. Bosk (dir.), *What Would You Do? Juggling Bioethics and Ethnography*. Chicago, IL: University of Chicago Press, 87–99.

Bourrier, M. (1996). Organizing maintenance work at two American nuclear power plants. *Journal of Contingencies and Crisis Management*, 4(2), 104–112.

Bourrier, M. (1999). *Le nucléaire à l'épreuve de l'organisation*. Paris: Presses Universitaires de France.

Bourrier, M. (2002). Bridging research and practice: The challenge of "normal operations" studies. *Journal of Contingencies and Crisis Management*, 10(4), 173–180.

Bourrier, M. (2011). The legacy of the high reliability organization project. *Journal of Contingencies and Crisis Management*, 19(1), 9–13.

Broadhead, R. S., & Rist, R. C. (1976). Gatekeepers and the social control of social research. *Social Problems*, 23(3), 325–336.

Bryman, A. (1988). *Doing Research in Organizations*. London, NY: Routledge.

Cifkin, M. (Ed.). (2010). *Ethnography and the Corporate Encounter, Reflections on Research in and of Corporations*. New York and Oxford: Berghahn Books.

Clarke, L. (2001). *Mission Improbable: Using Fantasy Documents to Tame Disaster*. Chicago, IL: University of Chicago Press, 1999. Paperback 2001.

Crompton, R., & Jones, G. (1988). Researching white collar organizations: Why sociologists should not stop doing case studies. In: Alan Bryman (Ed.), *Doing Research in Organizations*. London, NY: Routledge, 68–81.

Dekker, S. (2016). *into failure: From hunting broken components to understanding complex systems*. Boca Raton, Fl.: CRC Press.

Deeb, H. N., & Marcus, G. E. (2011). In the green room: An experiment in ethnographic method at the WTO. *PoLAR: Political and Legal Anthropology Review*, 34(1), 51–76.

Demchak, C. C. (1991). *Military Organizations, Complex Machines: Modernization in the US Armed Services*. Ithaca, NY: Cornell University Press.

Denny, R., & Sunderland, P. (Eds.). (2014). *Handbook of Anthropology in Business*. Walnut Creek, CA: Routledge.

Feldman, M. S., Bell, J., & Berger, M. T. (Eds.). (2003). *Gaining Access*. Walnut Creek, CA: Altamira Press.

Gherardi, S. (1998). A cultural approach to disasters. *Journal of Contingencies and Crisis Management, 6*(2), 80–83.

Goldman, M. (2006). *Imperial Nature: The World Bank and Struggles for Social Justice in the Age of Globalization*. New Haven, CT: Yale University Press.

Ho, K. (2009). *Liquidated: An Ethnography of Wall Street*. Durham and London: Duke University Press.

Hollnagel, E., Woods, D. D., & Leveson, N. (2007). *Resilience Engineering: Concepts and Precepts*. Farnham: Ashgate Publishing, Ltd.

La Porte, T. (1996). High reliability organizations: Unlikely, demanding and at risk. *Journal of Contingencies and Crisis Management*, 4(2, June), 60–72.

La Porte, T. R., Roberts, K., & Rochlin, G. S. (1987). *High Reliability Organizations: The Research Challenge*. Mimeo, working paper.

Lagadec, P. (1982). *Major Technological Risk*. Oxford: Pergamon Press.

Latour, B. (1987). *Science in Action: How to Follow Scientists and Engineers through Society*. Harvard University Press.

Le Coze, J. C., Pettersen, K., & Reiman, T. (2014). The foundations of safety science. *Safety Science*, 67, 1–5.

Macrae, C. (2014). *Close Calls: Managing Risk and Resilience in Airline Flight Safety*. Berlin: Springer.

Madsen, P., Desai, V., Roberts, K., & Wong, D. (2006). Mitigating hazards through continuing design: The birth and evolution of a pediatric intensive care unit. *Organization Science*, 17(2), 239–248.

Malinowski, B. (2013). *Argonauts of the Western Pacific: An Account of Native Enterprise and Adventure in the Archipelagoes of Melanesian New Guinea [1922/1994]*. London: Routledge.

Monahan, T., & Fisher, J. A. (2015). Strategies for obtaining access to secretive or guarded organizations. *Journal of Contemporary Ethnography*, 44(6), 709–736.

Patton, M. Q. (2002). *Qualitative Research and Evaluation Methods*, 3rd edition. Thousand Oaks, CA: Sage.

Perin, C. (2005). *Shouldering Risks: The Culture of Control in the Nuclear Power Industry*. Princeton, NJ: Princeton University Press.

Perrow, C. (1983). The organizational context of human factors engineering. *Administrative Science Quarterly*, 28(4) 521–541.

Perrow, C. (1984). *Normal Accidents: Living with High Risk Systems*. New York: Basic Books.

Perrow, C. (1991). A society of organizations. *Theory and Society*, 20(6), 725–762.

Pettersen, G. K. (2008). *The Social Production of Safety, Theorising the Human Role in Aircraft Line Maintenance* (Doctoral dissertation, Dissertation, University of Stavanger).

Pettersen, G. K. (2020). Organizational risk: "Muddling through" 40 years of research. *Risk Analysis*. https://doi.org/10.1111/risa.13460

Pidgeon, N. (1997). The limits to safety? Culture, politics, learning and man–made disasters. *Journal of Contingencies and crisis management*, 5(1), 1–14.

Power, M. (2004). *The Risk Management of Everything: Rethinking the Politics of Uncertainty*. London: Demos.

Reason, J. (1987). The Chernobyl errors. *Bulletin of the British Psychological Society*, 40(206), 1–20.

Rees, J. V. (1996). *Hostages of Each Other: The Transformation of Nuclear Safety Since Three Mile Island.* University of Chicago Press.

Rochlin, G. (1993). Defining high-reliability organizations in practice: A taxonomic Prologomena. In: K. H. Roberts (ed.), *New Challenges to Understanding Organizations.* New York: Macmillan, 11–32.

Rochlin, G. I. (1998). Essential friction: Error-control in organizational behavior. In: Nordal Åkerman (ed.), *The Necessity of Friction.* Boulder, CO: Westview Press, 196–232.

Rochlin, G. I. (1999). Safe operation as a social construct. *Ergonomics*, 42(11), 1549–1560.

Rochlin, G. I., & Meier, A. V. (1994). Nuclear power operations: A cross-cultural perspective. *Annual Review of Energy and the Environment*, 19(1), 153–187.

Roe, E., & Schulman, P. R. (2008). *High Reliability Management: Operating on the Edge* (Vol. 19). Palo Alto, CA: Stanford University Press.

Sagan, S. D. (1993). *The Limits of Safety: Organizations, Accidents, and Nuclear Weapons.* Princeton, NJ.: Princeton University Press.

Steffy, L. S. (2010). *Drowning in Oil: BP and the Reckless Pursuit of Profit.* London: McGraw-Hill.

Turner, B. A. (1976). The organizational and interorganizational development of disasters. *Administrative Science Quarterly*, 21(3), 378–397.

Turner, B. A. (1997). *Man-made disasters.[Facsimile].* Boston: Butterworth-Heinemann.

Vaughan, D. (1997). *The Challenger Launch Decision: Risky Technology, Culture, and Deviance at NASA.* Chicago, Ill.: University of Chicago Press.

Vaughan, D. (2006). The social shaping of commission reports. *Sociological Forum*, 21(2), 291–306.

Vidal, R., & Roberts, K. H. (2014). Observing elite firefighting teams: The triad effect. *Journal of Contingencies and Crisis Management*, 22(1), 18–28.

Weaver, C. (2008). *Hypocrisy Trap: The World Bank and the Poverty of Reform.* Princeton, NJ: Princeton University Press.

Weick, K. E. (1993). The collapse of sensemaking in organizations: The Mann Gulch disaster. *Administrative Science Quarterly*, 38(4), 628–652.

Weick, K. E. (1995). *Sensemaking in Organizations* (Vol. 3). Thousand Oaks, CA: Sage.

Weick, K. E. (1996). Drop your tools: An allegory for organizational studies. *Administrative Science Quarterly*, 41(2), 301–313.

Wiig, S., Aase, K., Johannessen, T., Holen-Rabbersvik, E., Thomsen, L. H., van de Bovenkamp, H., & Ree, E. (2019). How to deal with context? A context-mapping tool for quality and safety in nursing homes and homecare (SAFE-LEAD context). *BMC Research Notes*, 12(1), 259.

4 In the Footsteps of Turner

From Grounded Theory to Conceptual Ethnography in Safety

Jean-Christophe Le Coze

CONTENTS

INTRODUCTION

Studying the daily operations of safety-critical systems confronts the researcher with thorny methodological, empirical and theoretical issues. In a world shaped by digitalised and globalised flows of people, data, money, information and goods, our understanding of the interactions between practices, work, organisations, businesses, markets and regulations of an array of actors across countries and continents is a great challenge. Qualitative methods play an important role and are advocated by anthropologists and sociologists. Barry Turner is important in this respect, although this might not be the most obvious of his contributions. He is mainly known for his breakthrough contributions in organisational symbolism (culture) and the sociotechnical view of disasters. Yet, his early use and challenge of grounded theory, a method developed in the US in the 1960s, also makes Turner a pathbreaker in the qualitative study of safety-critical systems, although he mainly worked from secondary and retrospective data. I contend that his challenges to grounded theory can be turned into a new formulation which remains, I believe, true to his intention. This new formulation is conceptual ethnography. In the first section of this chapter, Turner's legacy is presented, underlying his interwoven interests in culture, disaster and grounded theory. His challenges to grounded theory are introduced and discussed. In the second section, more recent debates about grounded theory are presented, and linked to Turner's own contributions. In the last section, historical and desktop ethnographies of Vaughan and Hopkins are briefly mentioned to support, in the footsteps of Turner, the reformulation in safety research of grounded theory into conceptual ethnography.

TURNER'S RESEARCH: INDUSTRIAL SUBCULTURE, DISASTER AND GROUNDED THEORY

ORGANISATIONS, CULTURE AND DISASTERS

Reading Turner's writings and his personal retrospective (Turner, 1995), three research domains for which he made pathbreaking contributions appear in chronological order and are deeply interwoven: first, the issue of culture (or symbolism) in the study of organisations (Turner, 1971, 1986, 1990); second, the incubation model of disasters connected to topics such as risk, sociotechnical safety, safety culture and learning (Turner, 1978, 1989, 1990, 1992); and third, grounded theory (Turner, 1981, 1983, 1988).

In his research, Turner was not an isolated writer and collaborated over many years with a number of researchers in each of these different areas. Pasquale Gagliardi is one of them on the topic of symbolism and culture in organisations. He is one author among many others who met regularly in the 1980s through the activity of a European network of academics who shared an interest in culture, and which led to the publication of collective edited books (e.g. Gagliardi, 1990; Turner, 1990).

David Blockley, Nick Pidgeon and Brian Toft were close colleagues of Turner on disasters and safety (Pidgeon, 2010) and were part of a second active (UK-centred) network which also produced very rich material attesting to the relevance of Turner's incubation framing (e.g. Pidgeon et al., 1986, 1987, 1988, 1991; Turner et al., 1989, 1991). Finally, on the topic of grounded theory, Turner collaborated with Patricia Yancey Martin (Martin & Turner, 1986) but also Silvia Gherardi (Gherardi & Turner, 1987).

Beyond these close and direct collaborations, he has been an influential writer for a generation of researchers in these three different areas which have grown over the years to become broadly and loosely connected, fragmented and dispersed in the activity of different networks of researchers and publications, but which were tightly intertwined in Turner's practices, mind and writings.

Indeed, in safety research, Turner is surely known for his model of incubation of disasters, perhaps less so for his contribution to the study of organisational culture, and even less so for his research on grounded theory. Yet, it turns out that grounded theory is a very important ingredient of Turner's writings because it provided him with a structured way of innovating in areas with little prior publications (i.e. organisational culture, disaster incubation). Let's first introduce them and their context briefly.

Turner's interest in applying cultural lenses to organisations in the 1960s was a reaction to his lack of satisfaction, as a sociologist, with the orientation of the mainstream organisational studies which he was involved in at the time. In this period, mainstream structural views of organisations left little space for the study of symbolic realities. As a trained engineer who had worked in different companies before turning to the study of sociology, Turner felt that there was something of his experience of these places which was missing in the sociological approach of organisations.

> *My focus on culture was in part* (...) a way of compensating for some aspects of the disappointment with the work at Imperial College (...) I had been quite comfortable with the idea that we needed to investigate the sociotechnical aspects of industrial organizations, but also I wanted to set these in the context of the patterns of meaning in which people constructed and maintained in industrial organizations, those patterns which made the world of the factory feel distinctive when you enter it. (Turner, 1995, 290)

Published in 1971, *Exploring the Industrial Subculture* met little success, not before a decade when an international interest in organisational culture picked up during the 1980s (Turner, 1986), which propelled Turner to the heart of an active network of researchers on this topic, with dedicated symposiums then collective publications (Gagliardi, 1990; Turner, 1990). The study of culture has become a core thread of organisational studies since then.

His early inspiration for exploring the cultural dimension stemmed from a sensitivity to the writings of social anthropologists.

> It was natural that, as a sociologist, and particularly one with a passing acquaintance with the work of social anthropologists, I should have attended to rituals and ceremonies, to norms and social definitions, to distinctive organizational forms of language and communication, and to the cultural devices associated with hierarchy, with patterns of deference and with the manner in which power and dominance were articulated within industrial organizations. (Turner, 1986, 101–102)

The motivation for choosing the second research topic on disasters, which would become his thesis and then the book *Man-Made Disasters*, is not as clear as for culture in his autobiographical writings.

> I had become aware early in the 1970s that there was a potential store of information about administrative failures and shortcomings in reports of public inquiries into large-scale accidents and disasters and I concentrated this perception into my doctoral thesis which I wrote in Exeter and which subsequently formed the basis of a book, Man-made disasters. (Turner, 1995, 294)

This does not sound a particularly powerful motive. Why an interest in these kinds of administrative failures and shortcomings?

Several hypotheses can be formulated. I suggest four, each considering in turn educational, empirical, theoretical and historical elements of Turner's biography. A first one would be linked to his educational background in both engineering and sociology which offered highly relevant expertise to explore the mix of technological and organisational aspects behind such events (that is Vaughan's explanation in the foreword of the second edition of *Man-Made Disasters*, Vaughan, 1997).

This first hypothesis could be directly linked to a second, namely his familiarity with engineering incidents in the context of industrial organisations. Before turning to sociology, recall that Turner worked as an engineer in different businesses. It is more than likely that he had to deal with unexpected or unplanned events or incidents as part of these jobs, which needed strategies for problem resolutions requiring the coordination of various expertise from different departments of organisations.

An interesting quote in an article on grounded theory published after *Man-Made Disasters* gives credit to this assumption. He indicates, when explaining his theorising rationale that a

> common feature which formed the centre point of the analysis in each case was that each accident involved a large, complex problem, the limits and bounds of which were difficult to specify. This ill-structured problem (which provided the conceptual link with earlier batch production research already discussed in case B) was, in each case, being dealt with by a number of groups of individuals usually operating in separate organizations, and in separate departments within organizations. (Turner, 1983, 343)

This is a description that many engineers, if asked, could give of their own experience of sorting out the kind of typical production problems requiring the contribution of various employees situated in different functions (maintenance, production, process improvement, etc.). For an intellectual mind such as Turner's, it must have proven highly appealing to think about this, leading, perhaps, to a sustained interest about the ability for organisations to find solutions to their incidents, and perhaps, about their ability to better anticipate them.

A third hypothesis could be linked this time to UK history and the Aberfan national tragedy in 1966[1] in which 116 children and 28 adults perished buried in their school by a coal slide of the nearby tip. This event and the strong emotional response it triggered in British society must have made the news headlines the following weeks and months, with questions about the reasons for how it happened.[2] Turner studied sociology and organisations during this period of the late 60s, and it could have caught his attention, already sensitised by his own engineering practice (see above), with the intention to make sense of the tragedy based on his developing sociological insights.

The fourth hypothesis, which stresses a more theoretical side, relies on the indication that it is also a moment when Turner discovers an author who is for him important enough to be mentioned in his retrospective of the mid-1990s. Employed at Imperial College between 1966 and 1969 in the Woodward's research team, he writes, "Exploring resources at hand, I used the Science Museum library to inform myself about Kuhn and paradigms, and about developments in the area of general systems theory" (Turner, 1995, 287).

As for many social scientists in the 1960s, Kuhn was a major influence, and that is the reason why Turner mentions this author along with other sociologists such as Weber, Schutz, Berger and Luckman and Goffman (Turner, 1995), who shaped his sociological sensitivity. I have argued elsewhere and for this reason that Turner's framing of disasters was a Kuhnian one, a view which distinguishes him from the critical Marxist view and interpretation of disaster of Charles Perrow, among other (Le Coze, 2015).

Kuhn argued that shifts in paradigms best described scientific practices, when long-held theories had to be revised to account for observations which had so far been considered as anomalies to be explained away by slightly altering the theory (rather than abandoning it). The parallel with the incubation of disaster immediately strikes the reader. Indeed, in this model, anomalies are conceptualised as potentially accumulating until an accident shows that cultural and institutionalised worldviews should have been challenged beforehand to potentially prevent it. So, educational (his sociological and engineering background of the 1960s), empirical (his own engineering experience of industrial organisations), historical (the Aberfan tragedy of 1966) and theoretical (Kuhn's influential notion of paradigm shift published in 1962) contexts, taken together, provide some clues about Turner's research interest in an organisational analysis of catastrophes.

CONNECTING TURNER'S RESEARCH INTERESTS

A RESEARCHER AT THE VANGUARD

Now, one can find many commonalities, similarities and connections between the books *Exploring the Industrial Subculture* and *Man-Made Disasters*. I indicate three. One similarity is the rather long time it took, for each of his books, to gain recognition despite their ground-breaking insights.

When the book was published, it met a resounding silence. From my point of view, publication of the first book seemed to make no impact upon my life whatsoever until a decade or more later (...) Again, with Manmade disaster, I had the sense after publication that the book disappeared without a trace, but I have become aware in the past three or four years that it did reach an extensive if fairly specialized audience around the world. (Turner, 1995, 291, 295)

Why is this? As the impressionists in the painting art of the late 19th century, the novelty of his work did not fit in what was considered at the time to be central issues to the community of organisational theorists. He paid the price of the authors at the vanguard of research, who follow their genuine interest in intellectual development rather than sticking to existing domains.

> *I did not see myself as seeking to settle and colonize a piece of intellectual territory, but as someone setting out on a challenging intellectual adventure, someone for whom it was less important to seek intellectual security* (...) than to look for the interesting question to pursue. (Turner, 1995, 280)

Then, in the 1980s, the US management consultant firm McKinsey promoted business excellence through culture (Peters & Waterman, 1982). The huge success of their books made culture a hotly debated topic within and outside academia. More than ten years after its publication, *Exploring the Industrial Subculture* became an important resource of the organisational literature, and of these debates. In the 1980s again, Perrow's Normal Accidents book (Perrow, 1984) followed by the disasters of Bhopal (1984), Chernobyl (1986), *Challenger* (1986), Piper Alpha (1988) and Exxon Valdez (1989), supplied the material and created a momentum for core sociological controversies over the control of hazardous systems in the US (then beyond) in the 1990s (Sagan, 1993; Roberts, 1993), more than ten years after the release of *Man-Made Disasters*.

DISASTERS AS CULTURAL AND INSTITUTIONAL BELIEFS (AND POWER)

A second strong connection between the books *Exploring the Industrial Subculture* and *Man-Made Disasters* is the cultural lenses that Turner applies to his model of incubation. Although the rationale is as much technical and structural as cultural because of the multifaceted aspects of accidents, the concepts that he forged for his symbolic treatment of organisations in *Exploring the Industrial Subculture* is a very core ingredient of *Man-Made Disasters*. One finds many theoretical propositions directly imported from his previous book to his second, for instance, "Within work groups, patterns of shared values, norms and perceptions tend to emerge, leading to the development within organizations of cultures and subcultures which the members of the group hold in common" (Turner & Pidgeon, 1997, 101).

These cultures both allow and hinder perception and sensitivity to signals as they create what Turner characterised as institutional beliefs (not to say, paradigms) about the world; as the saying goes "a way of seeing is a way of not seeing." Moreover, he introduced and insisted on the problem of potentially diverging cultures within the horizontal and vertical differentiations of complex organisations. The consequence of this was to discuss the issue of power. "There will be competition and conflict between different elements of these organizational beliefs (...) the view of reality, and the decision premises which are likely to be dominant are the ones favoured by 'those with the bigger stick'" (Turner & Pidgeon, 1997, 139).

Power and culture are two central concepts of the social sciences, and it was part of Turner's reasoning even if the cultural interpretation dominated in the incubation model. But at times, power and incubation of disasters were explicitly combined and associated, as in the following quote.

> It becomes necessary, also, to start to take account of those other perennial concerns of the sociologist, the charting of the distribution of power, of the control of resources and of social reputation. The distribution of all these elements, together with the actions of those who seek to change their distribution, may lead to intentional or unintentional influences upon the nature of the information flow which characterise the incubation period, prior to the disclosure of an unsuspected and dangerous situation. (Turner & Pidgeon, 1997, 105)

Grounded Theory as the Third Research Topic

What this means concretely is that the exploration of the disaster topic from a sociological angle was very much influenced by his own previous developments on culture (and power) in organisations. This might sound a rather obvious comment, but it is one which will reveal its importance when discussing the next point because the third connection (which will be further explored in the rest of this chapter) is grounded theory. Published in 1967, grounded theory by Glaser and Strauss (1967) became an important resource for structuring the innovative path which Turner followed when venturing into his cultural treatment of organisations.

This is particularly clear in the preface of *Exploring the Industrial Subculture*. Talking about the importance that the Glaser and Strauss book had to support his work on organisational culture, he writes

> the method which they offered for dealing with qualitative data made it possible, without the necessity of a leap of faith, to expose and make explicit the conceptual framework inherent in a collection of field data, and then to examine the existing literature for concept and for theoretical linkages which related to the field observations. The literature identified is then unequivocally relevant because it fits those observations which have already been made. The present book is the product of the application of this process. (Turner, 1971, ix)

But he wrote just before, "I should hasten to add that the present book should not be taken as any way as an example of grounded theory: it lacks the rigorous concentrated analysis of a delimited area which Glaser and Strauss advocates" (Turner, 1971, viii). These sentences almost sound like a regret of not being able to apply fully the methodology because he discovered it too late in the process of writing his first book. However, the basic principles are already there in the above quote, namely this empirical movement which insists on beginning with the data to build a relevant conceptual framework to account for these data, and not the reverse.

Additionally, in the quotes above, one can also hypothesise how valuable the principles of grounded theory were for Turner to justify, against a mainstream structural view of organisational analysis in which he was involved, his inclination to study qualitative aspects of organisations in a systematic fashion that would match the rigour and objectivity expected by practitioners of surveys and quantitative types of social science research. With these points in mind, it appears that his second personal research topic, disasters, provided him with an opportunity, this time, to more fully embrace and apply the principles exposed in grounded theory by Glaser and Strauss.

In many places in *Man-Made Disasters*, one finds indeed references to the central proposition of grounded theory (which characterised its heuristic value for many), that data precede theory (not the other way around) and that theoretical production should be designed as a systematic and structured process of analysing data along the guidelines provided by grounded theory. Again, as for the cultural lens, many illustrative quotes can be extracted from *Man-Made Disasters* which directly relate to this concern. Here is an example.

> This book is concerned to examine some sets of data relating to disasters in order to
> try to provide terms and theories which will serve as a kind of "conceptual tool kit" for
> the examination of disasters. (Turner & Pidgeon, 1997, 24)

And here is another one.

> The categories constitute a rather unsatisfying list, because no clear relationship is
> apparent between the various elements on this list. The next stage of the investiga-
> tion, therefore, is to search for some helpful and acceptable framework which could
> bring some unity to the analysis without doing too much violence to the data. (Turner,
> Pidgeon, 1997, 66)

In fact, during the 1970s, Turner was one of the first users of grounded theory. Although not trained in the symbolic interactionist tradition of the Chicago school behind grounded theory, it suited his sociological preferences. His mentions and uses of phenomenological sociologist Albert Schutz and social constructionists Berger and Luckman in *Exploring the Industrial Subculture*, or of Weber (opposed to the style of Marx or Parsons) reveal a sensitivity for the direct empirical exploration of how humans create worlds as they ascribe meanings to their daily situations, a posture requiring close encounters with data through qualitative observations and interviews. "I preferred to be able to explore social reality without having to subscribe to a major system" (Turner, 1995, 281).

But Turner was not only a user of grounded theory and wished, based on his exten-sive fieldwork experience and theoretical developments, to discuss the value, limita-tions and possible avenues of improvements of the method. Three articles expose his ideas regarding grounded theory (Turner, 1981, 1983, 1988) plus an additional one in collaboration with Patricia Martin which puts together and expands some of Turner's earlier elements of discussion (Martin & Turner, 1986), and a report was also writ-ten with Silvia Gherardi (Gherardi & Turner, 1987). Fundamentally, Turner found grounded theory most interesting for its disciplined process of theorising, but he also had some concerns, which are contained in the following quote.

> In some ways, it is to be regretted that there has been no serious extended discussion
> of the principles underlying the generation of grounded theory, for there is an element
> of polemic in Glaser and Strauss's advocacy of grounded theory which leads them at
> times to overstress the extent to which existing theory can be completely ignored, and

to present their approach as a radical and novel one, rather than as a call for a reassertion of some of the more traditional principles of social inquiry. (Turner, 1981, 228)

What is particularly interesting is that again writing this sentence in 1981, Turner
anticipated, early, several strands of discussions and debates among practitioners
of grounded theory (and others more external to it) which led 20–30 years later, in
the 2000s, to the production of alternative versions along these lines, culminating
in what has been described as a second generation of grounded theory (Morse et
al., 2009) as well as quarrels with other fieldwork based sociological practices with
roots in ethnography (Burawoy, 2003). Let us come back therefore shortly to this
history, and see how this relates to Turner's concerns, and how he answered some of
them. This step is important because it provides the basis for the new formulation of
conceptual ethnography.

GROUNDED THEORY: EVOLUTION AND
CRITICS, A VERY SHORT OVERVIEW

THE ORIGINAL GROUNDED THEORY

The story of grounded theory translated the will of two authors, Barney Glaser and
Antelm Strauss, in the 1960s, to create a space in the US at the academic level for a
different view and status of qualitative fieldwork, for legitimising its scientific value
against dominant practices in sociology. Sociology at the time was indeed divided
in different schools but two were quite strong. Grounded theory targeted both. It
challenged on the one hand the a priori system wide theoretical orientation represented by a sociologist like Talcott Parsons and, on the other hand, the survey based-
quantitative studies represented by Paul Lazersfeld (criticised by Mills, 1959). In the
former orientation, qualitative fieldwork was subordinated to a system theory, and
in the latter, qualitative fieldwork was subordinated to the purpose of producing and
framing the content of survey questionnaires.

In both cases, qualitative modes of investigations were therefore considered of a
lower status than the other ways of practising sociology. The reverse proposition that
fieldwork qualitative data were first, that theory should emerge from data and that it was
possible to systematise or codify this process to a degree that would match the rigour
associated with the hypothetico-deductive approach which characterised the ideal of the
scientific method, was a significant move. This much-needed formulation for academics
(and their students), grounded theory, in valuing fieldwork over system theorising and
surveys, became a classic. But over the years, core assumptions of the book were challenged, turning first into diverging refinements (Strauss & Corbin, 1998) to finally new
alternatives endorsed by a new generation of authors (Morse et al., 2009).

A SECOND GENERATION OF GROUNDED THEORY

One centrally debated issue in the past decades is the relationship between data and theory
(Kelle, 2005). In the classic grounded theory statement, it is rather forcefully advocated

that no a priori theory should be brought into data collection, or at least it should be avoided as much as possible. This was precisely one of the core motives for grounded theory to be written in the first place, to reject armchair theorising characterising dominant US sociology. But when the two authors of the original version of grounded theory started publishing separately (Glaser, 1978; Strauss, 1987; Strauss & Corbin, 1998), their views drifted apart on this issue. Because of their differing backgrounds, Glaser maintained a strict inductive precept of no theory polluting data collection, while Strauss pushed the sensitising view (originating in the symbolic interactionism tradition, Blumer, 1969) that data could not be collected without a minimum of conceptual background.

This divide was then amplified and radicalised in the context of the growing influence during the 1980s and 1990s of powerful ideas in social sciences such as constructivism, postmodernism and post-humanism (culminating in the science wars, Hacking, 1999). These important ideas challenged the taken-for-granted discourse of the neutrality of science, of the objectivity of scientific theories, of the detached scientists, of the ideal of rationality, of the direction of progress and of the separation between material, natural and social entities.

They implied as a consequence, if considered seriously, a higher level of reflexivity on the part of researchers about the politics of (their) science and the implicit preconceptions (due to their own history, education, gender, experience, ethnicity, etc.) found behind their search for theoretical objectivity. These ideas also meant a higher degree of analytical sensitivity to additional entities (i.e. non-humans), and a more open-minded attitude regarding the limits and pretention of their expertise in relation to other forms of knowledge whether from other scientific disciplines or beyond the boundaries of academic specialities such as practitioners' experience.

This is what characterises this second generation of grounded theory. It imported these epistemological, philosophical and political controversies which challenged rationality, value neutrality of science, objectivity and the boundaries of the social. "A re-visioned grounded theory must take epistemological questions into account" (Charmaz, 2000, 522). Constructivist grounded theory (Charmaz, 2000) and situational analysis (Clarke, 2003, 2009) typically translate the implication of these ideas in renewed constructivist and postmodernist versions of grounded theory, as made explicitly by Charmaz in the following quote.

> A constructivist approach recognizes that the categories, concepts, and theoretical level of an analysis emerge from the researcher's interactions within the field and question about the data. In short, the narrowing of research questions, the creation of concepts and categories, and the integration of the constructed theoretical framework reflect what and how the researcher thinks and does about shaping and collecting the data. (Charmaz, 2000, 522)

THE EXTENDED CASE METHOD, AN ALTERNATIVE?

Another front challenged grounded theory as a method for fieldwork researchers (Tavory & Timmermans, 2009). It came this time from ethnographers and more precisely through the promotion of the extended case method developed initially in the context of the British social anthropology of Manchester (Gluckman, 1963), which was further exposed and argued by different authors in the past 50 years. In contemporary sociology, it is Michael Burawoy (Burawoy, 1998, 2003, 2009) who has popularised the method among fieldwork sociologists and led the quarrel with grounded theory. Its origin goes back to the anthropologist Bronislaw Malinowsky who back then set the standard of what would be known as ethnography, early in the 20th century. Practising ethnography meant for anthropologists to gain direct access to data by spending time talking, observing, interacting and living with studied populations (see also Kuran, Chapter 7, this volume).

As written by Gluckman about Malinowski, "he fought and won one important battle, during which he raised ethnographic fieldwork itself to a professional art. This battle was to establish that primitive life and primitive institutions were more complex, far more complex, than earlier theorists had thought" (Gluckman, 1963, 7). This was not the norm back then as many anthropologists would rely on secondary data for their research. Based on this ethnographic legacy of Malinowski, the extended case method as described by Gluckman relied on a mix of observations of the life of a community from a diversity of points of views combined with detailed events or incidents, in order to grasp social dynamics.

"We called these complex events social situations, and we used the actions of individuals and groups within these situations to exhibit the morphology of the social structure" (Gluckman, 1963). This approach finally challenged the structuro-functionalist anthropology which was a major trend back then, a theoretical orientation which favoured the description of communities as stability seeking and well-defined wholes in a synchronic way (a view also prominent in sociology in the US with Parsons' sociological grand theory). Another aspect of the extended case method is the continuous discussion among academics about their theory in relation to their data, the differing interpretation one can get from the same case when changing one theoretical orientation (see also Antonsen and Haavik, Chapter 5, this volume).

One could say that Burawoy has illustrated, refined, expanded and advocated this ethnographic posture from anthropology to sociology, while importing, just as the second generation of grounded theorists did, advances in social sciences (constructivism, postmodernism, etc.). Three points are worth emphasising. First, Burawoy explicitly targets grounded theory for its misguided inductive philosophy. Burawoy asserts that no data collection can be detached from an initial framing, without a priori theory. Second, he criticises the absence of historical and macro perspectives in the research produced by grounded theory, in relation to its roots in symbolic interactionism. Third, he acknowledges the presence of the ethnographer, not as objectively detached but as one fully immersed in their studies, part of a public ethnography which always connects the sociologist with a society's politics.

Some of these criticisms could probably now be reappraised in the light of the new versions of grounded theories introduced above (Charmaz, 2000; Clarke, 2003) because, as described, these new versions have completely embraced the inclusion of researchers in their multifaceted sources of subjectivity and philosophical conceptions, which both challenge the ideal of pure objectivity and the possibility of going in the field without prior knowledge. However, it is the historical and macro critics which are particularly interesting and challenging for grounded theory. Because of its Marxist view, Burawoy has always maintained the interpretive lenses of the system view in which historical and macro structures (e.g. capitalism) operate and affect micro situations described through ethnographic data collection.

Despite being immersed in ethnographic fieldwork, Burawoy does not wish to lose sight of the connections between the descriptions of the micro within their wider macro social structures. In the last two to three decades these structures have materialised in globalisation through flows of data, goods, people and money reshaping in very complex ways (Sassen, 2007) the morphology as well as the conditions of study of traditional units of sociological analysis (nations, companies, states, families, education, universities, etc.). This intellectual posture clearly differs from the origin of grounded theory and subsequent versions which are only loosely connected to the debates found in ethnography, a foundational difference of scope expressed explicitly in Charmaz (2000).

> Most grounded theory studies rely on detailed qualitative materials collected through field, or ethnographic, research but they are not ethnographies in the sense of total immersion into specific communities. Nor do grounded theorists attempt to study the social structures of whole communities. Instead, we tend to look at slices of social life. (Charmaz, 2000, 522)

The idea that one should perform ethnographies connecting observations of micro situations to social structures (the "whole") is therefore an important added value of the extended case methodology in comparison with the precepts of grounded theory (and the second generation), especially considering, today, the extent with which globalised forces have been concretely shaping the daily lives of people and institutions around the world.

Globalised processes have been shaping new operating landscapes of safety-critical systems through the liberalisation of trade and finance, deregulation and privatisation, and revolutions in IT and transport (maritime, aviation) which created powerful trends such as financialisation, self-regulation, digitalisation, standardisation and externalisation in the last three decades. Companies are now highly influenced by these trends and their profiles as a result have evolved toward what has been described as networked properties (Le Coze, 2017b).

BACK TO TURNER: THEORY–DATA CONNECTION, ETHNOGRAPHY, MACRO STRUCTURES, EXTERIORITY OF RESEARCHER

These short descriptions of how grounded theory evolved under the influence of powerful ideas (e.g. constructivism, postmodernism, post-humanism), and how other ethnographic methods such as the extended case method challenged some of its rationales and most importantly its lack of connection to macro (global) social structures (Tavory & Timmermans, 2009), provide a context to come back to Turner, and to justify the reformulation of conceptual ethnography. Framed this way, it appears that these debates completely relate to Turner's own lack of satisfaction with the original version of grounded theory, some explicitly explored (e.g. theory–data connection, ethnography), some less so and just mentioned in short sentences of articles without associated developments (e.g. macro structure, exteriority of researcher). Let me begin with the ones which were analysed.

THEORY–DATA CONNECTION

First, and as found in Turner's quote at the end of the previous section before the history of grounded theory, he identified two problems. One is the issue of collecting data without a theoretical background, "*Glaser and Strauss's advocacy of grounded theory* (…) leads them at times to overstress the extent to which existing theory can be completely ignored" (Turner, 1981, 228). As a reader of Kuhn and as someone who studied general system theory (and their related epistemological explorations, e.g. second order cybernetics), this precept seemed odd enough to be addressed.

Without rejecting the imperative that data collection should come first, Turner writes

> we become concerned with the researcher's perspective which must recognize an interplay between objectivity and subjectivity in the interactions of research. The personal style and mental set of the researcher are not irrelevant to the manner in which organizational research is carried out, especially, although not exclusively, in the stages of data analysis and theory construction. (Turner, 1988, 121)

I have made the assumption that his interest in disaster was linked to his reading of Kuhn, a hypothesis which introduces prior knowledge as a context for a choice of topic, emphasising one side of this subjectivity, and the presence of the researcher behind fieldwork. Back in the 1970s and 1980s, he was therefore anticipating the critics to come in the 1990s and 2000s as described above and he also wished to comment further on the source of the "creative theoretical imagination" (Turner, 1983, 335) which grounded theory encouraged through this cycle of confronting data, first, then concepts, second. To do so, he targets the limits of relying on induction as suggested in grounded theory.

"In fact this loose usage of the term 'induction' is of limited help, since all forms of research, qualitative and quantitative, are based upon a complex admixture of deductive and inductive procedures" (Turner, 1988, 111). Research on the cognitive processes behind theorising has subsequently added the notion of abduction alongside induction and deduction, precisely to stress the processes by which novel insights, the creative side of science, are gained (Reichertz, 2010).

ETHNOGRAPHY

Another is Glaser and Strauss and their inclination "to present their approach as a radical and novel one, rather than as a call for a reassertion of some of the more traditional principles of social inquiry" (Turner, 1981, 228). By this, he has in fact very specific ideas, articulated in the following quote.

> There is nothing novel about the relationship between the "grounded theoretical" researcher and his or her subjects. It is the kind of relationship which has been familiar

to anthropologists since the early studies of Evans-Pritchard and Malinowski, and to sociologist since Whyte's study of Street Corner Society. (Turner, 1983, 335)

Situated in the British context of social sciences, he might even have been aware of the post-war developments exposed in the extended case method (Gluckman, 1963), as we know that he described himself as "a sociologist, and particularly one with a passing acquaintance with the work of social anthropologists" (Turner, 1983, 333).

MACRO STRUCTURE

Now, among the more implicit topics, the themes of history and how macro social structures exert constraints on local, micro descriptions that are scrutinised through qualitative fieldwork were also acknowledged by Turner. These constraints can escape the reach of grounded methodology, as criticised in the extended case method. He writes that grounded theory "is of least use when dealing with large scale structural features if social phenomena, such as demographic or industrial trends or some aspects of macro organizational structures" (Turner, 1983, 335). This, of course, is not without creating a certain number of problems when one intends to delve into complex socio technical systems, as Turner did with the study of disasters.

This, again, is expressed in another quote.

> Glaser has suggested that formal theory can be best developed from grounded theory by concentration upon a single activity (...) I have not found myself wishing to take this approach up to now-perhaps I have a preference for the complex, multilevel explanation of existing systems rather than such a narrowly formal vision. (Turner, 1983, 347)

How did Turner therefore escape the grounded theory focus on such a restricted level of analysis? He could escape this through reports of public inquiries. Because these reports rely on a wide range of interviews of a diversity of actors linking them all to a specific event, he could cover a broad scope of society, from private companies to regulators and other institutions including industry or professions.

IMPLICATIONS OF RESEARCHERS' WORK

One can hypothesise that Burawoy's discussions of introducing macro structures in ethnographic fieldwork would have appealed to Turner[3] as well as the immersion of the researcher in his study. This latter aspect is about the implication and relevance of fieldwork conceptualisations for those studied, something which represents a first degree of reflexivity on the part of an ethnographer regarding the relationship between their work and the world. Writing about grounded theory, Turner says,

> I have also stayed with this approach because its properties make it more likely than many other forms of research that I can produce theoretical accounts which are understandable to those in the area studied and which are useful in giving them a superior understanding of the nature of their own situation. (Turner, 1983, 348)

This typically is the kind of question raised by constructivist, postmodernist developments of grounded theory or the public/reflexive ethnography that have been further addressed in the past decades and indicated previously. Even if he did not really develop much on these topics, I suspect that it seems possible that Turner was aware of these issues, although he could not pursue them further in the 1990s.

IN THE FOOTSTEPS OF TURNER: CONCEPTUAL ETHNOGRAPHY

I described in this chapter how Turner became an influential sociologist in organisational culture in the 1980s, then in disaster research in the 1990s, ten years after his leading publications (Turner, 1971, 1978). One can also find some traces of his reflections on grounded theory in manuals of qualitative research (e.g. Miles & Huberman, 1994), although this side of research never acquired the same academic impact as his two other main subjects did. More than 20 years after Turner writings, what about grounded theory, qualitative approaches and ethnography in safety and disaster research? He was one of the first sociologists who turned to public reports to extract data and theorise on their basis about accidents, but other followed in Turner's footsteps.

Two cases are Diane Vaughan (Vaughan, 1996, 1999, 2005) and Andrew Hopkins (1999, 2000, 2008, 2012), both have been highly influential authors in the past 20 years, too, with their similar grounded approach of analysing public documents, produced in the aftermath of disasters, as Turner did. Both refer to their celebrated predecessor, both embrace his sociological conceptualisation of accidents, and both reflect on their qualitative approach as he did.[4] Vaughan describes hers as historical ethnography (Vaughan, 2004, 2014) and Hopkins as desktop ethnography (Hopkins, 2006, 2016). These authors actually refine but also answer some of Turner's interrogations derived from his fieldwork and analytical practices. I briefly introduce Vaughan's work and do not introduce Hopkins because of space constraints (see Le Coze, 2017a, 2019, and also Hopkins, Chapter 2, this volume).

HISTORICAL ETHNOGRAPHY

Vaughan's analysis is well known to safety researchers, conceptualising the normalisation of deviance in both the loss of *Challenger* in 1986 and *Columbia* in 2003 (Vaughan, 1996, 1997, 1999, 2005), and based on a cultural rationale which connects it to the importance Turner granted to this dimension. What is probably less known, again as it is for Turner, is Vaughan's extensive reflection on case study, fieldwork, ethnography, grounded theory and analogical theorising associated with these empirical investigations (Vaughan, 1992, 2003, 2004, 2007, 2014). Because of her extensive and diverse empirical explorations of several topics (mistakes, misconduct, uncoupling, disaster) in several social contexts (private organisations, intimate relationships, public space agency), Vaughan explored the cognitive processes of theorising.

And she raised very similar concerns as Turner. "Grounded theory tied scholarship to the local, with no directions about pursuing the structural and the historical context of action. Also, Glaser and Strauss argued against that starting with a theory in mind invalidated the procedure" (Vaughan, 2004). If one retains, for the purpose of this chapter, these aspects treated or touched upon by Turner in his exploration of grounded theory (e.g. theory–data connection, ethnography, macro structure, exteriority), Vaughan's can be said to contribute to them all.

First, Vaughan goes beyond the ethnographic practice that Turner applied to accidents by searching outside official reports of commissions and by exploiting interview transcripts as well as engineering and administrative documents produced by the agency. Moreover, additional interviews with people of the organisation, NASA and its subcontractors were also performed (Vaughan, 2004). These ten years of ethnographic endeavour allowed her to reach empirical territories that Turner couldn't reach but pointed at, namely decision-making processes embedded in micro (artefacts, tasks, cognition), meso (structure, processes) and macro (political, professional, legal, demographic, political) layers of analysis.

This is a historical ethnography, for which theory matters before data collection (as a sensitising tool), and for which structural and institutional features of people's daily practices matter, and must be included in ethnographic work (Vaughan, 2004), consistently with Burawoy's plea (Burawoy, 2003). This historical ethnography turns out also to be a public ethnography (Burawoy, 2004), based on her rich experience of interacting with a diversity of actors beyond academia interested in her findings, following the publications of her books on *Challenger*, and her participation in the *Columbia* investigation board (Vaughan, 2003). So, further examination of and solutions to Turner's issues of theory–data connection, ethnography, macro social structure and researcher degree of exteriority are now available in safety and disaster research.

CONCEPTUAL ETHNOGRAPHY

Vaughan's and Hopkins' studies are retrospective ones. The description of historical ethnography refers directly to this backward-looking empirical exercise, and desktop ethnography refers to the situation of reading transcripts of hearings while seating in an office, getting immersed in organisations thanks to the amount of data available without going into the field (see Hopkins, Chapter 2, this volume). However, when one studies daily operations and spends time observing, interviewing, taking notes, etc., notions of historical and desktop ethnographies do not fit.

As already indicated several times, I suggest instead to use the notion of conceptual ethnography to translate the debates initiated by Turner on grounded theory which were amplified later under the influence of constructivist, postmodern or posthumanist discourses. Conceptual ethnography conveys this now shared idea that one never performs ethnographic work from nowhere. Our data collections, interpretations and interactions with different audiences are situated, value and theory laden – things that ethnographic safety research need to incorporate in its core assumptions (Le Coze, 2013). And, consistently with the extended case method, wider, macro and global processes have to be connected to our ethnographic observations (Le Coze, 2017a, 2018, 2020).

CONCLUSION

In this chapter, Turner's combination of work on organisational symbolism and culture, on disaster incubation and grounded theory, has been shown to be deeply intertwined. For each, Turner was at the vanguard of research, before these topics took off, and became hotly debated arenas. He therefore anticipated and contributed to debates on culture in organisation, on sociological interpretation of accidents and on the production of knowledge derived from fieldwork approaches. These debates have been strongly influenced by conceptual and empirical trends and debates of the past 20–30 years in social sciences among which include constructivism, postmodernism and post-humanism but also globalisation.

As a result, the themes which Turner started exploring about grounded theory (i.e. theory–data connection, ethnography, macro structures and exteriority of researcher) have been further investigated with new formulations (e.g. constructivist grounded theory, situational analysis, extended case method) outside the field of disaster and safety research, but also from within. In this respect, Vaughan's historical ethnography and Hopkins' desktop ethnography are two of these examples, and a suggestion of this chapter, when shifting from the study of past events to the study of daily operations in safety-critical systems, is to develop the notion of conceptual ethnography to incorporate all these refinements.

NOTES

1. http://www.bbc.co.uk/news/resources/idt-150d11df-c541-44a9-9332-560a19828c47
2. One is reminded of the emotional social uproar in the aftermath of the fire of the Grenfeld tower in 2017, almost 50 years later.
3. "I believe Barry saw the 'key insights' of Burawoy in his own work/efforts/analyses. He knew that whatever went on in a single organization (a school, a city, etc.) was linked to wider contexts, issues, dynamics, etc. So, he was right (in my view) in rejecting Glaser's "single activity" issue" (Patricia Yearly Martin, personal communication, June 2018).
4. This statement does not imply a sequential view of ideas which would simplistically introduce first Turner, then Vaughan or Hopkins. Relationships between authors, concepts and case studies are not that simple as they have their own historical trajectories, developing similar ideas at the same time without necessarily knowing each other works. It would require more space to explore this more thoroughly (see Hopkins' chapter in this book for some insights about this point).

REFERENCES

Blumer, H. (1969). *Symbolic Interactionism. Perspective and Method*. Los Angeles, CA: University California Press.

Burawoy M. (1998). The Extended Case Method. *Sociological Theory* 6(1), 4–33.

Burawoy, M. (2003). Manufacturing the global. *Ethnography*, 2(2), 147–159.

Burawoy M. (2009) Challenges for a Global Sociology. *Contexts* 8(4), 36–41.

Burawoy M. (2004). For public sociology. *American Review of Sociology*, 70(1), 4–28.

Charmaz, K. (2000). Constructivist and objectivist grounded theory. In: N. K. Denzin, Y. Lincoln (Eds.), *Handbook of Qualitative Research* Thousand Oaks, California (2nd ed., pp. 509–535).

Clarke, A. (2003). Situational analyses: Grounded theory after the postmodern turn. *Symbolic Interaction*, 26(4), 553–576.

Clarke, A. (2009). From grounded theory to situational analysis: What's new? Why? How? In: J. M. Morse, P. N. Stern, J. M. Cordin, K. C. Charmaz, B. Bowers, & A. E. Clarke (Eds.), *Developing Grounded Theory: The Second Generation*. Walnut Creek, CA: Left Coast Press, Inc., 194–233.

Gagliardi, P. (Ed.). (1990). *Symbols and Artifacts: Views of the Corporate Landscape*. Berlin: Walter de Gruyter.

Gherardi, S., & Turner, B. A. (1987). *Real Men Don't Collect Soft Data. Publisher*. Università di Trento, Dipartimento di Politica Sociale. http://eprints.biblio.unitn.it/4319/

Glaser, B. (1978). *Theoretical Sensitivity Advances in the Methodology of Grounded Theory*. Mill Valley, CA: Sociology Press.

Glaser, B., & Strauss, A. (1967). *The Discovery of Grounded Theory Strategies for Qualitative Research*. Mill Valley, CA: Sociology Press.

Gluckman, M. (1963). Gossip and scandal. *Current Anthropology*, 4(3), 307–316.

Hopkins, A. (1999). *Managing Major Hazards: The Lessons of the Moura Mine Disaster*. Sydney, NSW: Allen & Unwin.

Hopkins, A. (2000). *Lessons from Longford: The ESSO Gas Plant Explosion*. Sydney, NSW: CCH.

Hopkins, A. (2006). Studying organisational cultures and their effects on safety. *Safety Science*, 44(10), 875–899.

Hopkins, A. (2008). *Failure to Learn: The BP Texas City Refinery Disaster*. Sydney, NSW: CCH.

Hopkins, A. (2012). *Disastrous Decisions: The Human and Organisational Causes of the Gulf of Mexico Blowout*. Sydney, NSW: CCH.

Hopkins, A. (2016). *Quiet Outrage. The Way of a Sociologist*. Sydney, NSW: CCH Press.

Kelle, U. (2005). Emergence" vs. "forcing" of empirical data: A crucial problem of "grounded theory" reconsidered. *Forum: Qualitative Social Research*, 6(2). http://nbn-resolving .de/urn:nbn:de:0114-fqs0502275.

Le Coze, J. C. (2015). 1984–2014. Normal Accident. Was Charles Perrow right for the wrong reasons ? *Journal of Contingencies and Crisis Management*. 23(4), 275–286.

Le Coze, J. C. (2013). New models for new times. An anti-dualist move. *Safety Science*, 59, 200–218.

Le Coze, J. C. (2017a). Andrew Hopkins and the sociology of safety. *Journal of Contingencies and Crisis Management*, 25(1), 51–53.

Le Coze, J. C. (2017b). Globalisation and high risk systems. *Policy and Practice in Health and Safety*, 15(1), 57–81.

Le Coze, J. C. (2018). Resilience, reliability and safety: Multilevel research challenges. In: Wiig S., Fahlbruch B. (eds) *Exploring Resilience. SpringerBriefs in Applied Sciences and Technology*. Springer, Cham. https://doi.org/10.1007/978-3-030-03189-3_2

Le Coze, JC. 2019. Storytelling or theory building? Hopkins' sociology of safety. Safety Science. 120. 735–744. 126. 104660.

Le Coze, J. C. (2020). *Post Normal Accident. Revisiting Perrow's Classic*. Boca Raton, FL: CRC Press, Taylor and Francis.

Martin, P. Y., Turner, B. A. (1986) Grounded theory in Organizational Research. *The journal of behavioural applied science*. 13, 135–157.

Miles, M. B., & Huberman, A. M. (1994). *Qualitative data analysis: An expanded source-book*. (2nd ed.). Thousand Oaks, CA: Sage Publications.

Mills, C. W. (1959). *The Sociological Imagination*. New York: Oxford University Press.

Morse, J. M., Stern, P. N., Corbin, J., Bowers, B., Charmaz, K., & Clarke, A. E. (2009). *Developing Grounded Theory: The Second Generation*. New York: Left Coast Press.

Perrow, C. (1984). *Normal Accidents, Living with High-Risk Technologies.* 1st ed. Princeton, NJ: Princeton University Press.

Peters, T. J., & Waterman, R. H. (1982). *In Search of Excellence Lessons from America's Best-Run Companies.* New York: Harper & Row.

Pidgeon, N. F. (2010). Systems thinking, cultures of reliability and safety. *Civil Engineering and Environmental Systems,* 27(3), 211–217.

Pidgeon, N. F., Blockley, D. I., & Turner, B. A. (1986). Design practice and snow loading: Lessons from a roof collapse. *The Structural Engineer,* 64(A), 67–71.

Pidgeon, N. F., Blockley, D. I., & Turner, B. A. (1987). Reply to discussion of 'Design practice and snow loading'. *The Structural Engineer,* 65(A), 239–240.

Pidgeon, N. F., Blockley, D. I., & Turner, B. A. (1988). Site investigations: Lessons from a late discovery of hazardous waste. *The Structural Engineer,* 66(19), 311–315.

Pidgeon, N. F., Turner, B. A., Blockley, D. I., & Toft, B. (1991). Corporate Safety Culture: Improving the Management Contribution to System Reliability. In: *Reliability '91,* (Ed. Matthews, R. H.). Elsevier, 682–690.

Reichertz, J. (2010). Abduction: The logic of discovery of grounded theory. *Forum: Qualitative Social Research,* 11(1), Art. 13.

Roberts, K. (Ed.). (1993). *New Challenges to Understanding Organizations.* New York: Macmillan Publishing Company.

Sagan, S. (1993). *The Limits of Safety. Organizations, Accidents, and Nuclear Weapons.* Princeton, NJ: Princeton University Press.

Sassen, S. (2007). *A Sociology of Globalization.* New York: W. W. Norton & Company.

Strauss, A. L. (1987). *Qualitative Analysis for Social Scientists.* New York: Cambridge University Press.

Strauss, A. L, & Corbin, J. M. (1998). *Basics of Qualitative Research: Grounded Theory Procedures and Techniques.* Thousand Oaks, CA: Sage Publications, Inc.

Tavory, I., & Timmermans, S. (2009). Two cases of ethnography: Grounded theory and the extended case method. *Ethnography,* 10(3), 243–263.

Turner, B. (1983). The use of grounded theory for the qualitative analysis of organizational behaviour. *Journal of Management Studies,* 20(3), 333–348.

Turner, B. (1988). Connoisseurship in the study of organizational cultures. In: Bryman, A. (Ed.), *Doing Research in Organizations.* London: Routledge. 108–122.

Turner, B. (1995). A personal trajectory through organization studies. *Research in the Sociology of Organisations,* 13, 275–301.

Turner, B., & Pidgeon, N. (1997). *Man-Made Disaster. The Failure of Foresight,* 2nd ed. London: Butterworth-Heinmann.

Turner, B. A., Pidgeon, N., Blockley, D., & Toft, B. (1989). *Safety Culture: Its Importance in Future Risk Management. Position Paper for Second World Bank Workshop on Safety Control and Risk Management.* Karlstad, Sweden.

Turner, B. A., Pidgeon, N. F., & Blockley, D. I. (1991). The use of grounded theory for conceptual analysis in knowledge elicitation. *Int. J. Man-Machine Studies,* 35, 151–173.

Turner, B. A. (1971). *Exploring the Industrial Subcultures.* London: Macmillan Press.

Turner, B. A. (1978). *Man-Made Disaster. The Failure of Foresight.* London: Wykeham Science Press.

Turner, B. A. (1981). Some practical aspects of qualitative data analysis: One way of organizing the cognitive processes associated with the generation of grounded theory. *Quality and Quantity,* 15(3), 225–247.

Turner, B. A. (1986). Sociological Aspects of Organizational Symbolism. *Organization Studies.* 7(2):101–115.

Turner, B. A. 1989. How can we design a safe organisation? Paper presented at the second international conference on industrial organisation and crisis management, New York University, 3-4, November.

Turner, B. 1990. Failed artefacts. In: Gagliardi, P (ed). *Symbols and artifacts: views of the corporate landscape*. New York: Walter De Gruyter.

Turner, B. 1992. The sociology of safety. In: Blockley, DI. (ed) *Engineering Safety*, Maidenhead, Mc Graw-Hill.

Vaughan, D. (1992). Theory elaboration: The heuristic of case analysis. In: C. Ragin, & H. Becker (Eds.), *What Is a Case? Exploring the Foundations of Social Enquiry.* Cambridge: Cambridge University Press. 173–192.

Vaughan, D. (1996). *The Challenger Launch Decision: Risky Technology, Culture and Deviance At.* Chicago, IL: NASA, University of Chicago Press.

Vaughan, D. (1997). The Trickle-Down Effect: Policy decisions, risky work and the Challenger tragedy. *California Management Review*, 39(2), 80–102.

Vaughan, D. (1999). The dark side of organizations: Mistake, misconduct, and disaster. *Annual Review of Sociology*, 25(1), 271–305.

Vaughan, D. (2003). History as cause. Chapter 8. In: *Columbia Accident Investigation Board.* 195–201.

Vaughan, D. (2004). Theorizing disaster: Analogy, historical ethnography, and the challenger accident ethnography. *Ethnography*, 5(3), 315–347.

Vaughan, D. (2005). System effects: On slippery slopes, repeating negative patterns, and learning from mistakes? In: H. W. Starbuck, & M. Farjoun (Eds.), *Organization at the Limit. Lessons from the Columbia Disaster.* Oxford: Blackwell Publishing, 1997–1999.

Vaughan, D. (2007). Beyond macro- and micro-levels of analysis: Organizations and the cultural fix. In: Henry N. Pontell, & Gilbert Geis (Eds.), *The International Handbook of White-Collar and Corporate Crime.* New York: Springer, 3–24.

Vaughan, D. (2014). Theorizing: Analogy, cases, and comparative Social Organization." In: Richard Swedberg (Ed.) *Theorizing in Social Science.* Stanford, CA: Stanford University Press, 61–94.

5 Case Studies in Safety Research

Stian Antonsen and Torgeir Kolstø Haavik

CONTENTS

INTRODUCTION

Case studies have played and continue to play a key role in the development of safety science. This is not surprising, since major accidents represent singular, yet rare events that open a window into the processes that generate disaster, demonstrating how organisations perform under extreme strain. Nevertheless, the methodology underlying case studies is rarely discussed in the safety literature. Many of the most influencing theories and models that have explained what goes on when work in safety-critical organisations develops into failures or successes are based on extensive, in-depth studies of work and organising. Examples of theory development in a grounded theory–oriented fashion (Glaser & Strauss, 1967) are found in the works of Barry Turner (see Le Coze, Chapter 4, this volume) but also the foundational works within Normal Accident Theory (NAT) and High Reliability Organisations (HRO). Other case studies followed which contributed significantly to the vocabulary of safety research, such as Vaughan's study of the *Challenger* disaster (1996), Snook's study of the accidental shootdown of a US Army helicopter (2000) and Hopkins' study of the gas plant explosion in Longford (2000).

The aim of this chapter is to discuss the role of case studies in safety science and the role case studies can play in the future development of concepts, models and theories within safety research. By elaborating on the case study as a research activity, we highlight issues that do not always receive the attention they deserve, although they are decisive for the direction of research with respect to both design and outcome.

WHAT IS A CASE, AND A CASE OF WHAT?

Before embarking on the role of case studies in safety science, there are two more general questions that need to be answered: (1) What is a case, and (2) What are they cases of?

The first question may be answered with reference to the widely cited American social scientist Robert K. Yin, author of several books on case study methodology. His definition of a case study is the following:

> A case study is an empirical inquiry that investigates a contemporary phenomenon (the "case") in depth and within its real-world context, especially when the boundaries between phenomenon and context may not be clearly evident. (Yin, 2014: 16)

The word "case" refers to different things. Some use the term to refer to the empirical units in which scientific investigation is done, e.g. an organisation. Others refer to cases as theoretical constructs that emerge out of a social scientific analysis, e.g. the concept of an incubation period for man-made disasters (Turner, 1978). In the first meaning of the term, cases are entities that can be "found." In the second meaning, cases are constructed by invoking theory (see Ragin & Becker 1992 for a thorough discussion). The term "case" can thus refer to both context and phenomenon. In both meanings of the word, a remaining question is, what are the cases "cases of"? There is an, often implicit, assumption that the cases we study belong to a class of empirical units or phenomena that makes knowledge about them relevant outside the particular context where the study takes place. Within safety research, the cases studied share the trait that they are situated in a context where a source of hazard exists and some technological and organisational arrangements are in place to reduce the likelihood or magnitude of unwanted consequences. The link between the cases studied and the class of contexts or phenomena they represent is, however, rarely specified or discussed in studies of safety.

There are two important lessons to be drawn out of the general literature on case studies. One is Yin's emphasis on the relationship between the phenomenon and the context within which it is situated, and from which it is not always easily distinguishable. The other is the relationship between empirical observation and the development of concepts, models and theory. Both these have to do with the potential for generalisation across contexts, and how the strategy for selection (or construction[1]) of cases influences this potential. Not all case studies are based on ambitions of generalisation. Studies may be atheoretical, i.e. inductive descriptions of a phenomenon in itself, or use theory to describe and understand empirical observations. Others have ambitions of either concept or theory development, or even to test hypothesis by means of case studies. We focus our discussion on case studies with some level of ambition to create knowledge that is relevant outside the context of the phenomenon that is studied.

Strategic case selection is a crucial bridge between the particular context and more general applicability. Great opportunities lie in carefully delineating the phenomenon under study, selecting the context in which it is studied, and connecting the study to a clear theoretical potential. Strategic selection of cases provides a form of "variable control." One of the keys to unlocking this potential lies in a case strategy based on making deliberate choices about variation in the contexts of the phenomenon studied. Barry Turner's (1978) work, as the starting point for much organisational research in this field, is based on a deliberate case strategy. He included a variety of contexts where accidents had occurred in order to look for accident-inducing processes that were common across contexts. This allowed for a high degree of transferability of his findings. Other times, one wishes to reduce contextual variation. This means studying the same phenomenon many times with low variation in the context studied, which – when taken to the extreme – is the basis of experimental studies, such as investigations of human responses to audiovisual signals. Here, it will make sense to have as little contextual variation as possible (no distractions), but to have as much possible variation in the characteristics of the humans studied (e.g. gender, race, age, etc.).

In both case selection strategies – either maximising or minimising variation – there is an underlying ambition to *develop* concepts, models and theory rather than explaining phenomena with reference to *existing* concepts, models and theory. While the link to existing theory has been debated in the literature on qualitative methods (e.g. around the grounded theory approach), our view is that qualitative research is always theoretically informed but not theoretically determined. Equally, theory also needs reference points based on empirical observation in order to consider issues such as a theory's validity and the limits of its application.

The relationship and interplay between data and theory building thus lie at the core of case study research. This is particularly prominent in studies that are conducted from a grounded theory perspective (Glaser & Strauss, 1967, see also Le Coze, Chapter 4, this volume). This is an approach aiming at conceptualising patterns ("theories") in rich empirical data, consisting of a cycling between induction and deduction where the intent is that concepts, models or theories emerge from (that is, are grounded in) empirical observation. It consists of open coding of interview or other qualitative data, then organising data categories into more comprehensive concepts, which, in turn, are mirrored back to the empirical data to test their relevance and precision. Glaser and Strauss (1967) stress the need for constant theorising by writing short theoretical memos where the key concepts are identified and the relationship between them described. For instance, they studied how nurses responded to the death of patients and their handling of the next of kin. As they coded the descriptions of incidents, they generated the concept of "social loss" to denote the way nurses attached value to their patients by referring to age, education, beauty, talent and so on. This would be a topic in a dedicated theoretical memo, where the aim is to start the theorising process as early as possible in the analysis.

While case studies within safety research do not always make explicit reference to grounded theory, this way of developing concepts, models and theories on the basis of empirical observation has been influential in safety research.

THE CLASSICS OF SAFETY RESEARCH

As already indicated, we believe that case studies represent key building blocks in the foundation of organisational safety research. Barry Turner was a pioneer in this respect, not only by establishing a sociology of safety but also by introducing a grounded theory approach to safety research. Turner's analysis in *Man-Made Disasters* was based on qualitative analysis of 84 British accident investigation reports published between 1965 and 1975. Turner's motivation for doing this analysis was to identify common patterns that, in turn, could provide the basis of a model for understanding disasters. Summing up his own work, Turner describes his main contribution in the following way:

> I identified a pattern which suggests that large-scale accidents have many causes rather than a single cause and that their preconditions build up over a period of time, rather than springing into existence instantaneously. The model points to the way in which crises and disasters develop in a covert and unnoticed fashion during an "incubation period." (Turner, 1992: 192)

While there is much more to Turner's sociology of safety than is summed up in this quote, it illustrates key features in his research approach. First, his case strategy aims to maximise the number of units studied, and have as much variation between them as possible: a wide variety of disasters in a wide variety of contexts. If there are common traits that exist in highly different settings, then the findings will be more robust than if they were based only on one or a few specific contexts. Turner was aiming for generalisation, that is, moving from empirical description of particular observations to the formulation of models and theories with relevance outside the contexts of which the empirical analysis originated. Turner provided both models and theories. His accident model, ranging from a notionally normal situation, via an incubation period, a trigger event, onset of a crisis, rescue and salvage, to full cultural readjustment, has become a widely used graphical representation of disaster phases. His theoretical contributions gravitate around the concept of the incubation period and the observation that disasters are characterised by "the cultural disruption which is produced when anticipated patterns of information fail to materialize" (Turner & Pidgeon, 1997). From the empirical patterns found in the investigation reports, Turner (and Pidgeon) develops a vocabulary for understanding what goes on in the incubation phase, of which concepts like the "rigidities of perceptions and beliefs" and "decoy phenomenon" are probably the most well known. Importantly, this is done by developing the key concepts on the basis of in-depth studies of three of the cases, and "testing" the validity of these concepts on an additional ten cases. Moving between concept development and relevance testing in the empirical analysis is key to the grounded theory approach (Glaser & Strauss, 1967).

The point with this brief expose of Turner's work is not to go into details of Turner's vocabulary and theory but to emphasise that his contributions are the results of (1) a careful design of the case study, (2) an underlying ambition to create general concepts, models and theory, and (3) an ongoing interplay between theory and data. It is safe to say that Turner was a pioneer in utilising the strengths of case study methodology in safety research.

Several excellent studies followed in the 80s and 90s. Perrow's *Normal Accidents* (1984, 1999) can be seen as a case study, where "case" means the theoretical construct of the match between technological properties on the one hand and organisational control strategies on the other. While his main emphasis is on nuclear technology, he contrasts this to other industries. Perrow famously mapped different industries according to the axis of coupling and complexity, providing a comparative basis to his analysis, and also indicating the generalisable relevance of his key concepts and conclusions. While Perrow would perhaps not label *Normal Accidents* a case study, he introduces his book in the following way:

> In this book we will review some of these [high-risk] systems – nuclear power plants, chemical plants, aircraft and air traffic control, ships, dams, nuclear weapons, space missions and genetic engineering. Most of these risky enterprises have catastrophic potential, the ability to take the lives of hundreds of people in one blow (...) The good news is that if we can understand the nature of risky enterprises better, we may be able to reduce or even remove these dangers. (Perrow, 1999: 3)

The ambition is clear – to analyse a variety of instances where there is a relationship between high-risk technologies and organisational strategies to control the risks involved, with an ambition to develop general knowledge about "the nature of risky enterprises."

Research on High Reliability Organisations (HRO) (e.g. Rijpma, 1997) is also based on a deliberate case strategy – the units they study are cases of organisations or operations where theory would predict major accidents to happen regularly, but where things seem to go wrong less often that one would expect. Moreover, the case design is theoretically informed and positioned according to Perrow's concepts, but they still have clear theoretical and conceptual ambitions on their own. Once the cases are selected, they can go out and gather information about the properties of these organisations. The key to the quality of the study lies in their strategic selection of cases and their ability to build concepts out of empirical observations.

Another excellent case study within safety research is Scott Sagan's *The Limits of Safety* (1993). Sagan sets out to study the intersection between HRO and Normal accident research, with a design aiming to test the explanatory power of both theoretical frameworks. Based on historical documents he goes behind the surface of seemingly highly reliable organisations and studies mishaps and near-misses where things could have gone very wrong. What makes Sagan's work a classic is the way he positions his empirical work in the nexus of two theoretical frameworks. It is also a good example of the construction of a case. There is no specific organisation under study in Sagan's work, only documents describing what goes on behind the scenes of seemingly highly reliable organisations.

Vaughan's book *The Challenger Launch Decision* (1996) is a somewhat different case design. It is a single-case study of the way the safety margins of high-tech organisations can slowly degrade due to an interplay between external pressure, organisational and cultural factors and decision-making processes. What makes her study an important contribution to safety science is the impressive depth of the investigation, the way she uses multiple sources of data and the way she distils general concepts (e.g. the normalisation of deviance) out of particular historic trajectories in a high-risk organisation.

We have only singled out a few studies here. Others have already been mentioned, such as Snook's *Friendly Fire* (2000) and Hopkins' *Lessons from Longford* (2000). The point would still be the same – most of the literary canon in the domain of organisational safety stems from high-quality case studies combining rich empirical material with concept development and theory building.

CASE STRATEGIES IN SAFETY RESEARCH

Now we have stressed the importance of the strategic design of case studies, we will move on to describe some examples of such strategies for case studies with ambitions to either develop new concepts, models and theories, or to shed light on the validity of existing concepts, models and theories. We base these examples on a combination of the general literature on case studies and studies of organisational safety.

INDUCTIVE CASE STUDIES

Inductive case studies are based on an open, exploratory empirical approach, with limited theoretical influence at the onset of the study. This being said, the label "inductive" should not be taken too literally. All empirical observation is in some form theory laden in the sense that the choice of which phenomenon to study, and in which context, is usually influenced by existing knowledge. Turner's work is an example of this (see Le Coze, Chapter 4, this volume). The choice of phenomena (accidents) and their context (formal organisations) is far from incidental. Nevertheless, the approach starts out inductively in the sense that the theoretical hypotheses originate in the empirical material. Despite the inductive outset, the empirical analysis presented by Turner & Pidgeon (1997) also contains a more deductive side where findings are "tested" against the same empirical data, according to the grounded theory approach described above.

This approach to case studies makes it more akin to basic research than applied research. The strength of the approach is that it provides theoretical perspectives with a solid empirical foundation. It has a high potential of developing new concepts that can illuminate immature research areas and spur further theoretical development and empirical research. The downside is that, in established research areas at least, it comes with a risk of reinventing old insights under new names and brands. While there is usually more academic prestige in inventing new concepts rather than being able to apply old ones, the result could be a plethora of different concepts basically referring to the same phenomenon (see also Schulman, Chapter 10, this volume).

DEDUCTIVE CASE STUDIES – OF THE LEAST LIKELY CASE

Some studies use case strategies aimed at testing, refining or nuancing existing theoretical concepts. "Testing" is here not meant in the sense of quantitative hypothesis-testing, rather as an explicit ambition of finding the limits of applicability of existing theoretical frameworks. One such strategy is labelled "the least likely case" (Eckstein, 1975, Andersen, 2005) – trying to find a case where it is least likely that a theory has explanatory power. This is an approach that lies at the heart of social science. Durkheim's (1951) classic study of the suicide patterns in different religious, socio-economic and cultural contexts is an example of this. His theoretical ambition was to show that a purely psychological and atomistic view of human beings was only half the story, the other half being their social nature and context. In order to analyse his hypothesis, he chose to study the most individual decision there is – a person's decision to end his or her life. By studying this case, Durkheim put his hypothesis to the most challenging test. When he found the influence of social integration and social regulation on the frequency of suicide across different contexts, he was able to draw robust conclusions and create a general theory on social integration and individual action, and in the process established sociology as a scientific discipline.

In many ways, HRO research belongs to the same category, as their cases are portrayed as systems that should be exposed to the Perrowian dilemma between centralisation and decentralisation as control strategies for tightly coupled and interactively complex technology. If Perrow was right in his predictions, the nuclear-powered hangar ships should be prone to major accidents. When they are not, and the organisations seem to be able to reconfigure between centralisation and decentralisation, this can be seen as challenging the validity of Perrow's model.

The strength in the "least likely" approach lies in its ability to challenge existing theories and models. This does not have to be a question of proving theories right or wrong – it has to do with shedding light on their areas of application. For instance, HRO researchers rarely denounce the insights of Normal Accident Theory. It can just as well be seen as way of showing how things go right, even in complex systems operating in demanding conditions. The pitfalls of this approach lie in the possible construction or exaggeration of the antagonism between different theoretical frameworks, which for a long time was the case between HRO and NAT (Rijpma, 1997, 2003).

DEDUCTIVE CASE STUDIES – OF THE MOST LIKELY CASE

For studies using the "most likely case" approach (Eckstein, 1975), the aim is to find a case where it is the most likely that a theory should apply and put it to the test (usually based on a hypothesis that it does not apply). If it does not apply, then there is good reason to believe that it is flawed in one way or another. One example of this approach is Antonsen's (2009) comparison of the results of an investigation of a serious gas release on the offshore platform Snorre Alpha, with the results of a safety climate survey done on the same installation one year prior to the accident. The study originated in an interest to examine the links between the "before" and "after" descriptions of the same phenomenon (safety culture at the installation). The investigation painted a rather gloomy picture of the safety culture on Snorre Alpha. If safety climate surveys are to be able to serve as valid indicators for major accident risk, it should be possible to find some traces of a "poor safety culture" before the accident. This is what makes this an example of a "most likely case" approach. The "before" picture of safety culture was almost the diametric opposite of the "after" picture, and once other possible explanations were considered, Antonsen concluded that the difference could be related to an inadequacy of safety climate surveys to serve as a stand-alone method to describe safety cultures. Note that Antonsen's (2009) claim goes beyond stating that this particular survey was unable to show signals of danger related to the safety culture at Snorre Alpha. The claim is that the survey, which admittedly had its weaknesses, could be seen as a case of safety climate surveys in general, thereby making a generalisation.

One of the pitfalls of this approach, like the "least likely case" approach, is the construction of theoretical propositions that do not pay respect to the nuances in the theories being put under scrutiny. For instance, in the review process of Antonsen's paper, one of the reviewers rejected the paper, giving very insightful reasons for doing so. The reviewer felt that the paper's argument was based on a false premise, wrongfully attributing predictive ambitions to safety climate research in its entirety. One might agree or disagree with the reviewer's conclusion, but the discussion was an example of a foundational issue and debate that would be highly interesting to have out in the open, and not only in the closed review process. This is why we see value in making bolder theoretical claims through empirical research – it invites criticism, sometimes annoyance and often efforts to prove other researchers right or wrong.

HOW CAN SAFETY SCIENCE BENEFIT FROM FUTURE CASE STUDIES?

We believe there is great theoretical (and empirical) value in carefully designed case studies within the domain of safety science. There is no lack of possibilities in defining such case studies, but we will highlight some approaches which we believe can create important insights about safety. These approaches belong to two general categories – studies at the intersection between theories, and comparative case studies where context is treated as an independent variable.

STUDIES AT THE INTERSECTION BETWEEN THEORIES

Safety science is comprised of different communities and schools of thought, sometimes with limited communication between them. Often, there is an under-utilised potential lying at the intersection between these. One example is the relationship between Resilience Engineering (RE) and HRO research, which was the topic of a 2019 special issue of *Safety Science*.

RE (Hollnagel et al., 2006, 2008, 2011) emerged as a safety perspective continuing a tradition of focusing on the mechanisms behind successful outcomes – like HRO –

and the strategies of coping with and adapting to the unexpected – like Wildavsky (1988). It has later cultivated such perspectives and developed into a safety paradigm of its own with defining concepts such as Safety II (Hollnagel, 2018), functional resonance (Hollnagel, 2012) and efficiency-thoroughness trade-offs (Hollnagel, 2009).

In considering this field, Haavik et al. (2019) discussed two topics of relevance for case studies in safety research. The first is the generative potential that lies in "feeding off controversies" between different schools of thought. In the diversity of approaches to safety (Le Coze, 2019) lies a great potential for a richer understanding of safety, and a "silent truce" between proponents of RE and HRO is not the best way to trigger advances in our understanding of how safety or reliability is produced.

The second, and related, point is that studies in the intersection between theoretical frameworks are particularly useful when it comes to building theory from empirical data. In our 2019 paper on RE and HRO (Haavik et al., 2019), we tried to follow how the two research traditions "travelled" into new empirical domains, with the healthcare sector as an example. We found that RE was primarily used as a framework for close-up, situational studies of normal work operations, while HRO was a theoretical lens to study and improve organisational design. In this way, their differences (at least in their application) were more related to the level of analysis than their theoretical essence. If this contention is right, this intersection would be fertile ground for case studies of how resilience plays out in normal operations within the context of organisations designed according to HRO principles.

As underlined in Yin's definition of case studies, it is not always clear where to draw the boundary between the phenomenon under study and the context in which it is studied. Taking this seriously would imply that the understanding of a phenomenon requires detailed knowledge and description of the context. For instance, if successful resilient practices in normal operations are what is studied, this should include a consideration of the organisational context that makes successful outcomes more likely. The other way around – if you study organisations from an HRO perspective, the study should not be satisfied by understanding the overarching organisational principles but also zoom in until the practices become visible and the effects of redundancy, the preoccupation of failure, etc., manifest themselves. We acknowledge that constructing a case must involve some sort of framing, foregrounding some aspects and backgrounding others, and that it is impossible to combine extremes of depth and width in all studies. Nevertheless, we see potential for developing new insights from studies that aim to combine different theoretical lenses to shed light on how safety in organisations is created and recreated. When it comes to the case strategies described above, this would represent an inductive orientation to the phenomena studied, although the selection of the context for the study (the organisations) would be strongly informed by theory.

Another contemporary theoretical intersection in the safety domain is between what is termed Safety I and Safety II (Hollnagel, 2013, 2018) and the connected discussion of safety as a "dynamic non-event" (Weick, 1987) and safety as "dynamic events" (Hollnagel, 2018; Haavik, 2013; Macrae, 2014). A core question in this discussion is whether safety can best be understood by studying failure or success. This question touches upon the ontological and epistemological foundations of safety as a research object. The perspective that safety is a dynamic non-event (Weick, 1987), implying that much of the work that prevents accidents from happening is invisible, has recently been subject to criticism (Hollnagel, 2018; Haavik, 2013). One of the comments is that visibility (or invisibility) is not to do with the essence, or the ontology, of these phenomena, but rather with the methodology for investigating them. One possible explanation of why certain phenomena are not visible to safety researchers is that safety science has not sufficiently integrated theories and methods from other academic and research disciplines and traditions such as sociology and anthropology of work (e.g. Suchman, 1995; Goodwin, 1994; Star and Strauss, 1999; Latour, 1987), which for decades have explored and analysed (invisible) work. With the notion of dynamic events (Hollnagel, 2018; Haavik, 2013; Haavik et al., 2019), attention is led to safety research that builds on the orientation of HRO and RE towards successful and normal operations, and the theoretical and methodological repertoire from the sociology and anthropology of work. This equips us with a conceptual toolbox for strategically constructing and undertaking a type of case studies that, so far, has not been particularly accessible. For instance, Rosness et al. (2016) studied the way that the collaborative structures in a surgical team were instrumental in enabling a capacity to anticipate and prepare for undesired but foreseeable events. The study makes visible what *is* going on when unwanted consequences do *not* occur, and does this by delving into the micro-level patterns of prospective sensemaking in the surgical team.

Still, there are some peculiarities with the context of hazardous organisations that influences the very notion of success, and that is that failure can have devastating consequences. This would mean that the cases for studies of normal operation aimed at understanding why major accidents do not happen should be positioned against the possibility that something really bad might actually happen. This is essentially what the HRO researchers did when studying why the naval aircraft carriers experienced less than their fair share of failure. What makes the HRO studies so relevant for understanding success when it comes to reliability is the unacceptability of failure. Again, we see how central the selection of the case is for the robustness of the findings, and issues of case design should not be dealt with lightly. It may require a lot of theoretical and practical work, but when approaching the shared ambition of HRO and RE – of learning not only from that which goes wrong – it is necessary.

COMPARATIVE CASE STUDIES – CONTEXT AS AN INDEPENDENT VARIABLE

Comparative case studies provide opportunities to select cases that vary in some respect but are equal in other respects. This can provide a form of "variable control" that increases the basis for developing concepts, models and theories (Andersen, 2005). One option is to study the same phenomenon, e.g. improvisation, in contexts that are as similar as possible, looking for similarities and differences in the way improvisation occurs. Another is to study the phenomenon in contexts that vary as much as possible, e.g. to identify common patterns in safety-related improvisation that can be found in different contexts, or differences in the room available for improvisation in various contexts. In both strategies,[2] the aim is to develop a research design with a level of control over variation and similarities between the cases that are to be compared.

Some such studies already exist in safety science. Nielsen et al. (2008),[3] for instance, studied two "twin" plants from the same manufacturing company where one of them was subjected to a safety intervention, while the other was not. The plants made identical products, had the same production system, organisational structure, top management, policies and safety management system. While there may be several sources of variation between the plants that was beyond the control of researchers, their design provided a more robust basis for concluding that the intervention actually worked than would have been the case if they only had information about the plant where the intervention was made. This is a well-known "experimental" design where the strategic design and selection of cases helps to strengthen the knowledge claims produced by the case study – which represents a way of designing case studies that we believe is under-utilised in safety research (see also Martin, O'Hara and Waring, Chapter 11, this volume).

THE POLITICS AND PRAGMATICS OF CASE SELECTION

As should be clear by now, case selection is closely connected to the type of knowledge one sets out to build. Careful and strategic selection of cases and hence a deliberate foregrounding and backgrounding of phenomena can make the difference between an original and fruitful case study, and one that merely reproduces what is already known.

In an ideal world, choices about research and case study design could be made without too many constraints. In practice, a number of limitations apply. Many of these limitations fall outside the category of things we usually think of as purely scientific issues, and as convincingly demonstrated in a large body of literature from Science and Technology Studies (Sismondo, 2010) there is no such thing as "pure" science. Every choice made in science and research has political dimensions, just as there exists no (credible) politics without science (Latour, 2004). In organisations where production processes involve major accident risk, safety will be a sensitive topic in terms of organisational reputation, shareholder value and the licence to operate within their industries. For instance, for an airline to be part of a study of safety culture can be considered too risky in terms of the possibility of negative information reaching the media, passengers and shareholders. For studies with limited financial means, it may be hard to include enough participating companies to guarantee complete anonymity and we suspect that many safety researchers have experience with organisations choosing not to ask questions because the answers may be hard to handle (see also Bourrier, and Hayes and Maslen, Chapter 12, this volume). This means that not all cases will be available for "selection" when safety researchers design their case studies to gain insight into the way safety and accidents may be created in high-risk organisations.

Research – whether it is done within academic institutions, research institutes or industrial companies – is usually organised as projects, and the establishment of a specific project is often subject to competition with other project proposals. In that competition, the strategic priorities and programmes of the granting authorities – which are influenced by political processes in the university, the research council, company management, etc. – may be as important, or more important, than the scientific rigour in the research design. One example of this is the guidelines for European research programmes like H2020 that often require empirical studies in research projects to include locations that reflect the composition of nations in the EU. An important question in such circumstances is to what degree and how such requirements influence the formulation of research questions, and to what degree researchers are pragmatic and decide to "make the best of it" with a suboptimal case design.

Case design is also influenced by practical limitations. Case selection cannot be done without considering where one may get access, for practical or economic reasons (see also Bourrier, Chapter 3, this volume). For example, if one wants to study communication and collaboration in complex and technology-rich safety-critical settings, one may have a first choice case where one does not get access (for example, an emergency trauma unit), and ends up with a case at a surgical operations theatre where most of the procedures are planned in advance. Or, for example, one may have a research project studying offshore/onshore collaboration in the oil and gas industry, where one ideally would want to spend as much time at the offshore facilities as the onshore facilities, but for reasons of limited bedspace offshore must carry out the case study spending only 20 per cent of the time offshore and 80 per cent onshore, with corresponding implications for the data generated.

A last constraint that deserves some consideration is the room and opportunity for publishing case studies. The classics of safety science mentioned in the introduction of this chapter have a common feature in addition to being excellent case studies – they are all books. While there are also high-quality case studies published as journal articles (e.g. Schakel et al., 2016), the journal format has clear limitations regarding the level of detail in the description of rich empirical material and the way the relationship between theoretical constructs can be delineated. In the "publish or perish" era there are strong incentives towards publishing research results in journals, and this is likely to influence the way knowledge from case studies reaches the audience of both researchers and practitioners. In general, one may say that the framework conditions for doing extensive qualitative case studies, particularly basic research, is under pressure from increasing political and practical constraints, and this is something that is reflected in the type and quality of knowledge that is produced.

CONCLUSION

In this chapter, we have used previous approaches to case studies as a point of departure for considering the role of case studies in safety science. Our main argument is that case studies have played a vital role in creating key concepts, models and theories in this research field, and that it should continue to play a key role in further development.

At this point of the chapter, readers trained in qualitative methods may experience a slight after-taste connected to our use of terms like "variables," "deductive case studies" and "testing of theories." After all, the qualitative research traditions have strived for decades to be recognised as a research tradition equal to the quantitative paradigm. Are we now calling for a positivist turn in safety science? The short answer is that we are not. We see great value in qualitative research, be it ethnographic studies, storytelling approaches, or discourse analysis. What we do call for is more explicit links between theoretical perspectives and empirical analysis in order to utilise the full potential of our research efforts, irrespective of whether the studies are qualitative or quantitative. This is particularly important when the technology and organisation of hazardous systems is rapidly changing, at the same time as there are fundamental changes in their societal and global context. Making our case, the strength and importance of a strategic approach to case design comes before the choice of individual methods, and is central to the process of actively positioning empirical case studies in relation to theoretical ambitions. Such positioning is in our view a prerequisite for a more cumulative progression of safety science and the development of new perspectives to understanding safety in organisations.

NOTES

1. In the following, all references to *case selection* also includes *case construction*. The different epistemologies of selection and construction is important, but will not be elaborated further here.
2. In the case literature, these two strategies are labelled "most similar systems" and "most different systems" approaches (Andersen 2005).
3. This is a quantitative study, illustrating that case studies should not be treated as synonymous to qualitative research.

REFERENCES

Andersen, S. S. 2005. *Case-Studier og Generalisering* (In Norwegian. Case Studies and Generalization). Bergen: Fagbokforlaget.

Antonsen, S. 2009. Safety Culture Assessment: A Mission Impossible? *Journal of Contingencies and Crisis Management* 17(4): 242–254.

Durkheim, É. 1951. *Suicide: A Study in Sociology*. New York: Free Press.

Eckstein, H. 1975. Case Study and Theory in Political Science. In: I. Greenstein and N. W. Polsby (Eds.), *Handbook of Political Science*. Reading, MA: Addison-Wesley, pp. 79–133.

Glaser, B. G. and Strauss, A. 1967. *The Discovery of Grounded Theory: Strategies for Qualitative Research*. Chicago, IL: Aldine.

Goodwin, C. 1994. Professional Vision. *American Anthropologist* 96(3): 606–633.

Haavik, T. K. 2013. *New Tools, Old Tasks: Safety Implications of New Technologies and Work Processes for Integrated Operations in the Petroleum Industry*. Farnham: Ashgate.

Haavik, T. K., Antonsen, S., Rosness, R. and Hale, A. 2019. HRO and RE: A Pragmatic Perspective. *Safety Science* 117: 479–489.

Hollnagel, E. 2009. *The ETTO Principle: Efficiency-Thoroughness Trade-Off – Why Things That Go Right Sometimes Go Wrong*. Aldershot: Ashgate.

Hollnagel, E. 2012. *FRAM: The Functional Resonance Analysis Method: Modelling Complex Socio-Technical Systems*. Farnham: Ashgate.

Hollnagel, E. 2013. A Tale of Two Safeties. *Nuclear Safety and Simulation* 4(1): 1–9.

Hollnagel, E. 2018. *Safety-I and Safety-II: The Past and Future of Safety Management.* Boca Raton FL: CRC Press.

Hollnagel, E., Woods, D. D. and Leveson, N., eds. 2006. *Resilience Engineering: Concepts and Precepts.* Aldershot: Ashgate.

Hollnagel, E., Nemeth, C. P. and Dekker, S. 2008. *Resilience Engineering Perspectives: Remaining Sensitive to the Possibility of Failure.* Aldershot: Ashgate.

Hollnagel, E., Pariès, J., Woods, D. D. and Wreathall, J. 2011. *Resilience Engineering in Practice: A Guidebook.* Farnham: Ashgate.

Hopkins, A. 2000. *Lessons from Longford: The Esso Gas Plant Explosion.* Sydney: CCH Australia Ltd.

Latour, B. 1987. *Science in Action: How to Follow Scientists and Engineers Through Society.* Milton Keynes: Open University Press.

Latour, B. 2004. *Politics of Nature: How to Bring the Sciences into Democracy.* Cambridge, MA: Harvard University Press.

Le Coze, J. C. 2019. Vive la diversité!. High Reliability Organisation (HRO) and Resilience Engineering (RE). *Safety Science* 117: 469–478.

Macrae, C. 2014. *Close Calls. Managing Risk and Resilience in Airline Flight Safety.* London: Palgrave Macmillan.

Nielsen, K. J., Rasmussen, K., Glasscock, D. and Spangenberg, S. 2008. Changes in Safety Climate and Accidents at Two Identical Manufacturing Plants. *Safety Science* 46(3): 440–449.

Perrow, C. 1984. *Normal Accidents: Living with High-Risk Technologies.* Princeton, NJ: Princeton University Press.

Perrow, C. 1999. *Normal Accidents: Living with High-Risk Technologies.* Princeton, NJ: Princeton University Press.

Ragin, C. C. and Becker, H. S. 1992. *What is a case? Exploring the foundations of social inquiry.* Cambridge: Cambridge University Press.

Rijpma, J. A. 1997. Complexity, Tight-Coupling and Reliability: Connecting Normal Accidents Theory and High Reliability Theory. *Journal of Contingencies and Crisis Management* 5(1): 15–23.

Rijpma, J. A. 2003. From Deadlock to Dead End: The Normal Accidents-High Reliability Debate Revisited. *Journal of Contingencies and Crisis Management* 11(1): 37–45.

Rosness, R., Evjemo, T. E., Haavik, T. and Wærø, I. 2016. Prospective Sensemaking in the Operating Theatre. *Cognition, Technology and Work* 18(1): 53–69.

Sagan, S. D. 1993. *The Limits of Safety: Organizations, Accidents, and Nuclear Weapons.* Princeton, NJ: Princeton University Press.

Schakel, J.-K., Fenema, P. C. v. and Faraj, S. 2016. Shots Fired! Switching Between Practices in Police Work. *Organization Science* 27(2): 391–410.

Sismondo, S. 2010. *An Introduction to Science and Technology Studies*, Vol. 1. Chichester: Wiley-Blackwell.

Snook, S. A. 2000. *Friendly Fire: The Accidental Shootdown of U. S. Black Hawks over Northern Iraq.* Princeton, NJ: Princeton University Press.

Star, S. L. and Strauss, A. 1999. Layers of Silence, Arenas of Voice: The Ecology of Visible and Invisible Work. *Computer Supported Cooperative Work* 8(1–2): 9–30.

Suchman, L. 1995. Making Work Visible. *Communications of the ACM* 38(9): 56–64.

Turner, B. A. 1978. *Man-Made Disasters.* London: Wykeham Science Press.

Turner, B. A. 1992. The Sociology of Safety. In: D. Blockley (ed.) *Engineering Safety.* London: McGraw-Hill, pp. 187–205.

Turner, B. A. and Pidgeon, N. F. 1997. *Man-Made Disasters.* Boston: Butterworth-Heinemann.

Vaughan, D. 1996. *The Challenger Launch Decision.* Chicago: The University of Chicago Press.

Weick, K. E. 1987. Organizational culture as a source of high reliability. *California Management Review* 29(2): 112–127.
Wildavsky, A. B. 1988. *Searching for Safety*. Berkeley, CA: University of California Press.
Yin, R. K. 2014. *Case Study Research: Design and Methods*. Thousand Oaks, CA: Sage.

6 Actor Network Theory and Sensework in Safety Research

Torgeir Kolstø Haavik

CONTENTS

INTRODUCTION

This chapter can be read as an introduction to how Actor Network Theory (ANT) can be made relevant and powerful for safety research. Being as much an ontology and a methodology as a theory, ANT-inspired studies offer new perspectives and new takes on safety research in sociotechnical systems, and inspire searches for new themes and concepts to support the adaptation and relevance of safety research in a rapidly changing world. This becomes particularly apparent if we consider how the world and its sociotechnical systems are becoming increasingly tightly coupled in networks traversing geographical spheres and hierarchical levels, and where the proliferation of artificial intelligence artefacts blurs the borders between humans and non-humans; we can no longer easily define in advance the external borders of the systems we want to investigate, nor can we decide in advance what classes of actors to follow and study, since *who is acting* is an empirical question, and *where causes originate from and where effects are manifest* is highly uncertain. Thus, a risk analysis in a hospital or in an airline cannot focus on *either* technical *or* human aspects since the gestalt of failures and successes do not discriminate between such categories. Likewise, societal resilience cannot anymore be studied within the borders of nations alone, since risks travel so swiftly across the globe and produce new politics of the global, all with potentially different local implications – be it a new coronavirus, a carbon fuelled culture or large-scale migration.

Sensework will in this chapter be introduced both as an empirical phenomenon and as a research programme oriented towards studying a particular domain, that of sociotechnical work in safety-critical operations, where groups of professionals in technology-rich environments work to put together different sorts of representations to give meaning to, and work through, familiar and unfamiliar situations. To an increasing extent, safety-critical work – be it complex police operations, remote and virtual air traffic towers or the organisation of underwater drone activity in connection with offshore energy installations – is organised in contexts that resemble control rooms or operation centres, where digital information is abundant and continuous framing and sensemaking is necessary. Sensework is a label for such work and such contexts.

ANT as a theoretical and methodological approach to science and technology studies can be traced back to the early 1980s, to works by Latour and Woolgar (1979), Callon (1986) and Latour (1987). At the time, ANT contrasted the discipline of sociology of scientific knowledge (Bloor, 1976, 1999; Latour, 1999a) with a move away from mere social explanations of science, and in the years to come ANT would also contrast postmodern sociology with an explicit turn to materiality (see e.g. Latour, 1992). What ANT in particular may offer the field of safety is a methodology that supports inclusion of social aspects of construction in safety research – compared to the dominant positivist, linear cause-effect oriented approach to risk and safety. The reason why ANT may represent these somewhat opposite positions towards sociology on one hand and safety research on the other is not that ANT offers a particularly precise balance between the social and the material (or subject/object, or society/nature) but, to the contrary, that it offers an ontology and a methodology that allow us to describe and analyse any sociotechnical system without treating such aspects in a dualistic manner at all; the central (though admittedly somewhat unsophisticated) reference to *humans and non-humans* in ANT, and the insistence on treating these aspects symmetrically and without any preconception about agency, allows us to analyse contexts where the borders between social and material aspects are irrelevant, or at best unclear.

An ANT view on what the world is made of – an ANT ontology – assumes a fundamentally relational world where everything is constructed, and – without maintenance – nothing is permanent. A crucial implication of this is that nothing – not "facts," not "religion," not "culture," not "safety" – exist in their own essence and without influence of the sociomaterial networks they are weaved into and out of. Note that this is not the same as social construction, which gained considerable popularity in the wave of postmodern theory in the 1980s and 1990s, where so many established institutions could seemingly be readily deconstructed because they were viewed as purely human constructs; in a constructivist ontology in line with ANT, the more social and material elements and processes that enter into networked phenomena, the more solid the phenomena are – and any reference to weak and strong construction in an ANT context would relate to this. Hence, according to ANT, facts may be very solid, although we would hesitate to call them "true." The same goes for "safe" and "risk." The literature on ANT perspectives on safety is sparse, but such an ANT ontology has been operationalised by Hilgartner (1992) in an illuminating manner in his elaboration of risk objects and auto-based transportation, which shows how the construction – we could say sociotechnical construction – of risk may be addressed and understood in quite different ways in an ANT analysis than through more traditional and engineering-oriented risk analyses (more about Hilgartner's study later).

The sensework perspective acknowledges and operationalises the ANT ontology in a context of organisational and operational safety. At the intersection between Weick's works on sensemaking, the NAT/HRO controversy on technological determinism/social construction and the Safety I/Safety II debate, sensework operationalises an ontological and methodological stance influenced by ANT, with respect to high-tech, safety-critical sociotechnical systems. One of the conundrums of safety science that thus becomes more tangible and manageable is that of *safety as a dynamic non-event* (Sutcliffe and Weick, 2013; Wcick, 2011): as we shall see later, there are ways – well anchored in theory and empirical demonstrations – that allow us to investigate and analyse *safety as dynamic events*. One of the great advantages of this is that we do not any longer have to search in the dark for an invisible safety, since we have methods to approach safety as *visible* phenomena.

First, however, we shall introduce some general principles of ANT, before we describe the sensework perspective in some greater depth, some of its practical achievements and some future prospects.

ACTOR NETWORK THEORY: SOME CENTRAL PRINCIPLES

ANT is a body of empirical and theoretical writings where sociomaterial relations are treated as network effects. These networks are always heterogeneous, in the sense that social relations, including power and organisation, are deeply entangled with materiality. Accordingly, in the first instance, the different ingredients of organisations, such as agents, texts, technologies, plans, architectures and the like, will in an ANT account be analysed in the same terms (Law, 1992).

ANT has, since the first publications that can be considered ANT writings – e.g. Latour and Woolgar (1979), Callon and Latour (1981), Callon and Law (1982), Callon (1986) and Latour (1987) – shown an impressive (and to many, frustrating) ability to avoiding stabilisation, not the least through a vocabulary constantly in motion. However, although articulated in different manners, some foundational assertions are traceable in most ANT writings, and these assertions make up the backbone of ANT, and its ontology and methodology. The following account is by no means exhaustive, but it includes some central elements of ANT that are particularly relevant for the issues of safety science that are addressed in this chapter, as well as for the development of *sensework* (Haavik, 2017, 2016, 2014b).

A central principle in ANT is that of a generalised symmetry; this implies that descriptions of sociotechnical systems (e.g. operating theatres, airliners or whole industries) do not at the point of departure discriminate between different types of entities in those systems. Hence, social/cultural phenomena and material/technical phenomena are described in the same terms without presupposing particular differences. The whole idea of an ANT description is to find out what difference the different actors – be they individuals, simulators or IT infrastructures – make in the actual system, and which are mediators and which are mere intermediates. This point of a generalised symmetry tends to turn out differently in different scientific disciplines. In the social sciences and humanities, for example, whose intellectual history revolves predominantly around "the social," the symmetry point tends to take shape as a programmatic stance to focus particularly on materiality.[1] In more technically oriented scientific disciplines, on the other hand, the point of generalised symmetry would imply a stronger focus on immaterial – such as social or cultural – aspects.

Another principle of ANT is that in sociotechnical systems, relations are more foundational than components. This is strongly linked to the perspective that no actor acts alone, and that actors do not have essential properties, but derive their properties through interactions in the networks that they are woven into. That is why ANT researchers prefer to start their investigations "in the middle of things, in medias res. Circulation is first, the landscape 'in which' templates and agents of all sorts and colours circulate is second" (Latour, 2005, 260).

Kept together, these two principles pose a challenge at the departure of any research endeavour, since the boundaries of the sociotechnical system cannot be set in advance. Hence the main credo of ANT is to *follow the actors*, wherever this leads you. The consequences of this are not only practical but also economic and political, since budgets for this type of research, assumptions of what entities will be drawn into it (leaders? the brand new IT-system? decisions made on the other side of the globe?) and the political consequences of this are pregnant with uncertainty.

A reformulation of ANT into a sociology of uncertainty (Latour, 2005) is well suited as a passage to a more explicit discussion of the methodological implications of ANT for safety research. Doing ANT implies dealing methodologically with the five sources of uncertainty listed below.

First source of uncertainty: no group, only group formation. We should not assume in advance a fixed list of groups making up our field of study, telling us which individuals we should observe and talk to and which individuals are more powerful than others, since social aggregates are associated with processes and uncertainty.

Second source of uncertainty: action is overtaken. We never act alone; that is why locating an action is a demanding and sometimes exhausting task, since one is constantly being sent elsewhere. When observing and describing actions of different sorts, the uncertainty of their origin is the driving force for the researcher, and the best guarantor for an adequate analysis of the subject of interest.

Third source of uncertainty: objects too have agency. Agency is traditionally reserved for human subjects and groups of humans, possibly including organisations, institutions and the like. ANT increases the type of actors at work to include both humans and non-humans, that is, actors belonging both to the social and the non-social. This introduces uncertainty of each particular study and each particular case; who should count as an actor, and what difference does this actor make?

Fourth source of uncertainty: matters of fact vs. matters of concern. This uncertainty is a direct consequence of the three uncertainties referred to above. The actions and the actors at work – social, material and technical – that are producing, maintaining and re-shaping organisational arrangements and the environment into states of different degrees of durability, stability and accountability: these processes are better addressed as matters of concern rather than matters of fact. Treating the subjects of interest as matters of concern rather than matters of fact means actively deploying uncertainty about the stability of any seemingly stable configuration.

Fifth source of uncertainty: writing down risky accounts. The results of most investigations, and certainly those of researchers, are written accounts. Latour defines a good account as "one that traces a network" (Latour, 2005, 128). Network in this context means a narrative or a description of the mediators – those actors who make a difference, by definition in unpredictable ways – at work. Such explanations tend to be made through frames. But frames – be they formalistic representations of work, organisational models or accident models from the safety literature – tend to provide analyses by means of abstract, general social or technical models. (Note, the resistance towards frames does not mean a resistance towards theoretical preconceptions. However, theories come in many fashions, and the more a theory resembles a causal model, the less interesting it is for a student or a researcher that is seeking to produce new knowledge, or challenge old.)

In ANT, any analysis is intimately associated with the particular (Latour, 2005) rather than causal models. (Another clarification is necessary here: the often sharp distinction between theory and empirical data is misleading, since bodies of theory are also important empirical data in many studies, for example Le Coze (2019), Pettersen and Schulman (2019) and Haavik et al. (2019), which discuss the theories of High Reliability Organisations and Resilience Engineering in a recent special issue in *Safety Science* (Wears and Roberts, 2019).) Still, every study of safety involves challenging decisions regarding how far the networks should be traced, or, in practice, where the system borders are to be set. Where we set the boundaries for our systems has a profound effect on which aspects are included as relevant, and which are not. However, the definition of system boundaries is not merely scientifically reasoned (e.g. by aid of models of safety and accidents) – it is also a pragmatic choice that involves politics and finite resources (Antonsen and Haavik, Chapter 5, this volume). Being transparent about these inclusion criteria and stop criteria thus implies being more true to the empirical world than to models of the world.

To get an idea of the operationalisation of ANT ontology in the context of safety, consider Hilgartner's (1992) elaboration of risk in the road sector: Hilgartner demonstrates the problem of defining links between harm and risky objects since risk can always be attributed to multiple objects and far-reaching networks:

> the risk of motor vehicle accidents can be attributed to unsafe drivers, unsafe roads, or unsafe cars (…). The risks posed by unsafe drivers, in turn, can be attributed to inexperience, irresponsibility, fatigue, or alcohol consumption. Inexperience, in turn, can be attributed to inadequate driver education programs, which can be attributed to shortages of funds. (Hilgartner, 1992, 42)

Building on the work of Nader (1965), Hilgartner explains how the dominant risk object in US traffic policy in the decades after the Second World War gradually changed from "the driver" and "driver failure" to "federal regulation of automotive design" that allowed a range of construction choices to come unchecked into the market. Through this and other examples from the transportation sector, Hilgartner portrays risk as a highly relational phenomenon, resting in heterogeneous sociomaterial networks where the borders between technologies and politics may be unclear. As with risk, safety must be understood as a relational phenomenon: for example, while airmanship and seamanship traditionally have linked safe air and sea navigation to individual capabilities of pilots, who once had a large degree of autonomy, standardisation and regulatory developments within aviation and the maritime sector over the last decades have transformed such capabilities into systemic traits, making it necessary to search for safety in relational networks of pilots, operating companies, regulators, manufacturers and standardisation companies instead. And these networks do not perform airmanship or seamanship, but rather they perform what we may call *airlineship* (Haavik et al., 2017) and *distributed maritime capabilities* (Kongsvik et al., 2020).

SENSEWORK

Sensework addresses a type of sociotechnical work in safety-critical operations where groups of professionals in technology-rich environments work to put together different sorts of representations and pieces of digital sensor data to give meaning to, and work through, familiar and unfamiliar situations. As a theoretical perspective, sensework is situated in the intersection between safety research and ANT. Sensework pays the same attention to material and social aspects, and treats them in principle the same way, with *pragmatism* as a main value; the material and the social intertwine and interact in unpredictable ways, and how different sociotechnical systems work is treated as an empirical question.

Sensework is a perspective and an approach that has grown out of a series of empirical research studies in safety-critical, technology-rich contexts. This research includes studies in domains of petroleum, aviation, operating theatres and the maritime sector (Haavik et al., 2017; Haavik, 2017; 2013, 2016, 2014b). It draws inspiration from and builds on central themes from some of the most central theories and perspectives in safety science – Resilience Engineering, High Reliability Organisations and Normal Accident Theory – with a commitment to a constructivist ontology found in ANT, and its accompanying methodological credo. Some more theoretical and conceptual traces of this exploration along the borderline between safety research and Actor Network Theory have been developed in prior work (Haavik, 2014a, 2011, 2017).

This slightly unconventional combining of lenses in sensework studies has sometimes drawn attention to similarities between perspectives where the discourse has otherwise emphasised differences, such as in the interpretation of the common sociotechnical constructivist ontologies of Normal Accident Theory, High Reliability Organisations and Resilience Engineering (Haavik, 2014a). Other times, central differences have been pointed out where controversies have not been much explored explicitly by the safety science community, such as issues relating to the (in)visibility of the ingredients of safety – the notions of safety as a dynamic non-event and safety as dynamic events, respectively (Haavik et al., 2019).

Models have played an important role in the history of safety research (Le Coze, 2013). One strength of models lies in their ability to convey a message or a relationship in a short and concise manner. But models are also inclined to betray their originators, such as when models are deployed in a superficial manner. For example, a list of HRO capabilities or Resilience Engineering cornerstones can do more harm than good if they are deployed as over-simplified recipes for transforming risky organisations into reliable or resilient organisations, since unspoken ontological considerations of models may be disregarded (consider, for example, different definitions of safety as dynamic non-events or as a dynamic event – which will be explored more later). Instead of models or lists, sensework is first and foremost associated with empirical settings, and a research methodology that renders visible the ingredients of safety as investigated through empirical studies and case studies. Having said that, sensework studies are at their best when they *produce* models or lists or concepts that describe and explain situated sensework in meaningful ways. One such description of sensework was published in 2014 (Haavik, 2014b), portraying the work of combining, analysing and acting on data and information in offshore drilling operations in a context of Integrated Operations (Albrechtsen and Besnard, 2013; Haavik, 2013; Rosendahl and Hepsø, 2013). The description of sensework illustrates the highly pragmatic and situationally conditioned work in operation centres; theory, experience, pattern-recognition, workarounds, models and rule-based solutions – these are all different expressions of the range of formalised and non-formalised practices that drive operations forward and that enter into the work of dealing with the stream of minor and major contingencies of such operations. Importantly, this is an empirical model, with a particular temporal and geographical situatedness. It is not a universal model with generic validity, but efforts have been made for it to be a strong model portraying sensework in Integrated Operations in a particular company at the Norwegian Continental Shelf around the second decade of the 21st century.

The best way to explain sensework is not through definitions but through demonstrations of *what sensework (research) does*. One might argue likewise about ANT. In the following, we shall see how ANT perspectives can support the introduction of new themes and new framings of themes in safety research so as to progress from stalled debates or cold controversies, before elaborating on a case of sensework, and future challenges of sensework studies in contexts of societal resilience.

HOW TO MAKE ANT PRINCIPLES RELEVANT FOR SAFETY RESEARCH

A recurring theme and something of a conundrum in safety analysis is the tracing of interactions between social and technical elements of sociotechnical systems. Complex sociotechnical systems are dynamic and unfold over time, and are a source both of risk and safety. To fully understand the mechanisms and the uncertainties associated with these issues the safety community could profit much from drawing on insights from other fields of research that have tackled these issues for as long a time, but from different perspectives and with different methods. The safety community – interdisciplinary, but with an evident technological heel – have for a while struggled with reconciling "hard" and "soft" approaches to understanding the functioning and ensuring the robustness of sociotechnical systems. Although the safety field is indeed multidisciplinary, it is not so interdisciplinary as we may like to think; spillover and original amalgams between traditional fields of, say, engineering, physics, sociology and philosophy are not flourishing. It is thus worth noticing that ANT has been working on these issues for several decades, developing vocabularies and methodologies for approaching such interdisciplinary issues within sociotechnical systems – for example, biological laboratories (Latour and Woolgar, 1986) and aviation (Law, 2002).

SOCIAL AND TECHNICAL COMPONENTS AND RELATIONS

In the discourse on differences and similarities between Normal Accident Theory and High Reliability Organisations, technological determinism and social construction have been central parts of the repertoire (Hopkins, 2001; Rochlin, 1999; Clarke and Short Jr, 1993; Gherardi et al., 1998). Controversies over what kinds of entities – technological or social – influence the workings of sociotechnical systems are controversies in need of a relational perspective. That would imply approaching and describing sociotechnical systems with a keen eye for interactions and relations that shape parts of the systems and form their coexistence, instead of taking as a point of departure sociotechnical systems already populated with pre-defined components and boundaries (Latour, 2005; Haavik, 2011). This may sound abstract and theoretical at first, but it implies a highly concrete and practical-empirical device: put on some clothing[2] that you are not afraid will get dirty, situate yourself in medias res ("in the middle of things") with an open mind for what types of actors influence the sociotechnical system, follow the actors and make note of everything you see and hear; you never know, in advance, what or who are acting and making a difference.

The device to enter the empirical studies "in the middle of things" does not necessarily mean to start the investigations in the "operational environments" in the sharp end, although one should definitely be prepared to travel there also during the study. Actually, the advice is better understood as to start *anywhere* where things are going on and follow the actions from there, instead of planning the field trip using the boxes and arrows of organisational flow charts. In the study of past events, accident investigations must therefore be used with care since their dramaturgy may follow methodical principles that are not always well revealed. If that is the case, they must be read with the same critical eye as historians read their sources: under what circumstances did they come about, who shaped them, what power networks influenced them, what were the common understandings that framed them?

Adopting an ANT perspective and methodology in research on safety-critical sociotechnical systems enables us to break free from the polarised social/technology dualism, and to realise that who and what should be granted agency is an *empirical* question and not something that can be assumed a priori. An example of this approach in practice is Haavik's (2011) review of the Snorre A accident[3] (Schiefloe and Vikland, 2009; Wackers, 2006) studies. In the aftermath of this accident in 2004, Antonsen (2009) observed that two studies of safety culture on the rig – one before

(Kongsvik, 2003) and one after (Schiefloe and Vikland, 2009; Schiefloe et al., 2005) the accident – varied greatly; the first survey did not offer any warnings that something serious was brewing, while the more qualitative investigation revealed that the safety culture had not been satisfactory at all. While this isolated finding could point in many reasonable directions – that safety culture surveys have little predictive value (Antonsen, 2009) or simply towards the *hindsight is 20/20* argument[4] – an ANT perspective would take particular interest in the methodological approach in the two studies. While the first study approached safety culture as an isolated and pre-defined component, the second study was more oriented towards relations, and how the different human and non-human entities of the systems acquired their competences and characteristics through interaction.

In the post-hoc account (Schiefloe and Vikland, 2009) of the organisational context leading to the accident, the platform's technical condition was shown to have deteriorated due to "the platform's turbulent history, and in the problematic economic situation of the first operator of the field" (Schiefloe and Vikland, 2009, 11). The technical condition was transformed by organisational rearrangements and the economic situation. Concurrently, the deteriorating technical condition of the platform played a role in shaping the competences of the rig crew: the technical condition and the working situation "resulted in a well-developed ability to be versatile and to improvise. The Snorre crew could handle almost any technical challenge in a swift and competent way" (Schiefloe and Vikland, 2009, 10–11).

One conclusion is thus that the assessments of organisational culture undertaken *before* the accident did not account for the same culture as the assessments of organisational culture undertaken *after* the accident. This is because the investigation before the accident did not take into account the relational aspects that played such an important role in the investigation after the accident. Or, in the words of Latour from his generalised narrative of the social and technical aspects of scientific work:

No event can be accounted for before its conclusion (...) If such a list were made, the actors on it would not be endowed with the competence that they will acquire in the event. (...) This list of inputs does not have to be completed by drawing upon any stock of resources, since the stock drawn upon before the experimental event is not the same as the one drawn upon after it. (Latour, 1999b, 126)

A relational approach thus represents an insurance against both technological determinism and social construction; in a relational account being true to the principle of generalised symmetry, the realm to which entities belong (social/cultural or technological/material) is subordinated to how they shape, and are being shaped by, their surroundings.[5]

Visibility/Invisibility

The question of visibility is central in safety work, as so many accident investigations have shown us. Conditions that are believed to have led to – or could have provided warnings about – an accident are often seen only after the accident has occurred. And one can only evaluate and fix that which one sees. Hollnagel (2006) used the phrase *What you find is what you fix* to state this clearly, and it is hard to disagree. However, if one draws conclusions too quickly, this could be interpreted as emphasising a link between safety and invisibility. This link, despite its popularity in the field of safety, is highly problematic, and the reason is that invisibility is often attributed to *essence* instead of *method*.

It is frequently repeated (e.g. Almklov and Antonsen, 2014; Sutcliffe and Weick, 2013; Weick, 2011) that one of the difficulties associated with investigating and ana-lysing the ingredients of safety is that they tend to be invisible. Weick and Sutcliffe have portrayed safety (and reliability) as a *dynamic non-event* (Weick, 2011, 1987; Sutcliffe and Weick, 2013). It is dynamic, they say, in the sense that "it is an ongoing condition in which problems are momentarily under control due to compensating changes in components" (Weick, 1987, 118). Further, "safety is a non-event because successful outcomes rarely call attention to themselves. In other words, because safe outcomes are constant, there is nothing to pay attention to" (Sutcliffe and Weick, 2013, 152).

That safety is dynamic, and that it is made up (also) of compensations and adapta-tions is quite clear – and we could add articulation work, too: "Work that gets things back 'on track' in the face of the unexpected, and modifies action to accommodate unanticipated contingencies" (Star and Strauss, 1999). That successful outcomes rarely call attention to themselves, that safe outcomes are constant and that there is nothing to pay attention to and thus can be considered as a *non-event* is, however, a deeply problematic statement.

What you find is what you fix, says Hollnagel, but that is not the whole story. Lundberg, Rollenhagen and Hollnagel (2009) remind us of an even more important condition, namely that *what you look for is what you find*. This statement is about framing, and the tools used for looking. Importantly, it addresses not the essential characteristics of that which you are looking for but the *methods* adopted by those who are looking. Looking for something else also involves looking with different eyes, in different ways.

This is where ANT, as a method of looking, becomes useful. And instead of fram-ing before looking, ANT explicitly advises us to get rid of the frames:

> [I]t is usually when a "frame" is called in that the sociology of the social insinu-ates its redundant cause. As soon as a site is placed "into a framework," everything becomes rational much too fast and explanations begin to flow much too freely. (Latour, 2005, 137)

Visibility is a question of the method applied by the observer, not a question of the essence of that which is to be seen; if you have a method for observing infor-mal work, and a language for describing it, informal work will be visible. If not, it will stay invisible. Likewise, invisibility does not reside in organisations or practices themselves, but in our lack of methods for seeing how they are constructed and stabilised. Methods are folded into practices, and practices are folded into culture. That is why Goodwin (1994) refers to vision and seeing as a cultural practice. And when Suchman (1995, 56) quotes Wellman and writes about "invisible work" that "How people work is one of the best kept secrets in America," she articulates the observation that in organisations, work practices are often invisible to others than those who perform them. Managers, or for that matter researchers, may not be well aware of the shape of and rationale for situated practices. Those actually performing the work, on the other hand, be they Suchman's secretaries or the professionals on aircraft carriers portrayed in the High Reliability Organisations literature (Rochlin et al., 1987; Weick and Roberts, 1993), know very well what they are doing. For the researcher, then, it is a question of method to get the professionals to articulate it, and to get in a position to observe it themselves. ANT (e.g. Latour, 1992, 1987) provides much methodical advice to make visible and understand mundane work that usually does not grab the attention of many, other than those performing it. It would take too long to demonstrate this thoroughly here, but a condensate would be to follow the actors, and to follow them through the actual terrain where they work, and not through the maps or organisational charts or theoretical models of safety (see also Le Coze, Chapter 4, this volume).

These two aspects of sociotechnical systems and safety – the relation of social and technological elements, and the question of visibility – provide only examples, but important ones, of how ANT as a qualitative method has potential to inform and enrich safety research and some of its enduring, unresolved debates. In the following we shall see a practical example of adopting an ANT methodology in sensework research.

SENSEWORK IN THE OPERATING THEATRE

With the increased importance of images, visualisations and simulations in the operating theatre, the nature of surgery is gradually transforming from craftsmanship to sensework (Haavik, 2016). The list of representational artefacts playing central roles in, for example, neurosurgery includes X-ray, CT (Computed Tomography) scans, MRI (Magnetic Resonance Imaging) scans and ultrasound, in addition to the traditional microscope images and figures and graphs visualising the numerous sensor measurements made by the anaesthetic nurses in course of operations. While some of these representational methods have existed for a long time, the novelty, and the game-changing aspect, lies in the way they are integrated in real-time into the surgeons' work. When surgeons work as much on representations of the patient as on the patients themselves, this also changes the way researchers describe and analyse their work – and how we may discuss it in terms of safety.

While descriptions of sensework in offshore operations show the distributed cognitive-performative aspects of offshore work, a study of sensework in the operating theatre (Haavik, 2016) in particular draws attention to the *sensation of representation* and the *decisional nature of actions*. This study exemplifies the ambition of sensework studies to carve out topics and themes that may support safety inquiries in digitalised high-risk, high-tech environments.

In the forefront of neurosurgery, different forms of representation now enter into configurations that project an augmented reality, and it is not unrealistic that these representations in the future will be even more independent and powerful so as to produce a hyper-reality that we may confidently act upon. Sensational as this might sound, we saw traces of evidence in a case study of a series of neurosurgical operations that we undertook during two weeks in 2015 (Haavik, 2016). In this densely empirical study – as any sensework study is – we were able to follow at close range the actions and considerations of the surgical teams as they were confronted with brain tumours, disc herniations, neck fractures, aneurisms and many other conditions.

The way the representations not only guided actions but also framed the larger perspective and the horizon of understanding of the patient's condition demonstrated a power of representation that forced the empirical world – the state of the patient and associated decisions (only post-hoc traceable) – which followed. Such sensework analyses with ANT-inspired perspectives on representations and agency allow us to expand on the repertoire of safety research and to bring new perspectives on themes such as sensemaking (Rosness et al., 2016; Weick, 1995), naturalistic decision-making (Klein, 2008) and resilience (Hollnagel et al., 2013). Just like offshore drilling operations, image-guided surgery is a creative activity. Sense is *constructed* through hard work, aided by a range of practical-cognitive tools and processes, some of which are formalised, others more ad-hoc. Surgeons, or geologists, do not make sense with their minds, nor is sense "transported" to them from the objects they monitor and investigate through sophisticated instruments and visualisations; surgeons and other senseworkers co-construct sense together with colleagues, surgical instruments, patients, heuristics, checklists, theories, hypotheses – and the ultimate test is not how *true* it is but how well it *works*. Sensework studies take such pragmatics seriously, and seek to investigate them systematically.

THE NEXT CHALLENGE: SENSEWORK AND SOCIETAL RESILIENCE

An implicit, but undercommunicated, issue in sensework research so far is the understanding of sensework environments as *centres of calculation* (Latour, 1987) – locations, facilities or institutions that are capable of gathering, combining and making sense of immutable mobiles (Latour, 1986) (in this context: representations) in a powerful way that is not possible in other, more sparsely equipped locations. If we think of operating theatres and Integrated Operations support centres as centres of calculation that collect, combine and act upon all sorts of representations of data and information, we are in a very good position to address safety issues at a larger scale – for example, the burning issue of global climate change. Insights from studies of relatively local sensework, such as the work taking place in an operating theatre or in an onshore support centre, suggest arrangements for studies of those environments and institutions that gather, combine, analyse and act upon climate data and information from all corners of the world: which are the most powerful centres, what are their modi operandi, what do the functions of their resonance or dissonance look like? These are examples of questions that could inspire a new wave of safety-related studies of large-scale issues such as global climate change (Haavik, 2020).

NOTES

1. The discourse of sociomateriality (Leonardi, 2012) is an example of this.
2. This could be read both literally and metaphorically.
3. The Snorre A accident was a gas blowout on the Snorre A platform on the Norwegian Continental Shelf.
4. When eye doctors evaluate a patient's vision, 20/20 is a measurement result that implies perfect vision. To label something a hindsight 20/20 argument implies criticizing those behind it for looking back and being wise *after* something has happened, and that this insight was not available for those dealing with the event as it unfolded.
5. Note: To trace accidents through chains of events, spanning over heterogeneous issues such as financial strains, deteriorating equipment and work processes is not particularly original and not something to be associated with ANT alone. Analyses that do not refer to ANT at all can still be ANT analyses. And there are strong ANT analyses and weak ANT analyses. Strong ANT analyses would at least avoid linear causality, pre-defined system borders and pre-defined categories of actors, and allow for non-human agency.

REFERENCES

Albrechtsen, E., and D. Besnard, eds. 2013. *Oil and Gas, Technology and Humans: Assessing the Human Factors of Technological Change.* Farnham: Ashgate.

Almklov, Petter Grytten, and Stian Antonsen. 2014. "Making work invisible: New public management and operational work in critical infrastructure sectors." *Public Administration* 92(2):477–492.

Antonsen, Stian. 2009. "Safety culture assessment: A mission impossible?" *Journal of Contingencies and Crisis Management* 17(4):242–254.

Bloor, David. 1976. "The strong programme in the sociology of knowledge." *Knowledge and Social Imagery* 2:3–23.

Bloor, David. 1999. "Anti-Latour." *Studies in History and Philosophy of Science Part A* 30(1):81–112.

Callon, Michel. 1986. "Some elements of a sociology of translation: Domestication of the scallops and the fishermen of St Brieuc Bay." In: *Power, Action and Belief: A New Sociology of Knowledge*, edited by John Law, 196–223. London: Routledge.

Callon, Michel, and Bruno Latour. 1981. "Unscrewing the big Leviathan: How actors macro-structure reality and how sociologists help them to do so." In: *Advances in Social Theory and Methodology: Toward an Integration of Micro-and Macro-Sociologies* edited by Knorr-Cetina, K. D., and Cicourel, A. V., 277–303. Boston, Mass.: Routledge and Kegan Paul.

Callon, Michel, and John Law. 1982. "On interests and their transformation: Enrolment and counter-enrolment." *Social Studies of Science* 12(4):615–625.

Gherardi, Silvia, Davide Nicolini, and Francesca Odella. 1998. "What do you mean by safety? Conflicting perspectives on accident causation and safety management in a construction firm." *Journal of Contingencies and Crisis Management* 6(4):202–213.

Goodwin, C. 1994. "Professional vision." *American Anthropologist* 96(3):606–633.

Haavik, Torgeir Kolstø. 2011. "On components and relations in sociotechnical systems." *Journal of Contingencies and Crisis Management* 19(2):99–109.

Haavik, Torgeir Kolstø. 2013. *New Tools, Old Tasks: Safety Implications of New Technologies and Work Processes for Integrated Operations in the Petroleum Industry.* Farnham: Ashgate.

Haavik, Torgeir Kolstø. 2014a. "On the ontology of safety." *Safety Science* 67:37–43.

Haavik, Torgeir Kolstø. 2014b. "Sensework." *Computer Supported Cooperative Work (CSCW)* 23(3):269–298. doi: 10.1007/s10606-014-9199-9.

Haavik, Torgeir Kolstø. 2016. "Keep your coats on: Augmented reality and sensework in surgery and surgical telemedicine." *Cognition, Technology and Work* 18(1):175–191.

Haavik, Torgeir Kolstø. 2017. "Remoteness and sensework in harsh environments." *Safety Science* 95:150–158.

Haavik, Torgeir Kolstø 2020. "Societal resilience – Clarifying the concept and upscaling the scope." *Safety Science* 132. doi: 10.1016/j.ssci.2020.104964.

Haavik, Torgeir Kolstø, Stian Antonsen, Ragnar Rosness, and Andrew Hale. 2019. "HRO and RE: A pragmatic perspective." *Safety Science* 117:479–489.

Haavik, Torgeir Kolstø, Trond Kongsvik, Rolf Johan Bye, Jens Olgard Dalseth Røyrvik, and Petter Grytten Almklov. 2017. "Johnny was here: From airmanship to airlineship." *Applied Ergonomics* 59(A):191–202.

Hilgartner, Stephen. 1992. "The social construction of risk objects: Or, how to pry open networks of risk." In: *Organizations, Uncertainties, and Risk*, edited by James F Short and Lee Ben Clarke, 39–53. Boulder: Westview Press.

Hollnagel, E. 2006. "The myth of human error in risk analysis and safety management." *Keynote Speech at RISIT Programme Research Conference*. Lillehammer.

Hollnagel, E., J. Braithwaite, and R. L. Wears. 2013. *Resilient Health Care*. London: Ashgate.

Hopkins, A. 2001. "Was Three Mile Island a 'normal accident'?" *Journal of Contingencies and Crisis Management* 9(2):65–72.

Klein, G. 2008. "Naturalistic decision making." *Human Factors: The Journal of the Human Factors and Ergonomics Society* 50(3):456–460.

Kongsvik, T. 2003. *Hvilke Barrierer? Ansattes Vurdering av sider ved Sikkerhetskulturen – Snorre A (Which Barriers? Employees' Perception of Aspects of Safety Culture – Snorre Alpha)*. Trondheim: Studio Apertura.

Kongsvik, T., T. K. Haavik, R. J. Bye, and P. G. Almklov. 2020. "Re-boxing seamanship: From individual to systemic professional competence." *Safety Science* 130:104871.

Latour, B. 1999a. "For David Bloor… and beyond: A reply to David Bloor's 'anti-Latour'." *Studies in History and Philosophy of Science Part A* 30(1):113–129.

Latour, B. 1999b. *Pandora's Hope: Essays on the Reality of Science Studies*. Cambridge, MA: Harvard University Press.

Latour, B. 1986. "Visualization and cognition: Thinking with eyes and hands." *Knowledge and Society: Studies in the Sociology of Culture Past and Present* 6:1–40.

Latour, B. 1987. *Science in Action: How to Follow Scientists and Engineers through Society*. Milton Keynes: Open University Press.

Latour, B. 1992. "Where are the missing masses? The sociology of a few mundane artifacts." In: *Shaping Technology/Building Society: Studies in Sociotechnical Change*, edited by Wiebe E. Bijker and John Law, 225–264. Cambridge, MA: MIT Press.

Latour, B. 2005. *Reassembling the Social: An Introduction to Actor-Network-Theory, Clarendon Lectures in Management Studies*. Oxford: Oxford University Press.

Latour, B., and S. Woolgar. 1979. *Laboratory Life: The Social Construction of Scientific Facts*. Beverly Hills, CA: Sage Publications.

Latour, B., and S. Woolgar. 1986. *Laboratory Life: The Construction of Scientific Facts*. New Ed.] Princeton, NJ: Princeton University Press.

Law, John. 1992. "Notes on the theory of the actor-network: Ordering, strategy, and heterogeneity." *Systems Practice* 5(4):379–393.

Law, John. 2002. *Aircraft Stories: Decentering the Object in Technoscience*. Durham, North Carolina: Duke University Press.

Le Coze, Jean-Christophe. 2013. "New models for new times. An anti-dualist move." *Safety Science* 59:200–218.

Le Coze, Jean-Christophe. 2019. "Vive la diversité!. High Reliability Organisation (HRO) and Resilience Engineering (RE)." *Safety Science* 117:469–478.

Lee, Clarke, and James F Short. 1993. "Social organization and risk: Some current controversies." *Annual Review of Sociology* 19(1):375–399.

Leonardi, Paul M. 2012. "Materiality, sociomateriality, and socio-technical systems: What do these terms mean? How are they related? Do we need them?" In: *Materiality and Organizing: Social Interaction in a Technological World*, edited by Paul M. Leonardi, Bonnie A. Nardi and Jannis Kallinokos, 25–48. Oxford: Oxford University Press.

Lundberg, Jonas, Carl Rollenhagen, and Erik Hollnagel. 2009. "What-you-look-for-is-what-you-find-the consequences of underlying accident models in eight accident investigation manuals." *Safety Science* 47(10):1297–1311.

Nader, Ralph. 1965. *Unsafe at Any Speed. The Designed-In Dangers of the American Automobile*. New York: Grossman Publishers.

Pettersen, Kenneth A., and Paul R. Schulman. 2019. "Drift, adaptation, resilience and reliability: Toward an empirical clarification (special issue on HRO and RE)." *Safety Science* 117:460–468.

Rochlin, G. I. 1999. "Safe operation as a social construct." *Ergonomics* 42(11):1549–1560.

Rochlin, G. I., La Porte, T. R., and Roberts, K. H. 1999. "The self-designing high-reliability organization: Aircraft carrier flight operations at sea." *Naval War College review* 40(4):76–92.

Rosendahl, Tom, and Vidar Hepsø. 2013. *Integrated Operations in the Oil and Gas Industry: Sustainability and Capability Development*. Hershey, PA: IGI Global.

Rosness, R., T. E. Evjemo, T. K. Haavik, and I. Wærø. 2016. "Prospective sensemaking in the operating theatre." *Cognition, Technology and Work* 18(1):53–69.

Schiefloe, P. M., and K. M. Vikland. 2009. "Close to catastrophe. Lessons from the Snorre A gas blow-out." *25th European Group for Organizational Studies (EGOS)*. Barcelona, Spain, 3–4 July 2010.

Schiefloe, P. M., K. M. Vikland, E. B. Ytredal, A. Torsteinsbø, I. O. Moldskred, S. Heggen, D. H. Sleire, S. A. Førsund, and J. E. Syversen. 2005. *Årsaksanalyse etter Snorre A-hendelsen 28.11.2004*. Stavanger, Norway: Statoil.

Star, Susan Leigh, and Anselm Strauss. 1999. "Layers of silence, arenas of voice: The ecology of visible and invisible work." *Computer Supported Cooperative Work* 8(1–2):9–30.

Suchman, Lucy. 1995. "Making work visible." *Communications of the ACM* 38(9):56–64. doi: 10.1145/223248.223263.

Sutcliffe, K. M., and K. E. Weick. 2013. "Mindful organizing and resilient health care." In: *Resilient Health Care*, edited by E. Hollnagel, J. Braithwaite and R. L. Wears. London: Ashgate.

Wackers, G. 2006. "Vulnerability and robustness in a complex technological system: Loss of control and recovery in the 2004 Snorre A gas blow-out." *University of Maastricht, Research Report* 42:2006.

Wears, R. L., and K. H. Roberts. 2019. "Safety science, high reliability organizations and resilience engineering." *Safety Science* 117:458–459.

Weick, Karl E. 1987. "Organizational culture as a source of high reliability." *California Management Review* 29(2):112–128.

Weick, Karl E. 1995. *Sensemaking in Organizations*. Thousand Oaks, CA: Sage.

Weick, Karl E. 2011. "Organizing for transient reliability: The production of dynamic non-events." *Journal of Contingencies and Crisis Management* 19(1):21–27. doi: 10.1111/j.1468-5973.2010.00627.x.

Weick, Karl E., and Roberts, K. H. 1993. "Collective Mind in Organizations: Heedful Interrelating on Flight Decks." *Administrative Science Quarterly* 38(3):357–381.

7 Learning to Do Ethnography in Complex Systems

Christian Henrik Alexander Kuran

CONTENTS

INTRODUCTION

> From: driver256@truckeronline.no
> To: christian@norwegianacademic.no
> Subject: Your project
> Hi again,
> I stopped driving long haul truck regularly in 2009 and I see that none of the people who used to share the road with me are still there when I drive around today. I do have a cousin who is still on the road, and he has about 20-25 years of experience now. You should have a talk with him as well. I would very much like to join in one of your focus group interviews if you can manage to travel to Bergen or Trondheim.

I received the above email form one of my informants, as participants are commonly called in ethnographic research, while working on a project focused on rule bending and rule breaking in the road-based heavy goods transport sector. The project design was rooted in my ethnographic background as an anthropologist, and in my training as a risk and safety researcher.

Ethnographic methodology and methods that were developed as the staple and signature of the discipline of social and cultural anthropology have in recent years been introduced and adapted to various disciplines, such as healthcare and management studies, and have become of central interest in safety research. This importation of methods and methodology has not been without its challenges (Cupit et al., 2018). The major challenges are not related to whether ethnographic methodology and methods can bring valuable data and perspectives to other disciplines (Dixon-Woods, 2003) but involve more fundamental questions, such as: What is ethnography, when can a project be labelled as ethnographic, can ethnography as method and methodology be considered and adapted separately, or even as a whole be separated from the parent discipline of anthropology? This chapter attempts to clarify these uncertainties, and suggest solutions.

First, I give a short overview of ethnography as method and methodology and look at how ethnographic data is collected, what it is and how it can contribute to safety research. In the previously mentioned project on the heavy goods vehicle (HGV) sector I reconciled the fundamentals of ethnographic methodology with safety research, and here I will present some lessons that were learned in that project about using ethnography in a complex sociotechnical system. The following sections of the chapter address the questions of what ethnography is and how it is done in practice, using the study of rule bending and breaking in the HGV-sector as an example. Further, I discuss the potential benefits of the use of ethnography in risk and safety studies and review the challenges to the import of ethnographic methods and methodology into safety science studies. Finally, I provide a toolkit for ethnographic experimentation in complex sociotechnical systems.

WHAT IS ETHNOGRAPHY?

History

The description and analysis of cultures and ways of life is the original intention of traditional ethnographic practice (Bernard, 2006). Practitioners of ethnography have been concerned with such diverse topics as the nature of power, how people understand kinship, how people organise themselves along political and ethnic lines, the construction of gender, gendered practices and witchcraft (Salzman, 1996). If there is a historical foundation of ethnography it is undoubtedly the fieldwork tradition established by Bronislaw Malinowski. His fieldwork among the tribes in the Trobriand Islands in the Pacific, and subsequent monographs of life on the islands in the early decades of the last century, established the practices that are today the hallmark of ethnography. The hallmarks are fieldwork sustained over time, participant observation as the primary method and the monograph as its major medium for the dissemination of knowledge (Gobo, 2008; Malinowski, 1922).

Near the end of the last century many anthropologists practised ethnography with a sense of urgency as the world was being affected by the increased flow of wares, people and ideas, known collectively as globalisation. The idea behind urgent anthropology was to preserve descriptions of what were thought to be vanishing local cultures. Current anthropology on the other hand has long since revised its theoretical models, along with the realisation that there is no such thing as cultures that are not in flux, and that social change is one of the universal traits of human societies. The practice of ethnography is primarily associated with its parent discipline anthropology, but the practice no longer exclusively concerns itself with the study of small groups of indigenous peoples, and is not confined to the anthropology discipline as it is used in a variety of areas ranging from Public Health to the Humanities. The practice of ethnography has increasingly become a reflexive way of approaching any social field that might be of research interest.

ETHNOGRAPHIC FUNDAMENTALS

The terms *ethnography* and *social-* or *cultural anthropology* are sometimes used interchangeably in the literature. I will refer to the social science discipline of the study of human culture and societies as anthropology, while I will refer to ethnography as anthropology's primary *methodology*, and describe the various *methods* used in the practice of ethnography (Gobo, 2008). Participatory fieldwork sustained over time is the primary method used in the practice of ethnography, and the intention is to spend enough time doing observation, and ideally to participate in the day-to-day work and practice in the field setting as fully as possible. Time spent in the field doing participant observation will ideally allow the researcher to create friendships and trust between him/herself and the participants, often called informants in the literature.

The term *methodology* refers not to the practicalities of fieldwork and data gathering, but to the underlying philosophical approach to the subject data (Gobo, 2008). The foundational philosophical underpinnings of ethnography are that of *holism* and *cultural relativism* (Stewart, 1998). The concept of holism refers to how ethnographers seek to understand and describe phenomena observed in field, in relation to a comprehensive sociocultural context. The concept of cultural relativism refers to how individual actions as well as described meaning, norms and cultural expressions observed in the field should be taken and analysed in their context in order to reduce the cultural biases of the ethnographer as much as possible.

While adherence to the classical traditions of the discipline is still very much how ethnographers judge the value of any ethnographical undertaking (Marcus & Fischer, 1996), developments in the discipline have made its practitioners observant of alternative types of fieldwork, such as multisite fieldwork, and para-site, as well as fieldwork in western cultures, and settings such as cities or multinational organisations. For convenience, the classical traditions of ethnographic research, as relevant to work in the safety sciences, can be summed up as:

1. Performing participatory fieldwork in a geographically delimited field.

2. Having holistic methodical and theoretical aspirations. Not relying on a single method of data inclusion.
3. Writing thick or rich descriptions of the cultural and social environments, and peoples in the field.

Spending time in field, ideally several months or years in length, is time-consuming and exhausting, and anthropologists are trained not to rely solely on observation. The researcher is expected to take field notes several times a day to structure and retain memories, and if possible, to re-visit the field after a period of writing out of field. As an example, in my very first fieldwork in Tanzania I spent eight months in field living as a paying guest in a local family's house, observed, participated in activities, interviewed when it felt natural and took pictures. I even got a class of local school children to write school papers on a chosen topic. Field notes were taken by hand each evening, and small scratch notes where written when it was inconvenient to write longer notes. The data material was later used to produce a monograph, a detailed description of the place, people and the topics I was researching.

The Primary Research Tool

At a social dinner for truck instructors, the conversation went lively; I had earlier in the evening described my project on rule bending in the industry to the other dinner guests at my table. The theme of my research intrigued the truck instructors, and as the evening progressed, I was through the stories told around the table introduced to a colourful menagerie of characters my conversational partners had encountered in their careers as truck drivers and truck instructors. Politely asking if it was okay if I jotted down some notes, and getting a positive reply, in the later weeks, I would follow up the different stories and people in informal interviews.

Ethnography is developed in the field (Stewart, 1998), and as an ethnographer you approach the field with yourself. The primary research tool of the ethnographer is not the taking of field notes, a practice that just as often might get in the way of the social interaction the ethnographer is involved in at any one time. The primary research tool is the ethnographers' bodily presence in the field. It is the medium used in social interactions that take place, and gives the proximity to ask a lot of questions used to test the ethnographers' tacit theories about the situation he or she is in. The ethnographers' approach to the study of social interaction while him/herself is situated in ongoing sociocultural situations makes ethnographic research particularly open for critique. How can situated researchers be objective, how can the results of such approaches claim validity, and how can the results be reliable, are all fair questions. As Fox points out, this "reminds us that fieldworkers are not simple mechanisms for recording of data. The art of fieldwork is intimately connected to the body and behavior of the anthropologist" (Fox, 2006: 349). This valuable insight has repercussions for how to approach a complex sociotechnical system, not only as an ethnographer but also as a safety scientist.

The Physical Data

"You should be making maps." During my first fieldwork, I spent several months in a complex of three rural villages in Tanzania; the theme of the project was democratisation. While defining the villages as a complex socio-technical system would be considered quite a stretch, its population in excess of ten thousand, and its polytechnic and multi-religious composition made the community quite complex. During the fieldwork, I had arranged to call my supervisor in Norway, giving her an update and getting some much-needed moral support. She was a great sparring partner but also gave me one piece of advice that puzzled me. "You should be making maps." While this was before the advent of Google Maps it still felt very old fashioned, surely, I could petition the local authorities for maps or some such instead of pretending as if I was an explorer or some other nonsense. It was rural, yes, but the main road to touristy Serengeti and Masai Mara was only a short walk away from where I slept. Still I did as my supervisor advised me, made maps, and had maps made for me, adding to a growing and confusing pile of written, audial, and visual data.

After a period of fieldwork, the physical ethnographic data the researcher has in his or her hands might consist of a diverse collection of field notes. These might be field notes properly transcribed on a computer, diaries, small pocket books, but also scratch notes on whatever pieces of paper were available at the time, notes on your cellphone, some documents procured from a company director, pictures you were allowed to take, a small mountain of business cards, folders of email and other documentation (Marcus & Fischer, 1996: 37; Sanjek, 1996).

The Description

As described, the researchers' presence in the field over time is the key to ethnographic methodology; indeed it is hard to envision an ethnographic study without sustained fieldwork over time. While the description of the field as well as the continuous interaction with informants and writing of field notes is the central method, the collection of details is not the focus, but the iterative interpretive text of the researcher is. Details are undoubtedly important, but details alone are not as important as the description the researcher develops in the field, using social scientific theory to iteratively survey the most relevant data.

Of interest is reflexivity in the field, as informants may find it interesting to discuss particularities of the field with and alongside the ethnographer. The final product is the deep description (Geertz, 1973), which should have significant resonance with the researchers' experience (Wikan, 1990). A constant fear of "missing out" the "important details," is a common feeling among researchers doing ethnography, and the ideal of having to go home with your informants and talk to them there also makes ethnography truly a labour-intensive process.

WHY ETHNOGRAPHY IS APPLIED IN RISK AND SAFETY RESEARCH

In a discussion on the difference between the social science disciplines of anthropology and sociology, the Norwegian anthropologist Tian Sørhaug put it like this: What makes ethnographic data ethnographic? His answer to this question, and I agree, is the disciplines *respect for the real* as ethnographic writings always include people in its narratives (Sørhaug, 2004). The main benefits of applying ethnographic methodology and its myriad of methods in fieldwork setting should be readily apparent (M Dixon-Woods, 2003; Mary Dixon-Woods & Shojania, 2016). It includes the ability to approach sensitive subjects, such as the adherence to procedures in workspaces, such as in the HGV-study described next. Also, it allows the researcher, among other things,

- to be sensitive to the context the informants live in day-to-day
- to see if work is actuality performed as described
- to observe the many work-around solutions in the workspace
- to observe the expression of the status of various forms of knowledge among the informants
- to make descriptions of risk perception and risk judgements that resonate with the lifeworld of the informants

AN ETHNOGRAPHY OF RISK AND SAFETY IN THE TRANSPORT SECTOR

Actors in the Norwegian road-based commercial goods transport business hold that the high accident risk of international truck drivers on Norwegian roads is connected to international actors not being "serious" about safety standards (Bråten et al., 2013; Jensen et al., 2014; Nævestad et al., 2014). Among political and professional actors in Norway, the combination of the two factors of deregulation of the market and the increase of international competition leading to significant differences in drivers' salaries and working conditions has led to rising concerns about good conduct in the transport business.

The bending and breaking of safety-related rules are common occurrences in the transport sector. In this project the main aim was to shed light on how rules and regulations where bent and even broken by actors. The sensitivity of rule bending as a research theme presented both ethical and reliability challenges in data collection. The use of ethnographic fieldwork in a large complex system also presented challenges to the external validity of the project. Several times, I thought that perhaps my choice of topic was too socially sensitive and stigmatised. Would the truck drivers, their bosses, forwarders and other actors in the HGV-sector grant me access, and what would my status be?

Doing fieldwork in the commercial goods transport sector is not as physically and mentally exhausting as being in field 24 hours a day far from home in rural Tanzania, but it is perhaps more frustrating in that it took considerable time before I had situated myself in the social networks of the sector in such a way that observation was fruitful. Luckily, my research methods were themselves quite useful to gain access and create relationships with informants. In the project, I constantly wrote field notes after and during every encounter with informants; I gathered trade magazines, joined in on multiple online discussion groups and went to events such as trucker evenings. Interviews and focus groups were also widely used and often spontaneous and unstructured.

The divide between etic and emic descriptions and terms refers to the differences of understanding between the social science discourse (etic) of a phenomenon and the context where the same terms are used by informants in everyday conversation (emic), also described as the difference between insider and outsider research (Beals, Kidman & Funaki, 2020). Taking this distinction to heart, it is important to note that while there is a significant literature on noncompliance and rule violation in the safety science literature, I tried first and foremost not to be coloured by the etic, or theoretical terms and concepts describing rule violation. Instead, in line with ethnographic practice I sought out the informants' emic and personal descriptions of rule bending and breaking, finding a rich world of metaphors used to describe non-conforming behaviour as adaptions constructed in the day-to-day working environment.

Inspired by the idea of the "para-site" I participated in a university level course for people working in the transport business (Hadi & Marcus, 2011). Marcus describes the para-site as:

> "a bounded space of orchestrated interaction that is both within the activities of a particular fieldwork project and markedly outside or alongside it- or lateral to it (...) They are imaginaries of alternative forms within the leviathan so to speak, and in basic accordance with the expectations of the traditional anthropological research form" (Marcus, 2013: 206, 210).

In the para-site technique the informants are invited in as "epistemic partners" in the fieldwork and analysis of the data while also being connected to it as informants in the regular sense. "Collaborators in fieldwork cease to be simply informants, they go on, and social science fieldwork is no longer simply 'the discovery of new worlds and the translation of the exotic into the familiar'" (Fox, 2006: 16–17).

As suggested, following the ideals of ethnography it is not enough to observe and write well to be a good ethnographer, but you also should attempt to include yourself in the social settings that initially might not seem to be relevant to the research questions of your study (Geertz, 1973; Wikan, 1990). For example, I not only observed people in the HGV-sector in their daily work but also got to know them in other settings. Such as at the informants' homes. New settings introduce new social roles and statuses and can have the effect that informants perceive you differently, which can lead to new conversational topics and new insights. In the study of rule bending and rule breaking in a complex sociotechnical system such as the heavy goods transport sector this was not possible all the time, but gatherings and congresses and trucker evenings provided ample opportunity. Also, as most of us now have an online presence in our daily lives, so do the actors we study, and in this case, I cultivated an online presence as safety scientist in various fora where people somehow engaged with the transport sector, taking part or simply observing.

I tried to observe occurrences and frequencies of non-conforming behaviour over time (Barth, 1994: 33), such as ignoring sleep time regulation or not securing the load of a truck properly. This approach made it possible to create transactional models showing the flow of communication between different status and roles in the business, and how bending a rule could be done by actors for various reasons. The models or conceptual maps of behaviour were then checked, and cross-checked with actors themselves in field. In the third year of the study, the methods proved successful, with informants from all levels of the system – drivers, auditors, professional and transport managers – approaching me online and in real life wanting to discuss the challenges of rule bending and rule breaking in the industry, confirming and criticising the descriptions of life in the HGV-sector.   •

In summary, the project had many practical challenges, such as access to the field, time consumption and getting informed consent in ongoing social settings. I also found a challenge in how the differences of the implicit normative objective of safety science research can conflict with a core tradition in ethnographic methodology: to collect emic theory-free data from informants. Finally, there is the question on how to judge if the ethnography was successful or even of good quality. While it is quite possible for other researchers to do fieldwork in the HGV-sector, the exact fieldwork described above is not replicable in detail. This is due to how fieldwork is by its very nature linked to the individual researchers' social skills used to construct networks and get in touch with people (Bosk, 2001). Still, I find that challenges of a practical, normative or qualitative nature can be overcome in most projects, and the following toolkit can be a way of doing so.

ETHNOGRAPHERS TOOLKIT FOR RESEARCH IN COMPLEX SOCIOTECHNICAL ORGANISATIONS

Some of the practical, normative and quality criteria challenges described in this section are common to all ethnographic projects, but some become more apparent, and more challenging when the methodology is imported to the study of risk and safety in organisations.

WORKING THROUGH THE PRACTICAL CHALLENGES

The objects of study in safety research are often large and complex sociotechnical systems, such as nuclear power plants, the aviation industry, or entire cities with millions of inhabitants. However, if the creation and maintenance of meaning is of interest to the project concerned, then the ethnographic approach to complex sociotechnical systems may provide valuable insights.

Complexity can refer to the number of actors involved in the system, the number of processes in the system, the interconnections of human behaviour and technological processes, and how the system might be continually changing due to internal and external pressures (Perrow, 1984; Rasmussen & Svedung, 2000). Nancy Leveson defines different types of complexity in systems, referring to simple complexity that might be analytically reduced to comprehensible models, statistical complexity that might be described using statistics, and finally organised complexity where the use of system theory description is necessary (Leveson, 2011: 63). A system might also have a high degree of social complexity which is not limited to a quantitative analysis of the number of connections between the human actors in the system, but can also be understood as the interaction of the socially placed status of different actors, and their use of it in day-to-day work and interaction.

Nonetheless, gaining access and finding a way to delimit the field of research need not be especially difficult in ethnographic studies of risk and safety. Ethnographic methodology has become more self-reflexive over the years. According to Marcus and Fisher:

> Fieldwork should be recognized as a complex web of interaction in which anthropologist in collaboration with others, conventionally conceived of as informants and located in a variety of often contrasting settings track connections amid networks, narrative influences of cultural forces and changing social perspectives. (Marcus & Fischer, 1996: 18–19)

Using this definition and recognising that creating narratives from complex sociotechnical systems make it possible to use several methods of data collection during one period of fieldwork (Salzman, 1996). The ethnographer can simultaneously do participant observation in the field while collecting different types of information, such as written and online sources, and use methods such as focus groups and formal interviews.

In addition, there are techniques such as the para-site suggested by Hadi and Marcus (2011) as an addition to participant observation, which is especially suited for complex organisations where the stratification of the power and agency of actors makes access difficult. The para-site is a stage or event, such as a workshop where the collaborators are not just subjects in the study but are considered to be "'reflexive thinkers' which may (...) therefore be open to risking interpretations together with researchers about ideas fundamental to the political organization of their institutional contexts and functions, and how these ideas circulate, have effect and change" (Hadi & Marcus, 2011: 63). Using multiple stages in combination with documentation relating to the field and collected over a sustained period of time will allow the ethnographer to slowly build a holistic description of the system.

When doing ethnography in complex sociotechnical systems with the intent of being in dialogue with the safety science community, and perhaps only receiving access to the many sites that make up his or her field or system because the gatekeepers hope that there will be safety-related outcomes that can reduce risk or uncover the potential for unwanted incidents, I would argue that the ethnographer has a status in the field as a safety scientist. This status should be acknowledged in the narratives written down by the researcher. The informants who might also be gatekeepers in the field will also see you as such, and certain expectations will be made of you, which might even confer additional access (Goffman, 1959) (see also Bourrier, Chapter 3, this volume on the issue of access to the field).

In addition, there are the problems of access to the right people, the right areas and the right information. An ethnographer doing fieldwork in a hospital setting or in an international development organisation might have to jump through administrative hoops to gain the proper level of access, and be faced with the risk that access can been withdrawn if informants feel that the researcher has access to information that they deem to be sensitive or private, or if other concerns emerge that the ethnographer's access should be limited. Also, attaining informed consent from informants can be hard in any fieldwork, and even more so in large complex organisations, where the informants' status in the organisation can have both legal restraints and social exceptions connected to it. Striking a balance, where the ethnographer is always clear in his or her intentions with regard to his or her research project, as well as the boundaries of the project, alongside actively using the status of researcher to reduce any perceived threats associated with the presence in the field, can mitigate some of the challenges.

The universal problems of time and money are unavoidable in any ethnographic study, as it is both costly and time-consuming to use weeks or months in the field. This in itself cannot be remedied, but alternately spending time in the field and time outside the field can help with the researchers reflections on the activity observed in the field, and perhaps shorten the time needed in field as it makes it easier for the researcher to spend the time in the field more efficiently. It can be frustrating to both novice and senior ethnographers alike to participate in social interaction and observe the setting for many hours and weeks, sometimes without finding any observations that might be "used" in analysis. Taking regular trips outside of the field, to write down experiences and thoughts, can help in this regard, as it prepares the researcher for the next trip into the field, making it easier to focus on what to ask informants.

WORKING THROUGH THE NORMATIVE CHALLENGE

While there might be many reasons for a researcher to use fieldwork and participant observation as the preferred method in his or her project, the adoption of ethnographic methodology in projects and studies aimed at risk and safety is challenging. Safety studies in complex sociotechnical systems are normatively situated in a discipline, which includes the theoretical background of the safety science disciplines, including traditional safety science concepts such as barriers, accidents and resilience (Perrow, 1984; Reason, 1997). The safety scientist submits articles to journals such as *Safety Science* and is in dialogue with classic studies and founding theories of the discipline, such as those introduced in *Man-Made Disasters* (Turner & Pidgeon, 1997). As safety scientists, we have concerns such as how to communicate risks to stakeholders, or how safety is maintained in a system. These concerns are implicitly normative and, taken together with the theoretical package inherent in the discipline, constitute a package of information that the ethnographer brings with him or her into the field – one that might colour observations in various ways (see also Le Coze, Chapter 4, this volume). The formulation of the research project undertaken, and the intent to be in dialogue with practitioners in the safety science discipline, makes it necessary to have a grasp of the governing terms and theories of safety science, such as risk, barriers and high reliability organisations (Reason, 1997). Also, the ethnographer will be working in complex organisations, and participants will be aware that the overall goal of the ethnographers' project is one of safety, or of another goal, if it is not. Transparency in the intentions of the researcher is the key.

These are all long-standing challenges. Marcus asks, "How can we study values when considering that many ethnographic projects have normative concerns?" (Marcus, 2013: 198), and since *value* is a term connected to concepts such as safety adherence and non-conforming behaviour, this question would be a necessary part of any ethnographic project with safety science aspirations. Forgetting that you are a safety scientist with all its normative implications would be likened to forgetting that you wear a uniform that informants see and react to. This means that the researcher must always be aware of his or her status in the social setting.

A common way of doing social analysis in ethnographic research is to make the distinction between the individual informants' status and roles (Barth, 1994; Goffman, 1959). Status refers to the individual's position in his or her social context, such as a controller or manager. The concept of role on the other hand refers to how the individual uses and manages his or her status. This dual concept of social analysis can also be applied reflexively to the ethnographer in field. As previously mentioned, the ethnographer should be visible in their own narratives. The ethnography should then also make apparent his or her status when in field, and how he or she used agency to shape the role of his or her status.

This realisation is not necessarily problematic, and absolutely solvable in the field setting. First, being a safety scientist might confer some status upon the research project, and it might be considered harmless in the eyes of gatekeepers, and decision-makers in the system. Second, it is necessary to clarify the ethnographer's status as a safety scientist both in interpersonal relations in the field and in the narratives the researcher writes based on the observations done. The outcome of ethnographic study is mostly written text. Written texts can be considered translations from the native point of view, the emic, to the theoretical, ethic. But where the early-20th-century ethnographers translated the meaning of social and cultural behaviour from non-Western cultures to the academic and wider public at home, the translation work of the safety science anthropologist will be one from the ethnographic data to the field of safety science.

WORKING THROUGH QUALITY CRITERIA AND THEIR DIFFERENCES

The methodology of ethnography differs from other social science disciplines in that the common and transdisciplinary parameters of reliability, validity and (for quantitative research) generalisability might simply not be the best way to judge if ethnographic descriptions and narratives of social interaction are successful. Leslie, Paradise, Grooper, Reves and Kitto (2014) also consider the distance between safety and quality improvement in health care and ethnographic methodology to be imported into research projects. They suggest that pains should be taken to strive for more rigidity and transparency in ethnographic studies of health care. Also, they suggest that the collection of ethnographic data and ethnographic decryptions should be less concerned with critique of the human action and life worlds in field and instead move towards descriptions that can be useful to safety science (Leslie et al., 2014).

In a more structured solution to the challenge of quality criteria, Stewart suggests using criteria other than validity, reliability and generalisability to judge the quality or successfulness of an ethnographic contribution to the scientific discussion (Stewart, 1998). His criteria are:

- Does the ethnographic contribution demonstrate *veracity*: that is, is the contribution faithful to the truth, or at least successful in its description?
- Does the ethnographic contribution demonstrate *objectivity*: that is, does the contribution give conclusions that may be shown to transcend the perceptions of the individual research object?
- Does the ethnographic contribution demonstrate *perspicacity*: that is, does the contribution create understandings that are applicable to the study of human behavior in other research settings?

The use of these criteria can be considered a philosophical move away from the three standards of social research: reliability, validity and generalisation as established in mainstream social sciences (Bernard, 2006), with veracity replacing internal validity, objectivity replacing reliability, and perspicacity replacing generalisation or external validity (Table 7.1). Using these criteria as alternative approaches to the standard trio could provide a platform to open the dialogue between importers and exporters of ethnography in the risk and safety studies, and expand understanding of how to judge the quality of ethnographic contributions.

TABLE 7.1

Expectations for Ethnographic Method

Epistemic Value	Veracity	Objectivity	Perspicacity
Conventional equivalent	Validity (excluding external validity)	Reliability (excluding consistency)	Generalizability, external validity
Underlying question	Verisimilitude of depiction	Transcendence of perspectives	Applicability of insights elsewhere

Source: From Stewart 1998, modified by author.

CONCLUSION

From: driver256@truckeronline.no
To: christian @norwegianacademic.no
Subject: Your project

…

The subject you are researching is very important and contains a lot of information not fit for the light of day; many people just close their eyes to it. Another challenge is that few drivers are unionized and have to lower the safety demands in order to "survive in the business," and the unions are only preoccupied with increased tariffs. Having a slightly higher salary does not help when you daily risk your own and others' lives under the pressure of losing your license and livelihood.
Hear from you!

A common ground between safety science and ethnography can be found in how both ethnographers and safety science practitioners use organisational perspectives and are concerned with how meaningful practices are made, maintained and changed by people situated in groups and organisations.

Ethnography is best described as a methodology and not by its individual methods. But as a methodology, the ethnographic ideals of participatory fieldwork in a geographically delimited field, its holistic aspirations and the thick or rich descriptions of people and their cultural and social environments should not be written in stone in a way as to complicate the export of the methodology to less-than-ideal fields of study (Mary Dixon-Woods & Shojania, 2016).

In this chapter I have sought to show how ethnographic methodology can successfully be applied to the study of risk and safety. Challenges and differences in quality criteria can be overcome in dialogue between the importers and exporters of the methodology. This chapter has tried to suggest context-sensitive approaches to a context-sensitive methodology. Each new system, be it the HGV-sector, a nuclear plant or a healthcare institution will naturally demand a rethinking of the challenges and the methods used. In my example of studying rule bending in the heavy goods vehicle sector, a less-than-ideal field, the use of *safety researcher* as a participant status, the creative use of para-site method, multisite fieldwork as well as the cultivation of an online presence as a safety researcher both created access and established the trust over time that is necessary to developing thick descriptions.

REFERENCES

Barth, F. (1994). *Manifestasjon og Prosess*. Oslo: Universitetsforlaget.
Beals, F., Kidman, J., & Funaki, H. (2020). Insider and outsider research: Negotiating self at the edge of the emic/etic divide. *Qualitative Inquiry*, 26(6), 593–601. doi:10.1177/1077800419843950.
Bernard, H. R. (2006). *Research Methods in Anthropology* (5th ed.). Lanham: AltaMira Press.
Bosk, C. (2001). Irony, Ethnography, and Informed Consent. In: Hoffmaster, Barry, ed. *Bioethics in Social Context*. Philadelphia: Temple University Press, 199–220.

Bråten, M., Hovi, B., Jensen, R. S., Leiren, M. D., & Skollerud, K. H. (2013). *Arbeidsforhold i Vegsektoren. Et Forprosjekt.*

Cupit, C., Mackintosh, N., & Armstrong, N. (2018). Using ethnography to study improving healthcare: Reflections on the "ethnographic" label. *BMJ Quality and Safety,* 27(4), 258–260. doi:10.1136/bmjqs-2017-007599.

Dixon-Woods, M. (2003). What can ethnography do for quality and safety in health care? *Quality and Safety in Health Care, 12*(5), 326–327. doi:10.1136/qhc.12.5.326.

Dixon-Woods, M., & Shojania, K. G. (2016). Ethnography as a methodological descriptor: The editors' reply. *BMJ Quality and Safety,* 25(7), 555–556. doi:10.1136/bmjqs-2015-005117.

Fox, N. J. (2006). Postmodern field relations in health research. In: D. Hobbs & R. Wright (Eds.), *The Sage Handbook of Fieldwork.* London: Sage.

Geertz, C. (1973). *The Interpretation of Cultures: Selected Essays* (Vol. 5019). New York: Basic Books.

Gobo, G. (2008). *Doing Ethnography.* London: Sage.

Goffman, E. (1959). *The Presentation of Self in Everyday Life.* Garden City, NJ: Doubleday.

Hadi, D. N., & Marcus, G. E. (2011). In the green room: An experiment in ethnographic method at the WTO. *PoLAR: Political and Legal Anthropology Review, 34*(1), 51–76. doi:10.1111/j.1555-2934.2011.01138.x.

Jensen, R. S., Bråten, M., Jordfald, B., Leiren, M. D., Nævestad, T.-O., Skollerud, K. H., . . . Tranvi, T. (2014). *Arbeidsforhold i Gods og Turbil.*

Leslie, M., Paradis, E., Gropper, M. A., Reeves, S., & Kitto, S. (2014). Applying ethnography to the study of context in healthcare quality and safety. *BMJ Quality and Safety,* 23(2), 99–105. doi:10.1136/bmjqs-2013-002335.

Leveson, N. (2011). *Engineering a Safer World Systems Thinking Applied to Safety.* Cambridge, MA: MIT Press.

Malinowski, B. (1922). *Argonauts of the Western Pacific.* London: Routledge.

Marcus, G. E. (2013). Experimental forms for the expression of norms in the ethnography of the contemporary. *HAU: Journal of Ethnographic Theory, 3*(2), 197–217. doi:10.14318/hau3.2.011.

Marcus, G. E., & Fischer, M. M. J. (1996). *Anthropology as Cultural Critique. An Experimental Moment in the Human Sciences.* Chicago, IL: University of Chicago Press.

Nævestad, T.-O., Hovi, I. B., Caspersen, E., & Bjørnskau, T. (2014). *Ulykkesrisiko for tunge godsbiler på Norske veger: Sammenlikning av Norske og utenlandske aktører* (1327/2014).

Perrow, C. (1984). *Normal Accidents Living with High Risk Technologies.* Princeton, NJ: Princeton University Press.

Rasmussen, J., & Svedung, I. (2000). *Proactive Risk Management in a Dynamic Society.* Karlstad: Swedish Rescue Services Agency.

Reason, J. (1997). *Managing the Risks of Organizational Accidents.* Aldershot: Ashgate.

Salzman, P. C. (1996). Methodology. In: A. Barnard & J. Spencer (Eds.), *Encyclopedia of Social and Cultural Anthropology.* London: Routledge.

Sanjek, R. (1996). A vocabulary for Fieldnotes. In R. Sanjek (Ed.), *Fieldnotes the Makings of Anthropology.* London: Cornell University Press.

Sørhaug, T. (2004). *Managementaliet og autoirtetens forandring – Ledelse I en kunnskap- søkonomi.* Bergen: Fagbokforlaget.

Stewart, A. (1998). *The Ethnographer's Method* (Vol. 46). Thousand Oaks, CA: Sage.

Turner, B., & Pidgeon, N. (1997). *Man-Made Disasters.* Oxford: Butterworth-Heinemann.

Wikan, U. (1990). *Managing Turbulent Hearts: A Balinese Formula for Living.* Chicago, IL: University of Chicago Press.

8 Work as Planned, as Done and as Desired
A Framework for Exploring Everyday Safety-Critical Practice

Andrew J. Rae, David E. Weber
and Sidney W. A. Dekker

CONTENTS

WHY EXPLORE THE EVERYDAY?

One of the first and key methodological decisions in any safety research project is whether to focus on the exceptional or the mundane. Many researchers select, either explicitly or implicitly, accidents as the core objects of safety research (Ge et al., 2019). Safety research, under this approach, is about uncovering, describing and testing the processes and phenomena that lead to accidents (Aven, 2014; Hopkins, 2014). For example, well-known safety theories such as "Normal Accidents" (Perrow, 1999) and STAMP (Leveson, 2004) are essentially generalised explanations of accident causation. Such research struggles to rigorously establish and defend the cause–effect models it generates, because of the challenge of generalising from a sparse set of case studies. It can, however, at least in principle, generate testable hypotheses about accident prevention.

An alternative to studying accidents is to study normal situations. This has been expressed by Hollnagel (2018) as a "focus on how things go well," but this phrasing can be misleading. To study "success" is to adopt the exact same ontology as to study "failure." The difference is in the cases selected, not in the methodology. That is why we prefer the term "everyday work" rather than "successful work." The object of analysis is not the outcome, along with the processes and phenomena that cause the outcome, but the work itself, and the processes and phenomena that shape the work (Rae et al., 2020). Everyday work exploration can apply much more rigour in establishing descriptive models, because it does not need to make strong normative claims about what causes or prevents accidents.

Further discussion and debate about the choice between studying the exceptional and the mundane can be found throughout safety science literature (Havinga et al., 2017; Hopkins, 2007; La Porte & Consolini, 2008; Leveson et al., 2009). The remainder of this chapter assumes that the reader has decided upon, or is at least strongly interested in, everyday work as the object of investigation. The chapter will explain some of the methodological and practical challenges of studying work practices for the purpose of understanding safety, and will present a framework that may be helpful in the data collection and preliminary analysis phases of the research.

THE GAP BETWEEN WORK-AS-DONE AND WORK-AS-IMAGINED AS A RESEARCH CHALLENGE

There is nothing remotely original about the idea of studying work. Work is an important and interesting aspect of individual and social life, and almost every branch of social science has its own ways of thinking about and investigating work. Industrial and organisational psychology has investigated everyday work from the time of Mayo's Hawthorne experiments (Mayo, 1933) and Gilbreth and Taylor's time and motion studies (Nadworny, 1957). Ergonomists study work through methods such as cognitive task analysis and naturalistic decision-making (Hoffman & Militello, 2008), whilst sociologists examine workplaces as complex interactions of social structure and individual agency (Lawrence & Suddaby, 2006). Safety science has many methodologies and research practices that it can draw on, but also has difficulties particular to the field (see also Bourrier, Chapter 3; Le Coze, Chapter 4; Kuran, Chapter 7, all this volume).

"The challenge," as Havinga et al. (2017) explains, "is in talking about everyday work and safety at the same time." Accident-based models are difficult to validate but have straightforward suggestions for accident prevention. Descriptions of everyday work are easier to validate but seldom contain clear prescriptions for improving safety. How does describing the normal help prevent the exceptional? One response to this question focuses on the gap between Work-as-Done and Work-as-Imagined (Dekker, 2004; Hollnagel, 2017). Rules and procedures, the most common management tools for safety, are essentially prescriptive models of work. Any prescriptive model relies on an implicit descriptive model – a set of assumptions about how work happens in-between and around the rules and procedures. The accuracy of this descriptive model, with specific reference to its safety features, is vitally important for successful safety management. Closing the gap between models of work (Work-as-Imagined) and the reality of work (Work-as-Done) allows safety research to influence accident causation without directly investigating accidents.

While the Work-as-Done / Work-as-Imagined labels can be read as implying that there is a "true" picture of work, held by the workers, and an "imaginary" picture of work, held by managers, the reality is even more complex. The distinction between Work-as-Done and Work-as-Imagined originally comes from a distinction between "tâche" and "activité" (Leplat & Hoc, 1983), used to indicate that any formalised description of work is necessarily an abstraction placing emphasis on one particular viewpoint of work. Focussing on this viewpoint comes at the expense of other details which vary from instance to instance of the work, without changing the essential nature of the work. As a key example, a procedure always lists only some of the steps involved in a task. Someone is said to be following the procedure if they accomplish all of the steps in the prescribed order – the non-recorded steps are variable. A recipe asks for the addition of a cup of flour; it does not specify how the flour is to be obtained or measured. There is never a single objective description of work, even in the minds of the people who perform the work every day.

This, then, is the challenge and opportunity of researching everyday work for safety. There are many available representations of any work task. There are, as a representative but incomplete list:

- procedures telling people how to perform the task;
- rules that constrain the variability of the task;
- descriptions of the task made by observers;
- descriptions of the task made by participants;
- descriptions of the task made by non-participants whose work nevertheless interacts with or relies on the task;
- observations of specific instances of the task; and
- artefacts such as photos or videos recording parts of the task.

The role of a researcher of everyday work is to distil, synthesise and triangulate these representations to create a new description of the task which is authentic with respect to the other representations but still provides new useful insights. A key purpose is to identify the capacities that make work go well, and to seek the organisational, operational, procedural and design leverage points to enhance those capacities (Hollnagel, 2018), and to close gaps between planned and executed work (Borys, 2009; Dekker, 2012; Hollnagel, 2018).

METHODS AND METHODOLOGIES FOR RESEARCHING EVERYDAY WORK

It can be helpful to think of research design as a series of decisions. Each choice commits the researcher to a particular path of enquiry. We have already described the first of those commitments – focussing on the mundane rather than the exceptional. The second key decision is about the immediate goal of the research. Havinga et al. (2017) suggest that there are three main options here:

1. **Normative** research, which tries to find the best way to perform work, or at least to find opportunities to improve work;
2. **Descriptive** research, which tries to understand what is currently happening; and
3. **Formative** research, which searches for new ways to perform work.

Everyday work explorations are primarily descriptive. Whilst the ultimate goal of the researchers may be to make the world safer, the immediate goal of the research is to understand current reality.

In methodological terms, then, everyday work exploration for safety builds on, and is arguably a sub-field of, organisational ethnography. As Watson (2012) describes in the founding issue of the *Journal of Organizational Ethnography*, ethnography is not "simply a method of doing research." It is a

> "genre of social science writing which draws upon the writer's close observation of and involvement with people in a particular social setting and relates the words spoken and the practices observed or experienced to the overall cultural framework within which they occurred."

This methodology may embrace many practical investigative methods, including interviews, observation, participation, documentary analysis and even surveys. Ethnography is distinguished by the "metatheoretical assumptions" that influence "what we see as 'data,' how we collect and analyse the data, how we theorize, and how we write up our research accounts" (Cunliffe, 2011).

So what are the metatheoretical assumptions of organisational ethnography? The heart of ethnography is close, perhaps even "intense" (Watson, 2012) personal engagement and observation, whilst still recognising that the stories we hear and take part in as researchers are always only a part of larger stories told over longer periods of time or in wider contexts (Boje, 1995). We assume that each piece of data matters, whether it is a document, a conversation, a personal story or a muttered prophanity. However, we also assume that the way the data matters is not straightforward.

Common "traps" we are wary of include:

- believing that there is a single, knowable, objective reality that can be revealed through our data (the "Naïve realism" of Guba and Lincoln (1994));
- believing that what we notice is local and distinctive, when in fact it is a common feature of an industry or a culture (Martin, 2002); and
- accepting second-hand accounts about what other people believe (Dekker, 2014).

THE EDWE CODING FRAMEWORK

INTRODUCTION TO THE FRAMEWORK

Leading from the previous sections, the main methodological commitments of everyday work exploration (EDWE) are to focus on some type or aspect of everyday work as the object of the research, to aim to describe that work as it happens, to embrace all representations of that work as valid data that must be accounted for and to recognise that those representations are always part of a larger picture. These commitments drive the data collection in a very straightforward way. The researchers seek out existing representations of the work, and record new representations by observing the work or by conducting interviews and focus groups about the work.

As far as possible, EDWEs are conducted in the absence of a negative event such as an incident or accident. Whilst an accident report is technically a representation of how the work has performed, it contains data that has already been analysed with methodologies inconsistent with everyday work exploration – typically by trying to reconcile rather embrace a diversity of perspectives, and in doing so privileging normative representations of the work.

The method presented here is a framework for the coding of data. This framework is not intended to replace thematic coding, and is consistent with a number of different approaches for identifying themes. There is a long-standing debate in qualitative research about the role of existing theory in data gathering and interpretation. "Grounded Theory," for example, is sometimes interpreted to suggest that themes should emerge entirely from the data, ignoring existing theory (Howard-Payne, 2015). Other approaches emphasise the use of theory as a structure for analysing data. See Saldana (2015) for a thorough treatment of different approaches to coding.

The purpose of the EDWE coding is to supplement the primary coding method by identifying and labelling different representations of the same work. This supports the researchers in identifying missing representations, as well as in recognising when themes are consistent between representations, or arising as a result of particular representations of the work (see also Kuran, this volume about the collection of ethnographic data).

PRIMARY CODES

The framework to explore everyday work (Table 8.1) consists of codes for "subjects" and "modes." Each combination of subjects and modes (subject–mode tuples) relates to a particular actor and use of tense. For instance, discourse in present tense and with a reference to management exemplifies the tuple "management generalised." All subjects, modes and tuples are explained in more detail below.

Subjects

Each row in Table 8.1 refers to a different person or group. A typical workplace setting involves internal and external groups of subjects, often collectively called "stakeholders." Internal subjects are those directly employed by the organisation in which the EDWE is performed (subsequently referred to as "the organisation"),

TABLE 8.1

Framework to Explore Everyday Work

Subjects	Modes			
	Described	Generalised	Assumed	Desired
Worker/s (self)	[Reference to worker/s + verb in past] "We solved the problem: we made [the bar] of alloy."	[Reference to worker/s + verb in present] "I do not want to be promoted in this company."	[Reference to worker/s + assumed state] "I guess it feels like if something was to happen – you break your foot or something – you know you're going to be in trouble."	[Reference to worker/s + desired state] "I'd like to learn [name of terminal manager]'s job."
Colleague/s	[Reference to colleague/s + verb in past] "This [driver] physically noticed that [the truck] had an air leak."	[Reference to colleague/s + verb in present] "Current [safety] people haven't done the work."	[Reference to colleague/s + assumed state] "[Replacement drivers] probably wouldn't find half of my cages."	[Reference to colleague/s + desired state] "[Issues with dogs] should be classified as a near miss, sort of an incident."
Management	[Reference to management + verb in past] "There was a lot of management changes."	[Reference to management + verb in present] "[Management] have a problem with us and it gets brought up at the meeting. We need it solved there and then."	[Reference to management + assumed state] "There might be a little bit of chit-chat about it, but whether you get feedback that's probably a rarity."	[Reference to management + desired state] "Management needs to group and schedule [deliveries] a lot better so it doesn't put strain on the drivers."

(Continued)

TABLE 8.1 (CONTINUED)
Framework to Explore Everyday Work

| | | Modes | | |
Subjects	Described	Generalised	Assumed	Desired
Own Organisation	[Reference to own organisation + verb in past] *"All these important notes that we used to have on our systems, when [the company] took over, all got thrown to the dogs."*	[Reference to own organisation + verb in present] *"[Paperwork] just going through way too many hands. It becomes lost in the system."*	[Reference to own organisation + assumed state] *"[The company] would have spent all this money."*	[Reference to own organisation + desired state] *"[We] definitely [need] more collaboration within the organisation."*
Externals - Customers - Authorities - Other organisations	[Reference to externals + verb in past] *"[The agent] sold the business."*	[Reference to externals + verb in present] *"[Customers] don't understand things that go wrong on a site."*	[Reference to externals + assumed state] *"[Onsite,] you might forget there's a dangerous dog in there, and he's baled you up."*	[Reference to externals + desired state] *"If the customer had a better understanding of what LPG is and what it does, I think it would make a lot of things a lot easier."*

To exemplify the application of the framework to analyse discourse, we draw on an example of workers speaking about the consequences when experiencing injuries (Table 8.2). This conversation started when a worker reported to have injured their back when falling through a gap in a pallet used to transport materials. The worker's colleagues subsequently came with similar experiences and some workers decided not to report minor injuries in the future. Table 8.2 outlines the discourse as it relates to the subjects and modes. The initial conversation pertained to "work described" and gradually evolved to "work generalised" and "work assumed." No content directly related to "work desired." This can either be inferred from the existing data, or – ideally – result from following up with the respective participants and other relevant stakeholders.

whereas external subjects are other people, companies and institutions with which the organisation interacts. Examples of internal subjects are the workers, their colleagues or management, whereas external subjects are customers, authorities or other organisations. The framework should be extended with new subject codes if data refers to a person or group that doesn't match one of the existing subjects.

Workers are the people who directly and physically perform the work being discussed. This subject is implied when a research participant refers to themselves ("I"), or themselves and others ("we"). Examples are workers explaining:

> *"I won't hesitate to pull the pin on something if I feel it's unsafe."* or *"A very real fear that we do have here at the terminal."*

Colleagues are the people who the "workers" interact with within the same organisation. "Colleagues" can have the same or different roles as the "workers." Examples are call centre staff or replacement drivers. Participants refer to colleagues by using names (e.g. *"[Name] has got a lot of experience"*), company divisions (*"We're not supported from time to time by the call centre."*) or roles (*"A handful of really good, on-the-ball drivers"*).

Management involves those who administrate, organise and coordinate work activities in managerial roles. Examples of workers referring to management are:

> *"It'd be good to have backup from management,"* or
> *"[Name of manager] is higher than our manager."*

Own organisation relates to the company that "workers," "colleagues," and "management" work for. It makes sense to include "own organisation" as an individual group of subjects when people explicitly refer to the organisation they work for:

> *"The problem we have in this company at the moment is someone's turning up, 'Who are they axing?'"*

Externals are people, organisations or institutions outside the ("own") organisation. Examples are the customers, other organisations or government authorities. An example of people speaking about "externals":

> *We've got the fire brigade. We've got the SES. We've got other external people: Some of the plumbers would actually come in and help us.*

Modes

As well as relating to different stakeholders, data can refer to work in different ways, depending on whether it refers to specific experiences, generalisations, or aspirations about work. We refer to these different ways of talking about work as "modes." The framework consists of the following four modes (Table 8.1; columns).

Work described is the code assigned to a description of a specific situation or instance of work. These descriptions are typically in the past tense. An example is an illustration of how work was done in the past. For instance, workers spoke of a time when they replaced a heavy steel bar on a truck with an alloy bar to prevent injuries:

> *We solved the problem: we made the bar of alloy. So we went from a 10kg piece of steel that could knock you out to an alloy bar that's about 1 kilo. So the modifications didn't wreck the strength, but in case the bar did flip over, it didn't knock you out.*

Work generalised is assigned to a representations of a general state or lasting condition at work, for instance, how work usually takes place. People typically use present tense when speaking about work in general, for example:

> *Scheduling's bad. It's terrible actually. We do have a lot of problems with scheduling.*

Work assumed contains an element of uncertainty or speculation about work in the past, present, or future. The language is speculative and conditional. Words like "would," "might" or "probably" are typical. "Work assumed" often reflects people's concerns. An example is a worker who contemplated whether workers would receive any feedback about a concern raised in the form of lodging an observation for management:

> *I'd be surprised if you got any feedback or follow up when you put that observation through the system. There might be a little bit of chit-chat about the information provided, but whether you get feedback, that's probably a rarity.*

"Work assumed" not only reflects what people are unsure about but also reflects what makes them worry. For instance, workers can express concern that some colleagues might be unable to cope with the high workload during wintertime:

> *Winter brings its own challenges. We get a nightmare. Winter is different to summer. In winter we're flat out. We're exhausted because there's so much demand of work that physically, sometimes, we just cannot keep up. So many more trucks to do. So many more gas bottles to put on and off trucks. We can all predict come wintertime we're all going to be under the pump. That creates stress, fatigue on the road. I can see a point where maybe someone's going to crack.*

Work desired is assigned to normative accounts of how people would like work to take place, and what they recommend changing. People make suggestions, for instance, about what they would like to do or change, and how work processes could be improved. The language is characterised by wishful thinking and counterfactual expressions, such as "like to," "should be" or "needs to." For instance, a worker reported:

> *Whoever is in charge of making these policies or procedures should be making procedures specific to each site.*

"Work desired" offers rich descriptions of how people would like to improve work processes and the collaboration within the organisation. For instance, workers can report the absence of experienced colleagues in the company's call centre, trained and experienced enough to explain to customers about the issues at their sites when customers ring the call centre:

> *"We need experienced people who know about LPG in the call centre. Have experienced people on the phone that can explain to customers about their installation and what's wrong: 'Your site's non-compliant, but I'll put you to a person who will be able to explain to you.'"*

APPLYING THE FRAMEWORK

Whilst the framework is principally applied during preliminary data analysis, it can also be used to inform data collection and to clarify the findings and conclusions.

The subject codes provide guidance on the initial selection of participants, and on the expansion of the participant pool to fill gaps in the data. To effectively analyse everyday work it is important to avoid drawing conclusions based only on indirect representations such as workers talking about management or management talking about workers. It is most desirable to have multiple representations of the work from a variety of subjects.

The modes provide guidance about the questions that can be asked during data collection. (See Box 8.1 for an illustrative example.) Asking about "work described" is a good and easy way to initiate conversations, particularly when it is challenging to get people engaged. Questions about "work described" ask people to tell stories based on their own experiences. An example is to start by asking, *"Can you tell me of a time when you stopped work?"* when exploring workers' decisions to stop work for safety reasons (Weber et al., 2018). Then the questions are gradually shifted towards the other modes and eventually to "work desired." An example of how to explore "work generalised" is to ask people about the reasons they conduct work in the ways observed, for instance:

> *I noticed you are wearing a harness hooked up to a tripod in manholes that are less than a metre deep. Can you tell me about the reasons?*

The subject–mode tuples provide cases of everyday work. "Work described" is the most tangible mode, since it entails people's descriptions of situations as experienced in the past. "Work desired" is the most abstract mode, reflecting how work should and could be, but is not (yet). The following pairs of modes share certain characteristics: "work described" and "work generalised" are descriptions of work as it actually happens, and reflect work as it is experienced. In contrast, "work assumed" and "work desired" are hypothetical representations, characterising how work might, could, or should be (Cooperrider, 2018), and it remains uncertain if an assumption or reflection about work is true or will become reality.

BOX 8.1.   Illustrative Example of Using the EDWE Framework

Stopping and Non-Stopping: A Study of Everyday Work

"We can stop work, but then nothing gets done" (Weber et al. 2018) was a qualitative investigation of workers stopping tasks because of concerns about safety. Stopping unsafe work is a behavioural phenomenon that is highly relevant for safety. It is mentioned in many theories of safety and accidents, including behavioural safety (Abdelhamid & Everett, 2000), safety climate (Tharaldsen et al., 2008) and Drift (Dekker, 2011). And yet the act of stopping unsafe work, by its very nature, can only be considered as a hypothetical counterfactual when examining accident causes. To understand when, how and why workers stop unsafe work, the behaviour needs to be considered and examined as part of everyday work. It cannot even be usefully constructed as a "successful" or "correct" behaviour, because the outcome of not stopping (or of stopping) is almost always unknowable.

The work was funded by, and conducted within, an energy supply business. The particular branch of the business being studied supplies LPG gas bottles to residential and commercial customers.

The researchers made an explicit methodological choice to conduct an everyday work investigation. This choice drove the data collection. For an everyday work investigation, the data should consist of as many representations as possible of the work being studied – in this case, of the act of stopping work because of safety concerns. The researchers chose to use focus groups as the primary means of collecting the data. Focus groups were chosen ahead of direct observation because it was assumed that stopping work was an infrequent event. In fact, as this study hinted and a later study confirmed, stopping for safety is a very normal part of operational activity, and in hindsight direct observation could have been used. Focus groups were chosen ahead of one-on-one interviews because the researchers suspected that collective sensemaking about when it is appropriate to stop work was part of the phenomenon, and were hoping to collect responses from multiple participants about the same specific examples.

During the focus groups, the EDWE framework was used to guide the conversations to ensure a balance of work described, work generalised and work desired. In addition to the focus groups, the researchers also had access to normative representations of stopping work, in the form of company policies, procedures and senior management announcements.

During preliminary coding, the EDWE framework was used to label fragments, followed by the use of Corbin and Strauss' (1990) axial coding of themes. The following fragment illustrates the framework applied to differentiate between different subjects and modes:

"I complained about bottles being loaded on trucks without O-rings (WORKERS, DESCRIBED). *So you get to a delivery and you've got to replace*

O-rings (WORKERS, GENERALISED), *which should be something the bloke does that's filling the gas bottle* (WORKERS, DESIRED). *I probably filled out about three or four observations* (WORKER, ASSUMED). *It wasn't until [name of terminal manager] came out with me one day* (MANAGEMENT, DESCRIBED). *I showed him as we were unloading* (WORKER, DESCRIBED). *I've noticed that very rarely do I get a bottle without an O-ring or gauge now* (WORKER, GENERALISED). *So finally it got fixed* (OWN ORGANISATION, GENERALISED). *But that had to go to the point of somebody actually being on the truck with me, was in a position do something for the issue to get fixed"* (MANAGEMENT, DESCRIBED).

The EDWE coding does not provide the thematic analysis, but it helps the researchers steer a path between overly positivist reading of the passage, and a completely constructivist reading.

The work described can be matched with other similar work described as specific instances to support or contradict statements about work generalised – in this case to support a theme that perceived management inaction provides a disincentive for stopping work. The statements about management could not be cross-referenced with statements by management describing their own work, so the theme was labelled as "perceived management inaction" rather than "management inaction."

The EDWE framework provides a structure within which to explore all relevant subject–mode tuples during the collection of data on everyday work activities. When collecting data, it is important to get rich descriptions of the situations people choose to report. This can provide details and material that can be embellished by others, increasing the likelihood of other participants adding further experiences. Using an empty template of the framework for each conversation allows taking notes of what people have reported regarding each tuple, and helps to ensure that all areas of interest have been fully covered.

During analysis, the framework is applied mechanically, and independently of more researcher-interpreted coding. It should be possible to achieve very high inter-rater reliability when undertaking EDWE coding, with variance occurring only when there is syntactic ambiguity about the subject of a sentence, or there is irregular use of tenses, disguising whether a statement is intended as a description of a particular instance of work, a general statement about work or an aspiration about how work could be.

The EDWE codes are not intended to themselves form the basis of in-depth conceptual or empirical analysis, but they should assist in linking and prioritising thematic codes. For example, the following passage, once EDWE coding has been applied, contains a generalised worker description of a management behaviour.

"We keep reporting the same issues (WORKERS, GENERALISED), *and they go, 'We'll look into it'* (MANAGEMENT, GENERALISED). *And during the next toolbox we say, 'Did you do anything about it?'* (WORKERS, GENERALISED). *'Oh no,*

we're still looking into it.' As soon as they have a problem, it needs to be solved there and then (MANAGEMENT, GENERALISED). *Yet we constantly say to them what we need fixed. Scheduling: over and over* (WORKERS, GENERALISED). *So it really gets me how they can tell us what we need to fix and expect it to be sorted out on the day, yet they don't do the same for us. Anything we say to them just floats out the window"* (MANAGEMENT, GENERALISED).

To improve descriptions of work, researchers can seek out fragments on the same theme coded as DESCRIBED and DESIRED, along with fragments directly from the management point of view. In the absence of specific instances or alternate representations, this theme is a legitimate representation of a worker experience, but must be carefully labelled as only generalised experience from the workers' perspective.

The discourse relating to any particular theme (e.g. Table 8.2) can be further condensed for reporting as suggested in Table 8.3. This condensation allows company decision makers to reflect on everyday work in terms of the relevant subjects and modes as it relates to any theme of interest. The framework also helps to support the development of rich descriptions (Geertz, 2017), following a formula of moving from GENERALISED to DESCRIBED to DESIRED.

ISSUES AND FURTHER CONSIDERATIONS

Researchers will often need to choose between an open or a closed investigation of everyday work. An open investigation does not specify a particular aim and focus other than trying to understand the challenges of everyday work in general. People's experiences about work are explored from the "bottom up." Participants are free to touch on any subject relevant to work. A closed investigation is conducted to better understand a pre-identified challenge or issue. A closed EDWE represents a top-down approach since it explores people's opinions about a certain topic of interest. An open investigation creates the potential for interesting surprises (Davis, 1971), but the nature of a broad-ranging and open investigation means that in the EDWE coding, each theme is unlikely to be densely supported with a variety of subjects and modes.

In principle, the choice of an open or closed investigation is a practical question of scope, rather than a methodological issue. Both open and closed investigations can be inductive, with themes emerging from the data, or can be deductive, applying existing theory to shape the themes (Howard-Payne, 2015). However, a closed investigation risks the question being framed in terms of a single representation of work, most commonly "Why do workers not follow the rules?"(see also Le Coze, Chapter 4, this volume). The more the scope of an investigation is limited, the more likely it is to violate the methodological commitments of everyday work exploration, by building contrary assumptions into the research question.

A related consideration is whether to involve the perspective of a single or multiple groups of stakeholders. The more topics and groups of stakeholders are involved, the richer and more representative the EDWE, yet the larger the dataset and more time-consuming the analysis. Consistent with the literature behind "work-as-done,"

TABLE 8.2
Discourse of the Example "Experiences When Suffering Injuries"

Modes		
Described	Generalised	Assumed
Workers	**Workers**	**Workers**
(2) *which I didn't want to because I know what my back's like.*	(3) *But no, "You've got to do the right processes." You have no right.*	(14) *I guess it feels like if something was to happen – you break your foot or something – you know you're going to be in trouble,*
(5) *which I still did, crippled over ...*	(12) *If I go into a property and I know there is a dog there, and I go in and I get bitten, it's my fault.*	
(8) *while I was still crook. So I still had to somehow jump in the car – which killed me – to do an interview,*	(18) *But how else do I get new cylinders?*	
(19) *I've actually injured myself and been on light duties: my elbow, from physical labour.*	(23) *I've never heard this in my life. Like, [we] don't do fifteen-minute exercises or stretches before we start.*	**Management**
(21) *And I said, "What do you mean? I'm not trained to do any [exercises]."*	(25) *If I hurt myself and it's not bad, I'm not going to report it.*	(15) *because they'll just say, "You shouldn't have done that."*
(34) *[So] I just lied to them and said [the injury] was better.*	(27) *Yep, no dramas, but doesn't get reported.*	**Own Organisation**
Management	(30) *"Oh God, that's too hard. I won't report it." I honestly believe that.*	(32) *because the injuries will probably not be a true scale.*
(1) *They put me on workers' compensation,*	**Management**	
(4) *And while I'm on compo, they still dragged me in, wanted me to come in,*	(11) *"Why did you do it? You knew it wasn't safe. Why did you do it?"*	
(7) *Somebody who was the trauma manager at the time, he came down [to the terminal] to do an investigation on what happened ...*	(26) *"It's your fault. Just watch yourself."*	
(16) *The bloke that was management at that time, he said it was my fault that I hurt myself, not [the company]'s [fault], because I walked on those pallets.*	(29) *I think they know they've to make [reporting] so difficult that you'll go,*	
(20) *One of the head safety guys came down and paid a visit, and asked what I did and all that. And then [the safety person] said to me, "Did you do your exercises?"*	(31) *[A lower number of injuries] makes them look great,*	
(22) *And [the safety person] said, "Before you start work, you should have done all your stretches and your exercises." Just out of the blue!*	**Own Organisation**	
(24) *So [the safety person] turned [the injuries] straight back onto me: it was my fault [to get injured] for not doing the exercises.*	(10) *[Name of company] is very safety-oriented, but "Get the job done regardless." Then if you get hurt, the finger gets pointed at you,*	
Own Organisation	(13) *Even though everybody says, "You've got to deliver. You've got to deliver!" It comes back to the person who gets injured. It's normally their fault.*	
(6) *to have interviews with head people.*	(28) *There's just so much paperwork and everything involved just to report something.*	
(9) *because it all had to get done, had to get done.*	(33) *You have a light injury or a cut or a scratch and you're supposed to report it and go through all the hoops and everything. But it's just too hard.*	
(17) *No one told me not to walk on those pallets.*		

Note.   The numbers indicate the original sequence of the discourse. Only sections of the framework are depicted to which the example relates.

priority is typically given to views of the frontline workforce conducing work at the sharp operational end (Hollnagel, 2014). Of course, someone's sharp end is someone else's blunt end, and vice versa. An example is a principle contractor performing work on behalf of a local government. The principal contractor's sharp end involves its site supervisors who supervise the work of a subcontracted delivery partner. From

TABLE 8.3
Summary of Discourse of Example "Experiences When Suffering Injuries"

	Modes	
Described	**Generalised**	**Assumed**
Workers	Workers	Workers
A worker was injured while working on pallets.	Worker feels that reporting processes have to be followed meticulously.	Workers suspect to get into trouble in case they experience injuries in the future.
Another worker had injured their elbow and was put on light duties.	Workers question how they can perform their tasks without stepping onto pallets.	
The worker did not want to be put on workers' compensation.	At the time, it was normal practice to step onto and walk on pallets.	Management
The worker followed the management's instructions to come in for a meeting, yet experienced pain while travelling.	Everyone used to walk on pallets. Workers do not perform exercises and stretches before work.	In case of an injury in the future, workers expect to be told by management not to have done what went wrong.
One of the workers lied to management about their recovery stage to be allowed back at work.	Workers have never heard about the requirement to do exercises and stretches prior to commencing work.	
Workers explained to management that workers are not trained to do warm up exercises or stretches prior to commencing work.	Workers are unwilling to report light injuries to management in the future.	Own Organisation
Management	Management	Workers suspect numbers of injuries to be distorted.
Management put worker on workers' compensation.	Management is expecting workers to be more careful.	
HSE professionals conducted an investigation into the accident.	Injured workers are questioned why they did what they did despite they knew an activity was unsafe.	
HSE professionals enquired what had happened and whether the worker had done all their exercises prior to commencing work.	Own Organisation	
A manager expressed it was the worker's rather than the company's fault that the worker had injured themselves due to walking on pallets.	Light injuries, such as cuts or scratches, have to be reported as well. The company is very safety-oriented, yet expects everyone to complete their jobs, such as to complete deliveries.	
HSE professionals turned the injuries onto the injured worker, due to not having performed the necessary exercises.	Injuries are normally the fault of those who get injured (e.g. when getting bitten by a dog onsite).	
Own Organisation	Reporting involves much paperwork.	
The company requested the worker to come in for a meeting with management while the worker was on workers' compensation.	Workers believe the company is making reporting difficult to prevent workers from reporting.	
Nobody had instructed workers not to step onto pallets.	A lower number of reported injuries makes the company look good.	

Note. Only sections of the framework to which the example relates are depicted.

a subcontractor perspective, the principal contractor's site supervisors represent the blunt end since these are people in supervisory or managerial roles, whereas the sharp end are the workers who execute the physical, safety-critical work on behalf of the subcontractor. If resources are available, the perspective of each individual group can be explored complementarily.

The framework favours subjects over objects. This is because descriptions of work contain a vast variety of objects, many of which are irrelevant to the themes. For instance, in the following statement, workers address 12 specific objects (e.g. trucks, gauges) and two unspecific objects (e.g. something, it; as highlighted by asterisks). None of these are relevant to the overall point being made by the statement, which relates to management fixing issues.

> *"I've been here nearly seven *years. I complained about the *bottles being loaded on the *truck[s] without *gauges, *O-rings missing. So you get to a *delivery, pull the *bottle off [the *truck], and you've got to replace *O-rings, which should be *something that the bloke [does] that's filling the *gas bottle[s]. I probably filled out about three or four *observations. [Yet] it wasn't until [name of manager] started [his job at the *terminal] and came out with me one *day on the *truck, and I showed him on the *back of the *truck as we were unloading. I've noticed that very rarely do I get a *bottle without an *O-ring or a *gauge now. So finally *it got fixed. But that had to go to the point of somebody actually being on the *truck with me, [who] was in a *position [to] do something for *it to get fixed."*

If a particular object appears important, this can be captured in the thematic coding.

Extensions and adaptations to the framework are possible. For instance, the data analysis can distinguish between positive, negative and neutral statements. If confidentiality is not an issue, company names can replace "own organisation" or "other organisations." Additional subjects can be included or sub-groups of subjects defined. An example of the latter is to further distinguish between subjects' roles (e.g. drivers, maintenance, administration) or seniority (e.g. middle, senior, executive management).

The distinction between work-as-planned (or -imagined) and work-as-done (Hollnagel, 2017) is a valuable starting point when exploring everyday work. Yet none of the four modes of the framework directly relates to work-as-imagined or -done. Instead, work-as-imagined is mirrored in the views of "management" and partly "own organisation"; work-as-done is "work described" and "work generalised" by the "workers" and their "colleagues." As such, the framework indicates that tensions at work reach beyond the distinction between work-planned and work-done. Stakeholders, and particularly the workers, are generally pulling towards "work desired." On some occasions, work is executed as planned, yet "work desired" differs, whereas on other occasions work is executed as desired by diverting from work-planned.

SUMMARY AND CONCLUSIONS

Everyday work exploration, as an approach to qualitative fieldwork in safety-critical practice, can make visible opportunities, challenges and idiosyncrasies of work, over and above the distinction of work-planned and work-performed. This chapter has presented a framework to explore everyday work for the benefit of safety and productivity. The framework was developed on the basis of a study with an experienced workforce in the oil and energy industry. The framework consists of four descriptive modes that characterise how work is "described," "generalised," "assumed" and "desired" by different groups of stakeholders. The development and use of the framework has been illustrated with examples of workers' thoughts about work. Reflections on the framework and explorations of everyday work have been discussed.

Explorations of everyday work help identify organisational vulnerabilities and opportunities to learn from what is going well and what could be improved in daily operation. Exploring and understanding everyday work provides a means to gain valuable insight into actual work practices and idiosyncrasies at work.

REFERENCES

Abdelhamid, T. & Everett, J. (2000). Identifying root causes of construction accidents. *Journal of Construction Engineering and Management, 126*(1), 52–60. doi: 10.1061/(ASCE)0733-9364(2000)126:1(52).

Aven, T. (2014). What is safety science? *Safety Science, 67*, 15–20. doi:10.1016/j.ssci.2013.07.026

Boje, D. M. (1995). Stories of the storytelling organization: A postmodern analysis of Disney as "Tamara-Land." *Academy of Management Journal, 38*(4), 997–1035. doi:10.2307/256618

Borys, D. (2009). Exploring risk-awareness as a cultural approach to safety: Exposing the gap between work as imagined and work as actually performed. *Safety Science Monitor, 13*, 1–11.

Cooperrider, D. L. (2018). *The Appreciative Inquiry Handbook* (2nd edition). Berrett-Koehler.

Corbin, J. M., & Strauss, A. (1990). Grounded theory research: Procedures, canons, and evaluative criteria. *Qualitative Sociology, 13*(1), 3–21. https://doi.org/10.1007/BF00988593.

Cunliffe, A. L. (2011). Crafting qualitative research: Morgan and Smircich 30 years on. *Organizational Research Methods, 14*(4), 647–673. doi:10.1177/1094428110373658

Davis, M. S. (1971). That's interesting!: Towards a phenomenology of sociology and a sociology of phenomenology. *Philosophy of the Social Sciences, 1*(2), 309–344. doi:10.1177/004839317100100211

Dekker, S. (2004). *Ten Questions about Human Error: A New View of Human Factors and System Safety* (1st edition). CRC Press.

Dekker, S. (2012). Resilience engineering: Chronicling the emergence of confused consensus. In *Resilience Engineering: Concepts and Precepts* (pp. 77–92). Ashgate. https://research-repository.griffith.edu.au/handle/10072/63581

Dekker, S. (2014). Deferring to expertise versus the prima donna syndrome: A manager's dilemma. *Cognition, Technology & Work, 16*(4), 541–548. doi:10.1007/s10111-014-0284-0

Dekker, Sidney (2011). *Drift into Failure*. Farnham, Ashgate.

Ge, J., Xu, K., Wu, C., Xu, Q., Yao, X., Li, L., Xu, X., Sun, E., Li, J., & Li, X. (2019). What is the object of safety science? *Safety Science, 118*, 907–914. doi:10.1016/j.ssci.2019.06.029

Geertz, C. (2017). *The Interpretation of Cultures* (1st edition). Perseus.

Guba, E. G., & Lincoln, Y. S. (1994). Competing paradigms in qualitative research. In *Handbook of Qualitative Research* (pp. 105–117). Sage Publications, Inc.

Havinga, J., Dekker, S. W. A., & Rae, A. J. (2017). Everyday work investigations for safety. *Theoretical Issues in Ergonomics Science, 19*(2), 1–16. doi:10.1080/1463922X.2017.1356394

Hoffman, R. R., & Militello, L. G. (2008). *Perspectives on Cognitive Task Analysis: Historical Origins and Modern Communities of Practice*. Psychology Press.

Hollnagel, E. (2014). *Safety-I and Safety-II* (New edition). Ashgate.

Hollnagel, E. (2017, March 2). *Why is Work-as-Imagined Different from Work-as-Done?* Resilient Health Care, Volume 2. doi:10.1201/9781315605739-24

Hollnagel, E. (2018). *Safety-II in Practice: Developing the Resilience Potentials*. Taylor & Francis. doi:10.4324/9781315201023

Hopkins, A. (2007). The problem of defining high reliability organisations. National Research Center for Occupational Safety and Health Regulation. January.

Hopkins, A. (2014). Issues in safety science. *Safety Science*, *67*, 6–14. doi:10.1016/j.ssci.2013.01.007

Howard-Payne, L. (2015). Glaser or Strauss? Considerations for selecting a grounded theory study. *South African Journal of Psychology*. doi:10.1177/0081246315593071

La Porte, T. R., & Consolini, P. (2008). Working in practice but not in theory: Theoretical challenges of "high-reliability organizations." In *Crisis Management* (Vol. 2). Sage Publications. http://politicsir.cass.anu.edu.au/staff/hart/pubs/40%20Rosenthal,%20t%20Hart%20&%20Kouzmin.pdf#page=63

Lawrence, T., & Suddaby, R. (2006). Institutions and institutional work. In *Handbook of Organization Studies* (vol. 2, pp. 215–254). doi:10.4135/9781848608030.n7

Leplat, J., & Hoc, J.-M. (1983). Tache et Activite Dans l'Analyse Psyholoique Des Situations. *Cahiers de Psychologie Cognitive*, *3*(1), 49–63.

Leveson, N. (2004). A new accident model for engineering safer systems. *Safety Science*, *42*(4), 237–270. doi:10.1016/S0925-7535(03)00047-X

Leveson, N., Dulac, N., Marais, K., & Carroll, J. (2009). Moving beyond normal accidents and high reliability organizations: A systems approach to safety in complex systems. *Organization Studies*, *30*(2–3), 227–249.

Martin, J. (2002). *Organizational Culture: Mapping the Terrain*. doi:10.4135/9781483328478

Mayo, E. (1933). The Hawthorne experiment: Western Electric Company. In *The Human Problems of an Industrial Civilization* (pp. 55–76). Macmillan Co.

Nadworny, M. J. (1957). Frederick Taylor and Frank Gilbreth: Competition in Scientific Management. *Business History Review*, *31*(1), 23–34. doi:10.2307/3111727

Perrow, C. (1999). *Normal Accidents: Living with High-Risk Technologies*. Princeton University Press.

Rae, A., Provan, D., Aboelssaad, H., & Alexander, R. (2020). A manifesto for reality-based safety science. *Safety Science*, *126*, 104654. doi:10.1016/j.ssci.2020.104654

Saldana, J. (2015). *The Coding Manual for Qualitative Researchers* (3rd edition). SAGE Publications Ltd.

Tharaldsen, J. E., Olsen, E., & Rundmo, T. (2008). A longitudinal study of safety climate on the Norwegian continental shelf. *Safety Science*, *46*(3), 427–439. https://doi.org/10.1016/j.ssci.2007.05.006.

Watson, T. J. (2012). Making organisational ethnography. *Journal of Organizational Ethnography*, *1*(1), 15–22. doi:10.1108/20466741211220615

Weber, D., Macgregor, S., Provan, D. J., & Rae, A. J. (2018). "We can stop work, but then nothing gets done." Factors that support and hinder a workforce to discontinue work for safety. *Safety Science*, *108C*, 149–160.

9 Combining Lenses
Pragmatics and Action Research in Safety Science

Trond Kongsvik and Petter G. Almklov

CONTENTS

INTRODUCTION

The cultural turn in safety science led to increased use of qualitative methods of analysing complex social processes. This turn is often associated with the development of models representing a social phenomenon or process. Turner and Pidgeon's (1997) now classical illustration of disaster as a multi-stage process is one example of a model based on a qualitative approach.

In this chapter we expand beyond a qualitative focus and, by taking a pragmatic starting point, we address what kind of knowledge a *combination* of methods in safety science can produce. Does combining qualitative and quantitative methods have a simple additive effect or can combinations produce deeper understandings of the phenomena we study? Action research will be presented as an example of a pragmatic approach for method combination, strongly involving researchers in safety improvement processes in organisations.

Safety science has a pragmatic foundation – what we strive for in our research activities is better protection of people and assets. We are also pragmatists in another sense and in this chapter we use our multi-method experience from the petroleum industry to discuss how different methods bring different aspects of the phenomenon under study to the fore, and how combining methods is useful for understanding the world. Similarly, we also argue that researchers should hone their skill in interpreting diverse data sets together so as to gain a deeper understanding of the phenomena those data are drawn from. This might be accomplished at personal or a group level and in academic discourse. In this chapter, particular attention will be given to the combination of these two modalities of pragmatism that are found in the action research tradition: implementing organisational changes to improve safety, which may itself generate data that are highly useful to understanding safety.

We limit our discussion to methods relevant to organisational safety. There is, of course, important work being done on technical safety, such as risk analyses and the construction of technical barriers, and important work to be done in the interstices between technical and organisational measures, but they will only be touched on in passing here.

PRAGMATISMS IN SAFETY SCIENCE

If it works, we're right. If he dies, it was something else.
                                   —Dr House, in the "Honeymoon" episode
                                              of the TV medical drama *House*

Pragmatic philosophy is normally traced back to the turn of the 20th-century American philosophers such as Charles S. Pierce, John Dewey and William James. Among our main inspirations in this chapter are Latour (1999) and Batson (1972), whose orientations are clearly inspired by the early pragmatists in the sense that scientific data and models are viewed not primarily as representations of the world but as tools whereby scientists explore the world (see Almklov, 2008).[1] Science does not, then, "uncover," represent and mirror an underlying reality. Rather it produces constructs and concepts that are more or less useful for our interaction with it.

Being pragmatic in the everyday sense of the word is about being realistic and practical, rather than blindly following predefined ideology, ideas or procedures. It means solving problems in a way that takes into account the specific contextual factors that apply in any given situation. In some fields of research, especially in the social sciences, the choice of research method is not always a pragmatic issue but may be a matter of different dogma, leading to suspicion and mutual distrust. Conflicts are often based on divergent perspectives, especially in relation to positivism/constructivism debates. The controversy has a long history that was summed up by the organisational researcher Stablein some years ago (1999, p. 255) in the following way:

> Some (...) run well-controlled experiments to produce data which others claim "have little or nothing to say about the realities of organizational behavior." Some spend months "in the field" reporting their data as ethnographic tales that others dismiss as mere anecdotes. Some ask hundreds and thousands of people to answer carefully chosen questions producing data which others disparage as simplistic, distorted reflections of the respondents' organizational reality, unrelated to their organizational behavior.

In some fields of research the controversy is by no means resolved and seems to gain new impetus when technological advances provide foundations for new methods. New brain-imaging techniques, psychometric tools, new sensors and methods of analysing "big data," etc., are among developments that seem to fuel the view among at least some people that qualitative methods are unnecessary or even misleading.[2] For example, Pål Kraft, a psychologist and manager of a major research programme of the Norwegian Research Council, stated in an interview that qualitative methods have become less relevant and that new technologies are better suited to psychological research:

> By using qualitative methods you could ask people about their thoughts, feelings and opinions. But by using new technology you could measure what people think, feel and mean. With methods like EEG, FMRI and "eye tracking," you can see mental processes. (Meisingset, 2016, our translation)

This interview started a heated debate among researchers, demonstrating that the qualitative/ quantitative schism is still a reality, at least in some research areas.

Such a methodological controversy seems to be less evident in the safety research literature, even as the whole range of methods are used in the field. Why is this? At least one part of the answer is that safety researchers are often pragmatic in their approach, as their main concern is to generate knowledge that can be used directly to improve practice in a specific context (Le Coze, Pettersen, & Reiman, 2014). The pragmatic side of safety science is also evident in its multi-disciplinary approach to problem solving. For example, the huge improvement in safety in commercial aviation since the 1960s can be seen as an achievement made possible by the efforts of engineers, psychologists, sociologists and others, which have resulted in more technologically robust airplanes, more efficient surveillance systems, implementation of crew resource management training, improved regulations, etc. Similar safety improvements can be observed in other industries as well. If we take the onshore industries in Norway as a whole as an example, the fatality rate was 6.81 per 100 000 employees in 1974, declining steadily to 0.95 per 100 000 employees in 2016 (The Norwegian Labour Inspection Authority, 2018).

Such numbers can be seen as the result of efforts by researchers from various disciplines, technology developers, policy makers and practitioners, and they illustrate the effect of working towards the same goal. The pragmatic stance of many safety researchers might stem from the seriousness of the issues at hand, and from being occupied with what works, rather than with debates about methodology or the route to a workable solution to a problem.

Nevertheless, the generally pragmatic approach of safety science says nothing about the extent to which individual researchers or research groups or projects within the organisational safety field combine quantitative and qualitative research methods and so to get an indication of this, we reviewed and categorised articles published in *Safety Science* in 2018 (issues 101–110). Research articles with "organization" or "organisation" in the title, abstract or keywords were selected. Two articles clearly outside the field of organisational safety were omitted, giving a total of 63 articles.

TABLE 9.1
Methods Applied in Research Articles Published in *Safety Science* in 2018

Methods Applied	Number of Articles	Percentage (%)
Quantitative methods only	30	47,6
Qualitative methods only	21	33,3
Combination of qualitative and quantitative methods	3	4,8
Not able to categorise	9	14,3
Total	63	100,0

These were categorised as applying quantitative methods, qualitative methods or a combination of both. *Safety Science* was chosen because of its multi-disciplinary scope and the wide range of industries and sectors covered. Although this is a relatively simple analysis, it gives an indication of the prevalence of method combination (Table 9.1).

We can see that only around 5 per cent of the articles involved a combination of qualitative and quantitative methods, and that quantitative methods were most common (48 per cent).[3]

This illustrates that combining qualitative and quantitative methods is not very common on the individual or research group level in organisational safety research, in spite of the pragmatic foundation. We will elaborate on possible reasons for this later in the chapter.

THE RESTRICTIONS OF USING SINGLE METHODS

Each method has its pros and cons and produces a different kind of information. Arguably, methodology should always be considered alongside ontology and epistemology: the ideas one has about reality and how knowledge about it can be achieved. Each method produces certain forms of knowledge based on such positions.[4] Figure 9.1 is a simplified chart of key methods in organisational studies of safety, with the vertical axis representing proximity to the research object and the horizontal, researcher involvement. There is great variation in research practice within each of the categories, so we urge the reader to not take out their rulers to measure relative positions.

By proximity to the research subject (the vertical axis), we mean where the knowledge object or the categories of interpretation are constructed. Quantitative surveys are labelled as *distant*, for example, because survey categories are constructed outside the field of the concrete research projects or cases and the analysis is also based on de-contextualised data. In contrast we argue that *situated* methods to a larger extent produce interpretative constructs based on, or embedded in, situations that are experienced in the field. Interview studies may vary considerably in the degree of proximity, but they are typically based on preconceived ideas about what is relevant to the investigation.

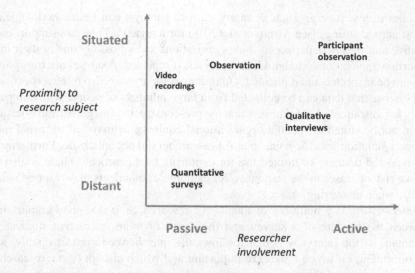

FIGURE 9.1 Schematic depiction of various methods showing the degree of researcher involvement and proximity to the research subject.

We use the term "researcher involvement" (the horizontal axis) to represent how different methods vary with respect to the researcher's involvement in the organisation's social life (passive/active), for example, whilst a researcher conducting observational research will aim to not intervene in social interactions, a participant observer gains information from trying to participate in social interactions and work practices. The active–passive dimension also denotes the degree to which the data are tested and discussed with informants. Surveys and observations, typically, fill data into predetermined categories whereas various ethnographic methods (such as participant observation or grounded theory) allow for an emergent development of analytical categories based on interaction in the field. As such, while some methods seek to remove the researcher from the data to observe the undisturbed work processes, other methods are based on an assumption that the researcher involvement in the field (with people and the material surroundings) is part of the process of producing knowledge (Almklov, 2008). It is worth briefly considering some of the key approaches in turn.

Observational studies can produce highly contextualised, rich data from a specific setting. Domeneghetti et al. (2018) provide an example from the nuclear industry, where decision-making processes were observed during emergency exercises. Observations are sometimes combined with video and audio recordings that are subjected to detailed analyses. Observational studies are resource-intensive and usually highly constrained in scope. It may, therefore, be difficult to apply the insights from one site to another, or to make generalisations that will apply outside the context in which the observations were made. Moreover, compared with participant observation research, such studies typically offer less insight into the perspective of the workers themselves and their explanations of why they do what they do.

Quantitative surveys, such as safety climate surveys, can be methodologically robust across settings (see Alruqi et al. (2018) for a review). They can support comparative analyses of departments and organisations and – possibly one of their most important applications – longitudinal analyses, if repeated. Analyses are transparent and can be inspected and replicated. Quantitative surveys are also relatively efficient in the sense that data can be collected from large numbers of people with comparatively few resources. A weakness is that the pre-constructed categories and questions might not be suited to specific organisational contexts, or may fail to reveal more intricate situation-specific issues that the researcher did not anticipate. Furthermore, self-reported data are of limited use for capturing tacit practices. There is also the known risk of biases in the responses, for example, respondents may respond strategically when answering surveys.

Interviewing is a mainstay of qualitative research, as it is a good compromise between the breadth of a survey and the depth of more interactive studies. An advantage of the interview is that it allows the interviewer to reflect openly, with the informant, on which issues are important and which are not (some researchers make greater use of this option than others). A downside, relative to surveys, is that interviews are more resource-intensive to conduct and harder to compare, as the interview process is more unpredictable, more influenced by the researcher's style and interests and the resulting data is not as structured as survey data. Some of the variations in the knowledge generated by interview studies are due to the analytical approach used: whether the researcher chooses to code and analyse data based on *a priori* hypotheses or use more emergent forms of coding such as grounded theory (Glaser & Strauss, 1999; see also Le Coze, Chapter 4, this volume).[5] An example of the latter is a study by Nixon and Braithwate (2018), who interviewed aircraft accident investigators and explored what kind of competencies made them good at their job. Compared with surveys, the results are more dependent on the analytical approach chosen and the researcher carrying out the analysis, and the analytical process is less transparent. Our own interview-based studies have largely been very interactive and open-ended, because we recognised that our preconceptions of the field needed to be challenged. On the other hand, some interview studies may use highly structured protocols and adopt an approach similar to a spoken survey (see also Hayes and Maslen, Chapter 12, this volume).

Participant observation is a method primarily associated with anthropological and micro-sociological studies. It is a very resource-intensive method as it typically involves prolonged research stays with a small group (see also Kuran, Chapter 7, this volume). An excellent example is a study by Oswald et al. (2018), where a researcher joined a health and safety management team in a construction project to study the reporting system. The results are heavily dependent on the skills and preferences of the researcher and are very hard, if not impossible, to validate. Successful projects may, however, result in important insights into situated work practices and workers' perspectives on their situation that have relevance outside the observation context. Extensive and prolonged interaction with the field may also give deep insights into implicit aspects of an issue, and discussions with the workers can be used to obtain their perspective on observations and to correct and elaborate them. The degree of participation varies considerably between researchers and work contexts, but in some cases (see e.g. Orr, 1996; and our own Almklov, 2008) participation may yield important insights into the workers' perspective, because watching is different from doing.

A general argument has been that using a single method might provide an oversimple and misleading picture of safety in an organisation. For example, Antonsen (2009) analysed data on a gas blow-out on the offshore petroleum facility Snorre Alpha in 2004. Although the direct consequences of the incident were restricted to economic losses and no injuries occurred, the incident had the clear potential to cause many fatalities, the loss of the entire platform and extensive environmental damage. The year before the blow-out employees on Snorre Alpha had completed a 20-item safety culture survey, giving a very positive evaluation of safety management on the facility, compliance, risk evaluation and the safety culture in general. The contrast with the findings from the accident investigation was huge; the latter revealed deficiencies in compliance, risk assessments, managerial involvement and governing documents. The discrepancy was attributed to the difference in methodology, with the argument that survey methods have clear limitations when it comes to providing valid descriptions of complex social processes and unconscious assumptions (see also Antonsen and Haavik, this volume).

COMBINING METHODS – WHAT CAN BE GAINED?

There are two types of researchers, those who are able to extrapolate from incomplete data.

Few would disagree that more data and diverse studies are a good thing. We believe, following Haavik (2014), that relying on a single method reduces the researcher's ability to provide a thorough account of organisations as sociotechnical systems. A key argument here is that the gains from combining methods go beyond simple additive effects. By exploiting the strengths of different methods, it is possible to perform interpretive extrapolation of data that can provide a more nuanced and detailed picture of sociotechnical systems, and also a basis for more meaningful and accurate intervention strategies. More specific advantages include:

- Cross-validations of results and findings. If analyses based on one method are supported by other types of study this is a clear strengthening of the findings.
- Hypothesis construction and testing. Participant observation is an excellent way of developing models and hypotheses that can then be tested by using other methods.
- Different forms of study address and can be better suited for different levels of the organisation. For example, broad quantitative studies of safety management systems may form an important context for detailed qualitative studies of practice at different levels in the organisation. Several important issues in safety science lie in the interstices between representations of work

and situated practice, the latter containing tacit, or less verbalised, elements of practice. In resilience engineering this is often discussed as the Work as imagined/Work as done distinction (see Ray, Weber and Dekker, Chapter 8, this volume). To go beyond viewing this as only a dichotomy, and to fully explore the relationships between the two, requires not only detailed studies of practice, but also higher-level studies on the systems that represent and manage those practices, as well as reflective analysis that combines them.

- Detailed, deeply situated studies can also be used to better understand findings in surveys, fleshing out mechanisms behind phenomena found in a survey. However, such studies are snap shots from one specific context, so such "fleshing out" is a matter of inference. For example, if a survey or interview study indicates poor compliance to safety rules in an organisation and a detailed observation of practice finds some reasons why this is the case, these findings cannot be blindly applied to the entire sample, but they can form the basis for inferences that go beyond what is found in the two data sets separately.

- Safety engagement of individuals and groups in the organisation. Any kind of safety research that involves data gathering from an organisation will create attention to the issue at hand, and combining methods will increase attention further. Although varying in impact, a safety survey or conducting interviews or having a researcher observe work have the potential to spur reflections, discussions, curiosity, resistance, etc. In our view, doing safety research on social systems is in itself a measure that can instigate change, similar to the so-called Hawthorne effect, referring to subjects changing behaviour just from being observed (Leslie et al., 2014).

Interpretation based on mixed forms of data is most often associated with research on social systems. It is however also found in interpretive scientific practices and applications in the natural sciences. In previous studies by the second author (Almklov, 2008; Almklov and Hepsø, 2011), similar pragmatics were studied in petroleum extraction. As geologists and engineers search for oil below the seabed off the coast of Norway, they must understand structures smaller than the relatively low resolution of their remote sensors (seismic data) and beyond the short range of the sensors that are lowered into existing wells. This necessitates a creative extrapolation between different data types. The detailed pinpricks of high-resolution measurements in the wells are stretched outwards inspired by the vague and coarse structural patterns seen in the "seismics." These extrapolations are based on data and imagination. Ideas about the geology in the areas below the seismic resolution emerge from a combination of the sources, aided by experience, theory and analogues. Seismics – which are essentially recorded reflections of sound waves – are, when corroborated by data and measurement taken from within the wells, said to represent geological structures, and the internal characteristics of these structures are filled in by extrapolating from the detailed measurements taken in nearby wells. This is a creative venture, or "educated guesswork" in the words of the experts. The subsurface professionals are able to make data-based inferences about the areas on which they have no direct data.

This example serves well to illustrate the pragmatics we aim for in method combination. First, individual data sources do not represent the reservoir, but they are employed as tools by which they are able to, by extrapolation, infer reference. Second, the geologists are pragmatic in the sense that their purpose is to find oil.

Like the detailed yet myopic well data, observations and interviews give detailed information that needs to be stretched in a qualified manner to be indicative of what goes on in a larger population. Survey data, like seismics, are greatly strengthened if one is able to interpret them in conjunction with the pinpricks provided by qualitative investigations. These combinations require skill, and their results will always be hard to validate scientifically, but skilled pragmatists find oil and we believe also that similar interpretative ventures are useful in safety research.

As our detour through petroleum extraction suggests, holistic analysis of different data sources may aid an interpretative extrapolation of data. For example, responses from a survey can be more meaningful and better understood when combined with knowledge gained from other methods. Like geologists' informed guesses about areas to which they have no direct access, such interpretations always contain an element of guesswork and are hard to validate directly, but they can prove very valuable. Some typical examples include how a detailed observation study can flesh out and give additional meaning to interviews and surveys. The observation study can only, strictly speaking, say something about the specific situation that is observed, but by skilful inference it can also give additional meaning to other data types from the same domain. Such combination of different data types requires that the individual researcher or research team possesses a profound understanding of the limitations and characteristics of all their methods and the data they produce and are able to take these into account in interpreting the data and draw inferences about patterns or processes that could explain the combined data set. Again, validation and corroboration of the data are the first steps. The next step is to delve deeper into underlying, unseen patterns that might exist in the data. Repeated surveys may, for example, reveal changes in workers' attitudes towards safety, but they apply simple metrics to often complicated processes. Interview data may help to put metaphorical flesh on the bones provided by these metrics by providing information about the dynamics of a change in attitude detected via a time series of surveys. Thus, interviews can not only supplement survey data but make them more meaningful. Survey data, which condense thoughts and feelings into numerical responses, are given more meaning when they are contextualised by interviews. Moreover, this combination may also lead to the discovery of completely new phenomena.

Even if there is a lot to gain, the combination of methods is not very common in safety research, as Table 9.1 suggests. There are several barriers. Factors that might hinder method combination include resources and economy (using one method is cheaper than using multiple methods), but more fundamental issues related to disciplinary world view, research traditions and publication outlets and career paths may also be relevant. The journal article format, typically short papers, makes it easier to publish single method studies. Careers and promotions tend to favour specialisation. In addition, combining methods necessarily requires skills in more than one methodological area, either by the individual or by a team that then has to build and hone a fruitful interdisciplinary discourse, sometimes including differences in fundamental epistemological issues and terminology. Further, in our experience one of the most critical challenges for multi-method safety researchers is access, and to convince organisations about the utility of such studies (see also Bourrier, Chapter 3, this volume). The opportunities are rare for holistic and open-ended systematic multi-method studies to be invited into organisations, especially through contracted research. Companies often seek more instrumental studies and answers to problems they themselves have identified. Initiatives such as Action Research, which we will turn to next, requires strong trust relations and commitment from the involved organisations.

ACTION RESEARCH

The proof of the pudding is in the eating.

–Old proverb

Action research (AR) involves a pragmatic combination of methods but also very active researcher involvement with the field of study. An AR project is typically a joint venture between researchers and "problem owners." The "combing of lenses" thus takes place in two dimensions. This will be elaborated by looking into general AR principles and by an example of an AR project.

ACTION RESEARCH PRINCIPLES

Kurt Lewin and his work in the 1940s is acknowledged as the starting point for action research (AR) (Willis & Edwards, 2014). AR has later been applied in social psychology, sociology, philosophy and critical theory (Kemmis, Nixon, & McTaggart, 2014). A unifying idea is that AR can be used to resolve social or organisational challenges (e.g. related to safety) by a joint venture between researchers and those who directly experience the challenges (Coghland & Brannick, 2010). In sociology and related disciplines, AR first emerged as a research tradition for empowering weak groups and communities. In organisational research it has been an important strand of applied research in organisational development.

AR is based on the principle that the external researchers facilitate collective learning processes by means of different research methods – both qualitative and quantitative (Greenwood & Levin, 1998). The main actors in an action research process are a research team and a group of "problem owners," for example, a local community or a company. The goal is to develop unified solutions to specific issues. The crux of the approach is that the researchers perform a mapping of the problem by using different methods and interpret and discuss the results and develop possible measures together with the "problem owners." The research team contributes to and studies the implementation process, and both parties evaluate the impact of the measures, and consider if they are appropriate to the issue at hand. Thus, the research team in an action research process is actively involved in a situated change process, and not merely "objective outsiders" who study a social phenomenon (Solem & Kongsvik, 2013).

AR has a pragmatic and situated starting point, as it is about solving particular problems in particular contexts. AR builds on researchers taking an active, participatory role and puts weight on joint reflection and learning processes. AR also involves a pragmatic approach to research methods, and different methodological approaches may be used in concert to map the nature and possible causes to a problem, and the effect of defined measures (Greenwood & Levin, 1998).

ACTION RESEARCH APPLIED: VALIDATION OF INTERVENTIONS

We have used AR to investigate safety challenges in the petroleum industry (Antonsen, Ramstad & Kongsvik, 2007; Kongsvik, Fenstad, & Wendelborg, 2012; Solem & Kongsvik, 2013). Over a three-year period a Norwegian petroleum company experienced a considerable increase in the number of collisions between offshore service vessels and offshore installations (Figure 9.3). Such collisions are considered a possible trigger for various major accident scenarios. During this three-year period there was also an increase in serious accidents and several fatalities related to vessel operations. The service vessels in the company fleet (approx. 35 in total) were contracted from several ship-owners and performed supply, anchor-handling and emergency preparedness tasks.

Our research group was engaged by the petroleum company to administer a safety improvement programme. Our first step was to distribute a quantitative survey to the vessel crews. This was complemented by observations and interviews carried out on the vessels. We concluded from an analysis of all the results that the deterioration in safety could be attributed to higher demands for efficiency, less rest and the development of symptoms of fatigue, which had decreased the vessel crews' trust in the petroleum company and contributed to work alienation amongst the crews (Kongsvik & Bye, 2004).

Using an AR approach, we then established a "Captains' Forum" involving the captains of all the service vessels and representatives from the relevant parts of the petroleum company (offshore installations, supply bases and traffic control). The Captains' Forum was held annually for 12 years (2001–2012) and was organised as a two-day search conference (Greenwood & Levin, 1998). The first day was dedicated to a broad search for causes for identified safety challenges, whilst the second day involved searching for effective improvement measures. The role of our research group was to present results from different research activities and act as a neutral facilitator of discussions throughout the two days. After the Forum, selected measures would be put into action and then evaluated the following year. Thus a "Plan-Do-Check-Act" quality management approach was inherent in the project (Figure 9.2).

Various research methods were used in the research projects that we performed: qualitative interviews, participant observations on board ships, document studies, quantitative surveys and quantitative analysis of accident statistics. Some examples of the research topics include:

- Safety and working environment surveys (every second year). Topics: Safety management, trust, safety climate, risk perceptions, communication, fatigue, etc.

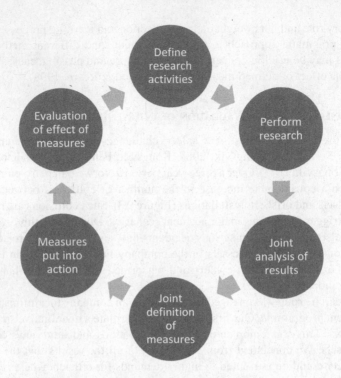

FIGURE 9.2   Research and improvement measures in the project.

- Quality and quantity of procedures and compliance
- Safety training of new sailors and navigators
- Evaluation of shift arrangements
- The captain as safety manager/role conflicts
- Short-term vessel contracts and safety
- Analysis of accident databases
- Communication and information flow in the logistics chain

The research was the starting point for the design and implementation of a wide range of measures, for example, revision of procedures, new training programmes, carrying an extra navigator to provide redundancy, establishment of new communication channels, technological and design changes, etc.

Safety improved considerably during the project period. The number of collisions declined significantly (Figure 9.3), and there were no fatalities after the safety development project started. In addition, safety and working environment surveys indicated that the level of trust between the contracted crews and the petroleum company was improving.

We argue that the combination of methods gave us, as a research group, a "high resolution" picture of the safety challenges and their possible causes as well as a knowledge base on which we could draw, in collaboration with the petroleum company and the captains, to develop measures that could remedy the problems identified.

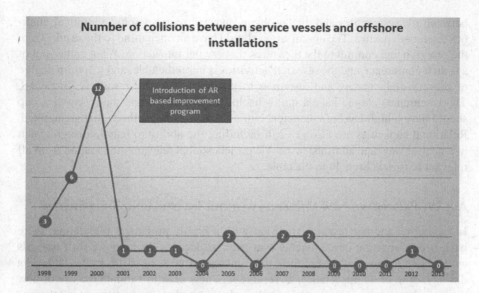

FIGURE 9.3 Development in the number of collisions between service vessels and offshore installations from 1998 to 2013 in a petroleum company.

In AR there is extensive interaction between the research team and actors within the organisation. As mentioned, AR projects employ a broad range of methods as a part of the analytical process. The results are typically presented to actors within the organisation in search conferences, facilitated workshops at which they are invited to contribute to analysis of the results and to propose and discuss measures that could be taken. In addition to being a powerful tool for delivering organisational change, AR may also supply important reflections on the implementation phase, for example, new organisational models, procedures and ways of working are evaluated not only for their academic merit but also for their feasibility and outcomes in practice. AR should not be seen simply as a way of organising reflexive implementation and change, but also as a method for gathering knowledge of what happens when safety science meets the practitioners, to paraphrase Almklov et al. (2014).

Thus, AR may in itself make an important contribution to improving researchers' understanding of organisational safety as a phenomenon, in addition to improving understanding of what can work in a particular context. To some extent the success of the change process validates the research. Also, meeting concrete, situational obstacles to the interventions gives additional understanding of the intervention. Rather than a passive analysis from afar, actually interfering with the organisational system is a learning process for the researchers and this goes beyond gaining knowledge about how to implement measures. For example, proposed interventions may be met by resistance based on tacit conventions that are hard to uncover before they are provoked by change.

There are of course some challenges related to applying AR. As the approach involves method combinations, the same challenges as described above are of relevance (resources, disciplinary world views, career paths, etc.). The close cooperation between the researchers and the "problem owners" requires some additional qualities from both parties. The organisation should have an open mind to the results from the research and commit to the measures that are put into action. What comes out of a search conference and the research activities is unpredictable, and a certain degree of bravery and capacity and resources to handle unexpected results is warranted. The researchers should have a multi-disciplinary mindset, and the research group should have a high level of competence in all the research methods that are applied. Relational capacities are also critical, including the ability to handle disagreement in a neutral way, and administer the AR project so that relevant information from all relevant parties is brought to the table.

ARE WE PRESCRIBING NEW MEDICINES WITHOUT INVESTIGATING THEIR EFFECTS?

In medicine, another research tradition with a highly pragmatic foundation, there have been important developments over the last decades, including a shift towards evidence-based methods. In a sense this represents the "end of theory," or at least a downgrading of its importance, as researchers now seek to validate the use of new drugs and clinical practices primarily based on statistical evaluations of outcomes.[6] It is impossible to emulate medical research's standardisation of cohorts in organisational studies of safety, as all organisations are unique (see also Martin, O'Hara and Waring, Chapter 11, this volume). It would, therefore, be highly unrealistic to expect similar forms of evidence-based science to emerge; however, one might ask if we are, in safety science, too far towards the other end of the spectrum. Are we forever prescribing new "drugs" without properly evaluating their effects? Comparative statistical evidence from multiple organisations is likely to have some value. If one is able to isolate Hawthorne effects then longitudinal survey studies can also provide useful metrics of the effects of interventions and qualitative and mixed-methods studies of change processes could also be useful. One potential problem is that academia incentivises the development of new improvement strategies and tools, rather than the evaluation of existing ones. As in medicine (see, e.g. Sismondo, 2008), we suspect there might be a bias against publication of negative results for safety interventions.

AR projects have shortcomings when it comes to evaluating effects of measures since the research team are heavily invested in the proposed measures, but AR is still one of the few perspectives to take the pragmatic foundations of safety research into account. If a measure works it continues to be used, but if it becomes clear that it is not working, we try something else. Some things work in practice and not in theory and vice versa. AR is one way to investigate different interventions, provided that researchers are able to combine extensive method triangulation with reflection on the implementation and change processes.

CONCLUSION

We have argued that good, descriptive representations of safety-related phenomena or processes are not dependent on use of particular (i.e. qualitative) methods. It may be fruitful to make use of the whole spectrum of methods, and combine them, especially when the aim is to solve safety problems. Qualitative approaches can provide rich information for describing phenomenon and problems, whilst quantitative approaches may be supplementary and provide information that can be used to design further qualitative investigations; supplementing qualitative research with quantitative research might have a more than additive effect on the knowledge that is produced.

We are not arguing that all organisational safety science researchers should become multi-method specialists or action researchers, but we do believe that the community as a whole and research groups and researchers involved in particular projects would benefit from exploring the added value of more extensive method combination. Combining lenses is a skill that needs to be honed by research groups and academic discourse to counter the institutional, economic and academic barriers to such an approach. In our own research, methodological pragmatism has been not only a matter of choosing the most suitable methods in each case but also an interpretative search for knowledge generated by combining diverse data types.

Based on this view of diverse data as a resource, a notion which will, of course, be familiar to those with knowledge of method triangulation, we also argue that AR offers added value, not only as an effective method of achieving effective and well anchored implementation of safety measures. Rather, we have argued, it produces knowledge about one of the most important challenges for organisational safety research: how to make a difference in practice. Consider the petroleum engineers from our example: when their wells are drilled and production has started they do not measure their success solely in terms of the volume of oil produced; rather they use the analyses and results to improve their understanding of the geology they have previously studied. In the same manner, AR produces unique data on the intervention process itself, which can further our understanding of the involved organisations as well as the safety interventions they have implemented.

The substance of this chapter can be seen as an argument for method triangulation in the broadest sense, but it is also a call to use triangulation as an exploratory method, rather than simply a means of validating or corroborating previous findings. Different methods provide different forms of information that can be interpreted together. We have focused on AR, because it provides us with data on one of the most important challenges for safety research, the validation of the knowledge it produces. Hence, our pragmatic perspective not only encompasses epistemic pragmatism, in that we see data as tools for exploration; we also argue for the use of organisational safety research methods that have the potential to improve safety directly (see also Schulman, this volume).

Working life is becoming more complex and intertwined. Through mergers and acquisitions, many companies have become giants and operate on a global scale. Complex cooperation between companies in project organisations, partly enabled by technological advances, has become more common and implies more organisational interfaces. These changes in working life give rise to new organisational safety challenges, and in order to address these as researchers, we need to make use of the whole repertoire of research methods. "Incubation periods" (Turner & Pidgeon, 1997) might become even harder to identify in a complex work environment (see also Pidgeon, Afterword, this volume). Thus, we should apply the principle of requisite variety (Ashby, 1958): We, as safety researchers, should meet increased complexity in working life by methodological variety to identify new safety challenges.

NOTES

1. See also Haavik (2014) for a related discussion of the ontology of safety science.
2. The promises seen in Big Data have even led some to pronounce the "end of theory," as science will be more driven by algorithmic analysis of big data sets. See e.g. Kitchin (2014) for a discussion of this.
3. A relatively large proportion of the papers (14 per cent) were actually discussion papers or proposed methods or conceptual frameworks without an empirical foundation.
4. See Haavik (2014) for a discussion of the ontology of different safety theories and traditions.
5. There are several variants of grounded theory and methods inspired by it. The main idea is to base the analytical categories on an analysis of the interview data that is as unbiased as possible, rather than fitting the data to pre-existing analytical categories.
6. This is the ideal type of EBM. In practice, the evidence base for several medical practices is less structured than these ideal types (see Knaapen, 2013).

REFERENCES

Almklov, P. G. (2008). Deltakelse og refleksjon. Praksis og tekst. [Participation and reflection. Practice and text.] *Norsk antropologisk tidsskrift*, 19(01), 38–52.

Almklov, P. G., & Hepsø, V. (2011). Between and beyond data: How analogue field experience informs the interpretation of remote data sources in petroleum reservoir geology. *Social Studies of Science*, 41(4), 539–561.

Almklov, P. G., Rosness, R., & Størkersen, K. (2014). When safety science meets the practitioners: Does safety science contribute to marginalization of practical knowledge? *Safety Science*, 67, 25–36.

Alruqi, W. M., Hallowell, M. R., & Techera, U. (2018). Safety climate dimensions and their relationship to construction safety performance: A meta-analytic review. *Safety Science*, 109, 165–173.

Antonsen, S. (2009). *Safety culture: Theory, method and improvement*. Farnham: Ashgate.

Antonsen, S., Ramstad, L. S., & Kongsvik, T. (2007). Unlocking the organization: Action research as a means of improving organizational safety. *Safety Science Monitor*, 11(1), Article 4.

Ashby, W. R. (1958). Requisite variety and its implications for the control of complex systems. *Cybernetica*, 1(2), 83–99.

Bateson, G. (1972). *Steps to an ecology of mind*. New York: Ballantine.

Coghland, D., & Brannick, T. (2010). *Doing action research in your own organization* (3rd edition). London: SAGE.

Domeneghetti, B., Benamrane, Y., & Wybo, J.-L. (2018). Analyzing nuclear expertise support to population protection decision making process during nuclear emergencies. *Safety Science*, 101, 155–163.

Glaser, B. G., & Strauss, A. L. (1999). *Discovery of grounded theory. Strategies for qualitative research*. New York: Routledge.

Greenwood, D. J., & Levin, M. (1998). *Introduction to action research: Social research for social change* (2nd edition). Thousand Oaks: SAGE.

Haavik, T. K. (2014). On the ontology of safety. *Safety Science*, 67, 37–43.

Kemmis, S., Nixon, R., & McTaggart, R. (2014). *The action research planner. Doing critical participatory action research*. London: Springer.

Kitchin, R. (2014). Big Data, new epistemologies and paradigm shifts. *Big Data & Society*, 1(1), 2053951714528481.

Knaapen, L. (2013). Being "evidence-based" in the absence of evidence: The management of non-evidence in guideline development. *Social Studies of Science*, 43(5), 681–706.

Kongsvik, T., & Bye, R. (2004). Alienation as an explanatory factor for increased risk on service vessels in the North Sea. In: Spitzer, C., Schmocker, U., & Dang, V. N. (eds) *Probabilistic safety assessment and management*. London: Springer.

Kongsvik, T. Ø., Fenstad, J., & Wendelborg, C. (2012). Between a rock and a hard place: Accident and near miss reporting on offshore vessels. *Safety Science*, 50, 1839–1846.

Latour, B. (1999). *Pandora's hope: Essays on the reality of science studies*. Harvard University Press.

Le Coze, J.-C., Pettersen, K., & Reiman, T. (2014). The foundations of safety science. *Safety Science*, 67, 1–5.

Leslie, M., Paradis, E., Gropper, M. A., Reeves, S., & Kitto, S. (2014). Applying ethnography to the study of context in healthcare quality and safety. *BMJ Quality and Safety*, 23, 99–105.

Meisingset, K. (2016). Rusforskning: Big-data og genetikk. [Intoxination research: Big data and genetics – in Norwegian only]. *Rus & Samfunn*, 9, Article 9.

Nixon, J., & Braithwaite, G. R. (2018). What do aircraft accident investigators do and what makes them good at it? Developing a competency framework for investigators using grounded theory. *Safety Science*, 103, 153–161.

NLIA (2018). Risiko for arbeidsskadedødsfall i det landbaserte arbeidslivet. En sammenligning av norske og utenlandske arbeidstakere. [Risk for work-related fatalities in onshore work. A comparison between Norwegian and foreign workers – in Norwegian only]. *KOMPASS Tema*. Trondheim: The Norwegian Labour Inspection Authority.

Orr, J. E. (1996). *Talking about machines: An ethnography of a modern job*. Ithaca: Cornell University Press.

Oswald, D., Sherratt, F., & Smith, S. (2018). Problems with safety observation reporting: A construction industry case study. *Safety Science*, 107, 35–45.

Sismondo, S. (2008). Pharmaceutical company funding and its consequences: A qualitative systematic review. *Contemporary Clinical Trials*, 29(2), 109–113.

Solem, A., & Kongsvik, T. (2013). Facilitating for cultural change: lessons learned from a 12-year safety improvement programme. *Safety Science Monitor*, 17(1), Article 4.

Stablein, R. (1999). Data in organization studies. In: Clegg, S. R., & Hardy, C. (eds) *Studying organizations. Theory & method*. London: SAGE Publications.

Turner, B. A., & Pidgeon, N. F. (1997). *Man-made disasters*. Oxford: Butterworth-Heinemann.

Willis, J. W., Edwards, C. L. (2014). The twist and turns of action research history. In Willig, J. W., & Edwards. C. L. (eds.): *Action research. Models, methods and examples*. Charlotte: Information age publishing, Inc.

10 Bridging an Unsafe Divide

In Search of Integrated Safety Science Research

Paul R. Schulman

CONTENTS

An Initial Premise. The argument here begins with a premise: the future success and impact of safety research depends on addressing two current failures in our research process itself:

A. *Failures of Integration.* The first is the failure to develop truly integrated and additive safety research methods and the resulting separation of social science research on sociotechnical systems (STS) from engineering research on these systems.[1] Safety science has been called by some "inter-disciplinary research." But while safety research might separately be multi-disciplinary or even cross-disciplinary within the social sciences, right now it isn't inter-disciplinary or even additive in findings between natural science, engineering and social science research on sociotechnical systems. Science and engineering research continues to conceive of "sociotechnical" systems largely as technical rather than human and organisational systems. Technical designs, system operation and even a great deal of risk assessment and regulation remain largely uninfluenced by social science findings (save for some of the physiological findings of human factors research).

B. *Failures of Specification.* A second deficiency is our inability to express key social science concepts, definitions and models pertaining to organisations with more specification than the ambiguous and sometimes contradictory

natural language expressions currently applied to them. Presently, the dominance of natural language expression of major concepts limits their precision and makes it difficult to forge agreement even among social scientists
on their usage. It also limits comparative measurements pertaining to safety
and its management across organisations and technologies.[2]

I believe the second deficiency has much to do with the first. But both are a major
barrier to the development of integrated social science, natural science and engineering safety research and its ultimate translation into safety design, safety regulation
and safety practice.

A Personal Reflection. I certainly do not exempt my own career of descriptive
research on organisations attempting to manage hazardous systems from the above
critique. With a group of colleagues from the University of California at Berkeley
I undertook researching over a period of years a set of organisations known to be
effective in achieving reliability and safety in the management of hazardous technical systems (LaPorte and Consolini, 1991; LaPorte 1996; Schulman, 1993; Roberts,
1993; Roe and Schulman, 2008; 2016). These organisations included a nuclear power
generation plant, an air traffic control centre and its national headquarters, a large and
complex high voltage electrical grid, and later, a number of interconnected infrastructures in electricity, gas, marine transport, telecommunications and water (including
dams, reservoirs and flood control systems). We termed these organisations "high reliability" organisations because they were managed reliably to prevent serious, if not
catastrophic system accidents, and also to provide continuous service delivery.

Our research basically consisted of extensive and extended (over months and in
one case years) on-site observations and interviews. We initially had no theory of
high reliability. We knew these were well-regarded organisations by experts and
the general public, in effect they were "institutions" in sociologist Philip Selznick's
sense of that term (Selznick, 1984). We wanted to see how they were different from
other organisations we had seen and read about in structure, behaviour and culture.
Our research findings consisted in describing what we had seen as these organisations having in common that differed from traditional cases and findings in organisation theory.[3]

We described our findings but did not have a formal theory of how, or if indeed,
the features we saw were the major cause of the reliability demonstrated in these
organisations. We also did not have a model of which organisational features might
be more fundamentally important than others. Further, we did not have an evolution-
ary model of how these organisations came to be high reliability organisations, nor
any projective assessment of whether they would remain high reliability organisations or how they might decay.

Finally, and perhaps most importantly, "high reliability" was for us a nominal
category of organisations based on their reputations and our selection of them as our
research set.[4] We did not have any interval measurement scale that could differentiate
comparatively "high reliability" from "higher" or "lower" reliability organisations.

Yet our research came to be characterised by others as "high reliability organizations (HRO) theory." To our surprise it was a short step to its transformation from

an empirical description of a set of organisations to a generalised prescription for all organisations. We saw many applications of our findings well beyond the context of our original research. We saw consultants offer our findings as a recipe for how to create HROs from organisations with quite different technologies from the mature and well-understood ones we described, as well as organisations in quite different social and political settings from the intensely regulated, public, non-competitive non-market organisations we looked at.

How did this happen? In retrospect, I believe one reason among many is the lack of formal theory specified in our research and the ambiguity of key concepts such as high reliability. Without a formal theory to bound the application of our findings and with no continuous variables by which to *test* applications – in other words to compare organisations and determine with correlation and regression analysis over a large enough sample to have significance, what contribution specific features were having on outputs and performance – most of the HRO prescriptions offered are in effect non-falsifiable hypotheses.

I and my colleagues have learned the hard way the price we paid for our conceptual ambiguity and our lack of clarity and precision in both theoretical causal formulation and organisational descriptions. Ambiguities in our social science concepts and findings were "clarified" not by us as researchers, but instead by consultants, lawyers and regulators.[5] In this process, significant distortions have occurred between our research and its practical application (see also Bourrier, Chapter 3, this volume). I will return to this danger later in this chapter.

THE DIMENSIONS OF CONCEPTUAL AMBIGUITY: THE SPECIFICATION OF "SAFETY"

It is revealing to consider that with all of the public concern, research and regulation focused on it we still do not have a clear or agreed-upon definition of "safety." Is it a continuous or discrete property? Are there only "safe" and "unsafe" conditions or can safety be expressed as a variable with a scale of more or less? If so, can we agree on what constitutes more or less safety? How would we actually measure it? Can safety be prospectively determined or only measured retrospectively by lagging indicators of "unsafety" in accident rates?[6]

One specific example of this confusion and ambiguity is found in the use of the term "near miss." A near miss is generally thought to be a reflection of unsafety. But would a "hit" have been better? A near miss could in fact be a good thing – it could reflect a robustness regarding safety. Or as some argue, if it is meant to refer to some aspect of unsafety the term should really be "near hit."

Also, what is the relationship between safety and risk? Are they opposites? Is more safety achieved proportionately through the mitigation of defined risks? An international study group composed of aviation regulators has denied this, asserting that:

Safety is more than the absence of risk; it requires specific systemic *enablers of safety* (emphasis added) to be maintained at all times to cope with the known risks, [and] to be well prepared to cope with those risks that are not yet known. (Aviation Safety Management International Collaboration Group, 2013; p.2)

Risk is about loss but safety seems to be about assurance. While a number of accidents or failures can occur within a given period without invalidating a probabilistic risk assessment, a single failure can invalidate the assumption of safety. Yet we talk routinely about "safety management systems" and "risk management." Are they the same thing? How exactly are they different?

These ambiguities of concept translate into deficiencies in measurement. And these deficiencies in measurement, I contend, underlie the present separation in analysis and practice between social science and engineering research on the safety of complex sociotechnical systems (see also Kongsvik and Almklov, Chapter 9, this volume).

Some Significant Disciplinary Contrasts. Concepts and definitions of major variables in the natural sciences (such as energy, mass, organic and inorganic molecules) and in engineering (such as stress, hardness, speed and voltage) are largely standardised and agreed upon and formally expressed through stipulated meanings often in artificial language that are physical descriptions or mathematical ones. Most are measured along interval scales.

Social and organisational factors such as leadership, authority, centralisation, decision-making, motivation, mindfulness, stress and resilience are grounded in concepts expressed in natural language with all of its ambiguities and imprecision. These concepts are then difficult to translate into measurable variables. (For example, compare engineering with social science concepts of "stress" in clarity and measurement.)

We frequently describe organisational and managerial "variables" as nominal categories (resilient or "high reliability" organisations) or treat them as opposites in binary pairs (e.g. flexibility/rigidity or centralisation/ decentralisation, or well or ill-structured problems), not as interval or even ordinal scales of measurement. They are treated as "factors" but not really as variables.

Currently, many organisational and managerial "pathologies" have been described in case studies comprising much of accident research (e.g. Medvedev, 1991; Vaughan, 2016; Perrow, 1999; Reason, 2016). But these pathologies – such as the "normalisation of deviance"; rigidification; "brittleness" (as opposed to "resilience"); practical drift; a culture of punishment and blame; ambiguity in authority or accountability; decision bias; "groupthink"; goal-displacement; communication blockages; information silos, etc. – are expressed (often as metaphors) in natural language with all of its ambiguities and imprecision (Petersen and Schulman, 2016). As insightful and useful as the case studies of accidents have been, these concepts have not been standardised in their definitions and descriptions, let alone been translated into measurable variables.

That means that those management or organisational conditions tied to accidents are described in terms that do not allow careful aggregate or comparative measurement across cases. It is then difficult to learn about the impact of organisational and managerial factors, both positive and negative with respect to safety, across organisations without standardised definitions and without some variables that can describe *more or less* of them.

Couldn't more careful specification of the meaning of our basic concepts, as well as their measurement, help us to build up a cumulative knowledge base in organisational and safety research? Couldn't this greater specification also help us as researchers to resist conceptual appropriation by lawyers, consultants, legislators and regulators for normative purposes well beyond their analytic roots?

Why Seek Inter-Disciplinarity in Safety Research? Currently, because of these disciplinary differences in specification, we have two separate and largely unconnected pictures of complex sociotechnical systems. Consider the following argument offered by a human factors engineer:

> If you open the plates of a circuit breaker, you will eventually have an arc. You don't want the electrons to arc, but no engineer would say that the electrons that formed the arc were lazy or complacent: if you don't want the arc, you engineer the system around the constraint. Human factors engineering operates according to the same principle; identify the constraints in the interactions between the employees and the workspaces, tools, and technology, and engineer around it.

This perspective may be fine for human factors and their use in ergonomic designs by engineers. I once worked on a project with an engineer who told me that he tried to design systems that were not only "fool proof" but "damned fool proof" so that even a damned fool couldn't screw them up. But we know that organisational and managerial factors can be a support for and not only a constraint on reliability and safety. For example, engineers can make design errors that humans can identify and organisations can correct (e.g. in training, formal procedures or in "work arounds"). But, given that technologies are sociotechnical systems, shouldn't we expect engineers to incorporate organisational and managerial factors within designs and not simply design "around them"?

At the same time, we know also that organisational and managerial behaviour may have more variance and be less predictable than the physical laws and principles engineers design within. But there is quite a significant predictability in human behaviour over large numbers and cases and if we had measures that could aggregate comparable data to larger scales we might discover more regularities in organisational behaviour relative to specific input variables.

Integrating organisational variables into technical design processes poses many challenges, not the least of which are basic technical system definitions themselves. The latest version of safety management standards applying to pipelines written by the American Petroleum Institute (API) defines a pipeline as:

> that which includes physical facilities through which hazardous liquids or gas moves in pipeline transportation, including pipes, valves, fittings, flanges (including bolting and gaskets), regulators, pressure vessels, pulsation dampeners, relief equipment and other appurtenances attached to pipes, pumps and compressor units, metering stations, regulator stations and fabricated assemblies. (API 1173, 2015)

Nowhere in this definition is there any mention of people or organisations – including installers, maintainers, operators – or requirements to keep the operation of the pipeline within limited temperatures and pressures. How likely is it, with this definition, that the API standards will encourage the inclusion of organisational and managerial factors into technical designs? The API report does define (separately) a "pipeline operator" as follows: "an organization that operates a pipeline." Not an encouraging description given that these two definitions are part of a document claiming to lay out a model for a pipeline safety management system.

Right now there are increasingly important opportunities for closer integration between the two research approaches to safety. The recent stress on safety management systems (SMS's) by industry groups and regulators has created growing demand for careful analysis, including metrics and measurement, of the implementation of these systems and, ultimately, the measurement of their impact on organisational behaviour, not simply on rates of incidents and accidents.

But there are risks. How can we take up these measurement challenges without simply sweeping aside the complexity and ambiguity we know to be part of the organisational phenomena we study? How can we develop measures and metrics without at the same time pursuing more precision at the expense of *accuracy* – two quite different properties.

One climate scientist recently remarked, for example, that climate change models are now more accurate than in earlier times but less precise. They are more accurate because there are more variables included. They are less precise because these new variables, now including green house gas emission predictions and possible policy responses, have wider fluctuation and ranges of impact (Martin, 2015). In other cases, however, accuracy can easily be sacrificed in what amounts to a *false* precision. The rest of this chapter describes a possible process for pursuing an opportunity, with appropriate safeguards, for adding organisational variables to risk assessments, and offers an example of how it might work.

A STRATEGY FOR THE MEASUREMENT OF ORGANISATIONAL AND MANAGERIAL VARIABLES AND THEIR INTEGRATION INTO THE ANALYSIS OF SAFETY AND RISK

Two Perspectives on Safety. Consider the following statements regarding "safety":

(1) Safety is defined and measured more by its absence than its presence (James Reason, 2016).

In this perspective, safety as a positive condition is indeterminate; it's hard to establish definitively that things are "safe." It's much easier to recognise "unsafety" or danger in the face of accidents.

But here is another perspective:

(2) Safety is the continuous production of "dynamic non-events" (Karl Weick, 2011).

Here safety is the consequence of positive actions – identifying potential sources and consequences of accidents, acting to prevent them, constantly monitoring for precursor conditions, training and planning for the containment of consequences of accidents if they do happen – in short *safety* as a managed property.

Under this concept we can offer an *operational definition* of safety as: those non-events which stem from the presence and operation of effective management in relation to the prevention of errors and precursor conditions that can lead to errors and, ultimately, to accidents. But then what constitutes effective safety management?

As part of this operational definition it is important to understand the distinction between safety as "dynamic non-events" and non-events in systems without careful management that have so far simply "failed to fail." Unfortunately, there is at present significant confusion about this conceptual difference.

An example is the following statement. It was part of a pre-trial motion from the Pacific Gas and Electric Company to a federal court in which it was being prosecuted for criminal negligence in the fatal explosion of a gas pipeline:

> The San Bruno pipeline explosion was a terrible accident which devastated many people and harmed an entire community. A pipe with a faulty weld was placed in service in 1956, where it performed safely for 54 years. Suddenly, it failed catastrophically. (Pacific Gas and Electric Company as quoted in *S.F. Chronicle* 9/4/14)

This is not a description of "safety." Instead, this 54-year period of non-events is better defined as *the failure to fail*. Understanding and measuring this distinction is an important analytic and measurement challenge. *How can we distinguish non-events that are simply "failing to fail" from those dynamic non-events that reflect safety, that is, effective safety management, without having to wait for an accident?*

To frame this challenge, consider that there is a growing interest in safety management systems on the part of system operators and their regulatory agencies. Yet surprisingly there is an inconsistent statistical impact in many organisations between their efforts at implementing safety management systems and actual reductions in their incident and accident rates (Kaspers, et al., 2017; Australian Transport Safety Bureau, 2011).

In reality, the absence of a consistent correlation is actually instructive. It highlights what organisational factors are really important in promoting dynamic non-events leading to safety, as opposed to what are all-too-often limited implementations of formalised safety management elements.

Propositions about the Implementation of Safety Management Systems (SMS). There is an important difference between implementing the structural features of an SMS in an organisation – "safety" officers; safety plans; formal meetings; safety budgets; formal accountability and reporting relationships – and

- achieving a widely distributed acceptance of safety management as an integral part of actual jobs in the organisation,
- a collectively shared set of assumptions and values concerning safety (a "safety culture") and
- commitment to safety as part of the individual identity of personnel in an organisation.

Without wide and deep employee engagement, an SMS will simply be an administrative artefact without a strong connection to actual behaviours that connect to safety-promoting performance and safer outcomes. Further, it takes time, persistent effort, adaptive behaviour, *continuous monitoring* (with metrics) and corrections to implement and maintain an effective SMS.

It is worth describing in general what metrics addressing safety management issues could look like. They can vary by the formality and specificity of the measurement scale applied (Stevens, 1946).

Different Types of Metrics. Some metrics may be only *nominal binary elements.* For example, only "Yes or No" answers are given to a question such as: is a safety management element, e.g. regular procedural reviews or a work planning process, present or not? Yes/no elements at least allow the calculation of relative numbers of yes-or-no answers per element.

Some metrics can be *ordinal* across some dimensions running from higher to lower. For example, maturity or developmental models can specify stages in the development of an SMS based on how many nominal elements (e.g. a corrective action programme, emergency response training, or root-cause analysis of all incidents) are present in an organisation (e.g. Fleming, 2000). Of course, the stages in the model are arbitrarily asserted and there might be many alternate paths to reaching "maturity".

Also "Likert scales" in opinion surveys are ordinal measurements with the ordering applied to given statements describing attitudes or elements of safety culture (such as "people here are encouraged to voice concerns about safety risks"). The ordering often runs from 1 ("strongly disagree" with given statements) to 5 ("strongly agree"), with intermediate positions (2–4) between these extremes. However, this ordering (as it is in a maturity model measurement), though it may run from 1 to 5, does not really establish equal interval distances between these numbers. An ordinal Likert scale does not mean that the attitudinal difference between a "strong disagree" and "disagree" answer is the same as that between a "disagree" and "no opinion" answer. Ordinal distributions can allow the calculation of a modal (most frequent) or median (the number that separates the higher from lower half of the answers or elements) measured, but the calculation and use of means must be viewed with scepticism, especially if means are used to summarize and compare attitudes across units or entire organizations.

Some safety metrics, however, can actually be based on *interval* variables – with a scale of standardized values running from higher to lower along equal intervals (e.g. number of incidents or accidents with injuries in a given year or the backlog of uncleared corrective actions reports in a given time period). These metrics can establish "less" or "more" of a property or quality in a way that allows calculation of a mode, median and mean for a distribution of values measured along the scale.

Finally, interval scales with natural zero points, such as injury or accident rates noted above, allow ratio measures to be calculated to compare two measurements (e.g. one organisation's annual accident rate or training hours may be twice or 50 per cent that of another).

There are, of course, different approaches to gathering information for metrics, some can be based on observational instances and/ or interviews by inspectors or other skilled visitors. Some (e.g. "safety culture") can be assessed based on survey results. Some safety culture metrics have even been derived from factor analysis of "narratives," for example, underlying themes e.g. "cost containment always trumps safety," derived from content analysis of employee comments in survey question-naires. In some cases, *index* metrics can be compiled from a combination of different types of metrics with equivalent or normalised scales and possibly weights applied to each.

A STRATEGY FOR SMS METRICS DEVELOPMENT

Safety management system metrics should be focused on *leading* indicators. Measures already exist for incidents and accidents, many required by law and reg-ulation. But these are lagging indicators collected after the fact. The strategy of SMS metrics is to provide *leading* indicators so that the integrity of an SMS can be assessed (and differentiated from an organization that is so far "failing to fail") *before* an accident occurs.

One metrics strategy for leading indicators employed by high reliability organisa-tions (HROs) is to first identify *precursor* conditions or states – those that make an accident or failure more likely. These precursor conditions include not only physical conditions (e.g. excessive temperatures and pressures, loss of back-up equipment) that exceed a bandwidth of acceptable operating conditions, but also organisational conditions (e.g. communication blockages, high cognitive loads or time pressures on operators, excessive noise in control rooms) that can diminish skills and invite error.

Within this strategy high reliability managers strive to keep operations out of precursor zones. In the case of nuclear power plants, control operators may shut down reactors if they believe operations have moved into precursor conditions, com-mercial pilots may refuse to fly if they believe equipment or weather conditions are "hazardous."

Precursor management includes careful error or failure analysis to identify cor-related precursor states, sensitivity to signals that indicate the approach of precursor conditions and a quick response to move away from or out of precursor zones. These activities can be observed, documented and measured through communication links and reaction speeds. Through effective measurement, a process of precursor man-agement can be detected and evaluated in an organisation. It is a process that can impart a special kind of "precursor resilience" to organisations (Roe and Schulman, 2016), allowing them to move back from designated precursor zones quickly and maintain a robustness in safe performance and reliable outputs.

Effective metrics as leading indicators should reflect *models of causation* pertain-ing to safety. It should be agreed-upon why they are important as safety metrics. The identification and analysis of precursors through experience, observation or simu-lated potential connections to accidents provides validity to them as metrics.

Further, *multiple metrics* and large amounts of data if possible should be devel-oped to cover each element of a safety management system to be assessed. More, metrics with more data can improve the overall reliability of the assessment of the management element. Single, high-value metrics offer perverse incentives to "manage to the measure" or to distort the measurement process itself. As one nuclear power plant official once conceded, "organizations will do what you inspect but not necessarily what you expect!"

Finally, safety management metrics should not in themselves define *punishable offences*. They should be widely accepted in an organisation as important tools for learning, not as instruments of control and punishment. Otherwise there will be counter-motives arising to bias their measurement (Moller et al., 2018).

Some Safety Management System Elements and Their Potential Metrics. Many SMS elements and metrics have been suggested in the US by the Institute of Nuclear Power Operators (INPO, 2012), Underwriters Laboratories (UL, 2013) and the Center for Chemical Process Safety (CCPS, 2010). Some examples are:

- the clarity, coverage and consistency of procedures: whether procedures are revised to reflect task learning or improvement: e.g. number or procedural revisions in last five years
- the existence and performance of a corrective action programme: e.g., its backlog and closure rate
- regulatory non-compliance reports: number per year, number outstanding
- incident and accident reporting: what's the median time lag between events and final reports? Are root-cause investigations of accidents undertaken? On what percentage of incidents are they undertaken?
- does the organisation have in place emergency response and crisis management frameworks to handle critical events should they occur: e.g. are there clearly defined roles and responsibilities in place for emergencies? How many employee hours of training or simulations are provided per year?
- is there an encouragement of the reporting of mistakes and error? (Likert-type opinion survey)
- does the organisation create and maintain complete, accurate and up-to-date documentation on operations, maintenance, incidents, etc.? (This can be determined by audits)
- is there participation in meetings concerning safety and risk across levels and units of an organisation: e.g. specific attendance rosters with signatures could be kept officially for all work planning and other safety meetings
- are proposed actions analysed with respect to safety in order to proceed, or do they have to be demonstrated as unsafe in order to be stopped: is there a safety culture that supports this practice? (opinion surveys)

In addition to these there are also "proxy" measures for the presence and effectiveness of an organisation's safety management system. The Institute for Nuclear Power Operators (INPO) does close inspections of work spaces in its visits to nuclear power plants to see if they are clean and if tools are in assigned places. They believe these are indicators of care and systematic management in miniature.

In some HROs managers attempt to maintain a "well-orderedness" in a variety of operations, even those unrelated to safety – in such things as cafeteria operations, in the proper functioning of all technical systems, including, lights, copiers and plumbing.

In short all of these organisational and managerial features can be important in the promotion of safety. Their presence, as indicated by extensive measurements, supported by careful and systematic observations, are what allows us to distinguish "dynamic non-events" from "failing to fail" in relation to safety. Importantly, the presence of a strong safety culture (itself subject to metrics) can reinforce the reliability of SMS measurements.

Adding Organisational and Management Metrics to Risk Assessment. A huge potential arena for social science research integration with technical and engineering perspectives now lies in the analytic processes of risk analysis and risk assessment widely practised in commercial, regulatory and other governmental organisations. Currently, these analyses tend to focus primarily on physical variables. Factors of organisation and management are typically neglected – in fact, often resisted as subjective, ambiguous, arbitrary and subject to unreliable measurement (Danner and Schulman, 2018).

Organisational factors that are included are often lumped into a broad category of "incorrect operations" closely tied to specific physical risks, such as valve settings or switching errors or operational failures in specific tasks. But "incorrect operations" can encompass a wide variety of actions and inactions. Further, many of these incorrect operations, such as not following procedures, or a skill-based operational error, such as turning the wrong valve, are actually *output* effects – lagging rather than leading indicators. What organisational or managerial conditions can lead to procedural lapses or an eventual operational error? Risk assessments typically do not parse out particular elements of organisation and management. But risk assessments could be important *receptors* for social science findings on organisations if we can put those findings in a usable format. Organisational metrics could be factored into the analysis and calculation of both probabilities and the consequences of failures and accidents.

A Strategy to Protect against False Precision in Risk Calculation. Adding organisational and managerial metrics can address some measurement problems that arise in many risk assessments focused only on physical dimensions of failure. But one obstacle is the current effort in many risk assessments to calculate risk and risk mitigation down to degrees of precision that actually undermine accuracy. An example is the effort to calculate "risk/spend optimizations" focused on the comparative yearly risk reduction per dollar spent. This effort at extreme analytic specificity can lead to a false and misleading "precision" that defies empirical meaning in any managerial frame of reference.

Consider the following example:

In one risk analysis (of potential helicopter accidents) a hazard scale of 3 ranges of severe injuries and fatalities (SIFs) was created: Extreme (2.5–12.5 SIFs per event), High (0.5–2.5 SIFs) and Moderate (0.1–0.5 SIFs). Then mid-points of the range for each category were computed and multiplied by an estimated likelihood for that category.

These separate risks were summed to get the total safety risk for the failure event (SIFs) under consideration during a year's worth of operations. In this helicopter risk analysis "on average 0.0424 SIFs are expected to occur per year" and the risk after one proposed mitigation is reduced to .0180 SIFs/yr. Replacing a single engine with a twin engine helicopter at $3M yields a normalised *risk/spend* efficiency of an 0.0081 reduction in SIFs/yr per $1 million spent.

But what is the actual meaning of 0.0424 of an injury or fatality per year? What is the meaning of a risk/spend efficiency of 0.0081 of a SIF/yr per $1 million spent?

This level of granularity has no meaning in any managerial frame of reference. It is one thing to divide probability over a multiple set of years to get a single year's probability. It's another thing to divide a *hazard* in this way. This division loses information about the hazard itself, for example, about how many deaths *could* occur in a single helicopter accident in a given year and the managerial utility of an 0.0244 annual risk reduction per year. This is a granularity and presumed precision out of proportion to both management and the concept of safety.

In short, a "risk/spend" efficiency is likely to be a highly misleading calculation in relation to both risk and safety management. Let's consider the "riskiness" of this calculation from another perspective (admittedly a parody on my part).

The California Public Utilities Commission (CPUC) has recently approved a requirement for utilities to use a new quantitative risk assessment methodology to calculate risk/spend efficiencies in order to compare potential utility risk reduction investment options for funding by means of approved utility rate increases. The method allows the calculation of the amount of risk reduction per year per dollar spent for each option. The utility can then select the most cost efficient option and seek offsets for its costs through Commission rate increases.

This method is offered as a way to answer a question all-too-often asked in American regulatory circles: "How much safety are we willing to pay for?" – sometimes asked with these added words: "given the value of a human life."

I propose a test of the wisdom of this approach through offering metaphorically what may seem like a mundane comparison: namely, the price of tickets to symphony concerts. My wife and I enjoy going to the symphony but symphony tickets have become expensive. This is becoming a problem for us.

Perhaps this problem has its roots in a failure of us and other symphony goers as consumers to ask ourselves the question: "How much music are we willing to pay for given the value to us of the individual notes?"

Currently, full-time orchestra members are typically paid a fixed annual salary. But wouldn't it make more sense to pay them for their actual production? In fact, a more precise valuation of their actual contribution to the orchestra would be the number of actual notes they play per piece per concert and the value attached to these individual notes.

The value per note could be negotiated at the start of the playing season and precisely pro-rated down from full notes to half notes, quarter notes, eighth and sixteenth notes. (Isn't it an additional effort, after all, to hold a full note as opposed to a shorter eighth or sixteenth note?) Of course, rest notes shouldn't have a value and would be free.

This opens the door for many calculations of note/spend efficiency on the part of consumers as well. How much do audiences wish to invest in tickets for a given concert in relation to different pieces by different composers, on a note-by-note cost?

The symphony could with this added precision offer a range of ticket options relative to consumer preferences. Why not try out, for example, a half-price Beethoven night based on playing only half the notes?

Beethoven symphonies, after all, are full of crescendos. Take the 5th symphony. Why does the orchestra have to play all the notes of its many crescendos? I and many other concert goers know this work well. We don't have to hear all the notes in a crescendo – after all we know where those notes are heading. The Orchestra could play only the first and the last crescendo notes, with free rest notes in between.

With this approach the symphony and its concert goers could finally and precisely answer the question: "How much Beethoven are we willing to pay for?" A note/spend efficiency could be achieved and optimal musical investments made by the symphony and its consumers.

This may all sound foolish – and it is. But it's foolish for reasons not all that far removed from the current approach to pursuing safety through risk/spend efficiency calculations approved by the CPUC.

The fact is that safety, like a symphony performance, has properties that cannot be divided and parsed out in individuated units. You cannot buy safety by the pound, nor Beethoven by the note.

Safety is a managed property. It takes an unwavering commitment by high level officials in utilities, regulatory agencies and, ultimately, the public. In infrastructure organisations, it takes a safety culture that permeates through all organisational levels, motivating personnel to include thinking about and acting for safety as part of their job and their personal identity. It takes a continual push towards improvement on many organisational fronts; a set of formal procedures carefully followed but also revised when flaws or better ideas for work performance are uncovered. Safety also takes a constant watchfulness for problems and error at all levels of an infrastructure and a communication system that allows information about error to flow freely across levels and units of the organisation. It also takes a supporting safety culture and effective efforts by regulators.

All of these elements have to reinforce one another. They form an ensemble, like the instruments in a symphony orchestra. A loss or degradation in one will affect the others. A competent conductor doesn't say: "we don't really need the double bases. They don't play that many notes and we can leave those out or distribute them to the oboes and the tuba." Nor would an effective safety manager say to subordinates: "You just do your job as I tell you and I'll worry about safety. We're not paying for you for any safety related conversations beyond 5 minutes, and you're limited to only one corrective action suggestion per year."

Instead of asking "how much safety are we willing to pay for?" a much better question would be "what level of safety do we actually know how to manage and regulate for?"

Unfortunately, this question is rarely asked in safety regulatory processes. Instead we seek to regulate safety in the midst of a sea of competing adversarial interests and pretend that we can "monetise" safety through the calculation of "risk/spend" efficiencies with an illusory precision. The risk/spent calculation methodology approved by the California Public Utility Commission, by the way, doesn't even include any safety management factors. Will new helicopters be inspected, maintained and operated properly? There are currently no metrics collected to shed light on that question.

Many may see a foray by social science into more efforts at variable specification and measurement as an invitation for social science researchers to themselves participate in exercises in false precision. But will social scientists really descend to this level? Aren't we, in fact, the best analysts to detect, highlight, question and ultimately help to correct errors of false precision?

At the same time, our participation in metrics development can bring organisational and management factors directly into risk analyses and assessments. We can increase the resolution and granularity of precursor management and risk assessments down to but not beyond their meaning at the level of actual task performance.

Safety management metrics can also reveal the potential uncertainty associated with physical risk mitigation investments relative to the probabilities and consequences that will surround these mitigations. How much risk mitigation will a new gas pipeline segment actually add if it isn't clear whether it will be installed or maintained properly or operated according to recommended pressures? With social science participation SMS metrics can offer a range of estimates that reflect key uncertainties and thus provide a more *accurate* picture of risk, even if it may be less precise. Social science research can in this way replace the hubris of single value risk calculations, including risk/spend efficiency calculations, with a more realistic *range* of likelihood and consequence estimates.

Further, targeted organisational and managerial metrics can help us define more reliably the margins of best and worst-case estimates for risks of failure of physical systems by giving us information about whether these systems are carefully and attentively operated and maintained by well trained and motivated workers, whether possible failures are anticipated and planned for, and whether resources, roles and skills are available for rapid and effective responses to these failures.

With careful safety management measurement it should ultimately be possible to combine key both physical and organisational variables into more comprehensive risk analyses and assessments. One way might be through the use of safety management *index* scores (SMI's) for given organisations. These index scores could be computed for individual organisations as the measured maturity of many of its overall safety management system metrics including its safety culture. The score would run as a percentage from 0 to 100 per cent, 1 for an immature, unsafe organisation through 100 per cent for an organisation with a mature and effective SMS. It could be employed by regulatory analysts as a discount factor, a multiplier percentage applied to any risk-reduction calculations offered by utilities in support of large-scale physical investments for which rate or price increases are requested. If an organisation's calculated risk reductions rested in a strong safety management foundation reflected in a high index score (say 90 per cent) it would receive regulatory credit for 90 per cent of the specific reductions calculated. If it has a weak safety management system index (say 10 per cent) if would have its estimated risk reductions discounted down to 10 per cent by its safety management index score.

Finally, it should be recognised that measured improvements in safety management can in themselves be important risk mitigations, often far cheaper than some new physical investments sought by organisations.

CONCLUSION

It will take a large and persistent R&D effort to achieve the integration of organisational and managerial variables as safety management metrics into the engineering analysis of technical systems. Metrics development specifically should also include a variety of participants beyond social scientists and engineers – including regulators as well as personnel familiar with operations at many levels within regulated organisations. Only with this wider participation can validity and reliability in metrics be developed to support a more integrated and holistic safety management perspective. This integrated research base for the analysis of sociotechnical systems, could significantly add to our understanding of how to manage and ultimately design them for increased safety.

Improved conceptual specification and modes of measurement may well be a means for social scientists to have their work penetrate more deeply into policy, regulation, management practice and even technical designs. Then we might truly come to understand, design, manage and regulate our critical technologies and infrastructures for safety as sociotechnical systems.

NOTES

1. For a classic statement of the sociotechnical systems perspective see Emery (1959).
2. See Gerring (1999) and Gerring and Baressi (2003) for insights on the pros and cons of natural language expression of concepts.
3. See, for a description of our findings and the differences, see T. LaPorte and P. Consolini (1991).
4. Todd LaPorte did try to estimate the number of operations that occurred over a fixed time period in air traffic organizations where theoretically a collision could occur and didn't. But it wasn't clear whether this alone was a variable of reliability and what constituted "high" in this measure. It also wasn't clear how this could be a comparative measure across different organizations and what would constitute a given countable "operation."
5. A very insightful discussion of how this process can occur can be found in Johan Bergstrom (2020).
6. An insightful exploration of these ambiguities in the understanding of safety can be found in Carl Macrae, *Close Calls* (2014).

REFERENCES

American Petroleum Institute (2015). *Pipeline Safety Management Systems: Recommended Practice 1173*. (https://www.api.org/~/media/files/publications/whats%20new/1173_e1 %20pa.pdf)

Australian Transport Safety Bureau (2011). "A Systematic Review of the Effectiveness of Safety Management Systems." (https://www.atsb.gov.au/media/4053559/xr2011002_final.pdf)

Bergstrom, J. (2020). "The Discursive Effects of Safety Science." In J.-C. Le Coze (ed.), *Safety Science Research: Evolution, Challenges and New Directions*. Boca Raton, FL: CRC Press, 173–186.

Danner, C. and Schulman, P. (2018). "Rethinking Risk Assessment for Public Utility Safety Regulation." *Risk Analysis*, 39(5) (November), 1044–1059.

Emery, F.E. (1959). "Characteristics of Socio-Technical Systems." London: Tavistock Documents, no. 527.

Fleming, M. (2000). *Safety Culture Maturity Model*. HSE Offshore Technology Report. Sudbury, UK, 2000.

Gerring, J. (1999). "What Makes a Concept Good. A Criterial Framework for Understanding Concept Development in the Social Sciences." *Polity*, 31(3) (Spring), 357–393.

Gerring, J. and Baressi, P. (2003). "Putting Ordinary Language to Work: A Mini-Max Strategy for Concept Formation in the Social Sciences." *Journal of Theoretical Politics*, 15(2), 201–232.

INPO (2012). "Traits of a Healthy Nuclear Safety Culture." (https://www.nrc.gov/docs/ML1303/ML13031A707.pdf)

Kaspers, S., et al. (2017). "Measuring Safety in Aviation: Empirical Results about the Relation between Safety Outcomes and Safety Management System Processes, Operational Activities and Demographic Data." *Presario Conference.* (https://pdfs.semantic-scholar.org/a672/aed3cee7a253cd5b1fe3e266de9b384f5ded.pdf?_ga=2.176329765.1417911500.1561933416-1398805658.1561933416)

LaPorte, T. (1996). "High Reliability Organizations: Unlikely, Demanding and at Risk." *Journal of Contingencies and Crisis Management*, 4(2), 60–71.

LaPorte, T. and Consolini, P. (1991). "Working in Practice But Not in Theory: Theoretical Challenges of High Reliability Organizations." *Public Administration Research and Theory*, 1(1), 19–47.

Macrae, C. (2014). *Close Calls*. London: Palgrave Macmillan.

Martin, R. (2015). "Why Climate Models Aren't Better." *Technology Review* (November). (https://www.technologyreview.com/s/543546/why-climate-models-arent-better/)

Medvedev, Z. (1991). *The Truth about Chernobyl*. New York: Basic Books.

Moller, N., et al. (eds.) (2018). *Handbook of Safety Principles*. Hoboken, NJ: John Wiley & Sons.

Perrow, C. (1999). *Normal Accidents*. Princeton: Princeton University Press.

Petersen, K. and Schulman, P. (2016). "Drift, Adaptation, Resilience and Reliability." *Safety Science*, 19(August), 460–468. (http://dx.doi.org/10.1016/j.ssci.2016.03.004)

Reason, J. (2016). *Organizational Accidents Revisited*. Boca Raton, FL: CRC Press.

Roberts, K. (ed.) (1993). *New Challenges to Understanding Organizations*. New York: Macmillan.

Roe, E. and Schulman, P. (2008). *High Reliability Management*. Stanford, CA: Stanford University Press.

Roe, E. and Schulman, P. (2016). *Reliability and Risk: The Challenge of Managing Interconnected Critical Infrastructures*. Stanford, CA: Stanford University Press.

Safety Management International Collaboration Group (2013). *Safety Management System Evaluation Tool*. (https://www.skybrary.aero/bookshelf/books/1774.pdf)

Schulman, P. (1993). "The Negotiated Order of Organizational Reliability." *Administration and Society*, 25(3), 353–372.

Selznick, P. (1984). *Leadership in Administration*. Oakland, CA: University of California Press.

Stevens, S. (1946). "On the Theory of Scales of Measurement." *Science*, 103, 677–680.

U.L. (2013). "Using Leading and Lagging Safety Indicators to Manage Workplace Health and Safety Risk." *Underwriters Laboratories*. (https://legacy-uploads.ul.com/wp-content/uploads/sites/40/2015/02/UL_WP_Final_Using-Leading-and-Lagging-Safety-Indicators-to-Manage-Workplace-Health-and-Safety-Risk_V7-LR1.pdf)

Vaughan, D. (2016). *The Challenger Launch Decision*. Chicago: University of Chicago Press.

Weick, K. (2011). "Organizing for Transient Reliability: The Production of Dynamic Non-Events." *Journal of Contingencies and Crisis Management*, 19, 21–27.

11 Large-Scale Mixed-Methods Evaluation of Safety Programmes and Interventions

Graham Martin, Jane O'Hara and Justin Waring

CONTENTS

INTRODUCTION

In this chapter we explore the terrain of research and evaluation within safety in healthcare. In contrast to many of the fields explored in this volume, healthcare is distinguished by a long history of methodologically robust research. Indeed research in health and healthcare has often been at the vanguard of methodological development to increase the validity and reliability of results. Both pharmaceutical and non-pharmaceutical interventions to relieve disease or improve health are typically (though not always) subject to an extensive array of evaluation processes before being incorporated into routine healthcare delivery, including for their safety, and they are often relatively well monitored for their continuing safety and effectiveness after adoption. However, this tradition of evaluation, we argue, brings challenges as well as advantages to the field of research and evaluation around quality and safety in healthcare. In particular, the dominance of methods associated with epidemiology and pharmaceutical development has meant that it is only relatively recently that key institutions in the field of health services research have come to recognise the contribution that qualitative inquiry, including ethnography, can make. More lately, however, the field has begun to enthusiastically adopt mixed-methods research, acknowledging the complexity both of safety interventions and the contexts in which they are expected to work, and of the role of qualitative insights in understanding how interventions lead to better safety – and the extent to which a seemingly effective intervention can be expected to work consistently through time and space.

Understanding of the prevalence and nature of issues of patient safety and quality in healthcare is a relatively recent phenomenon. It is only since the 1980s and 1990s that quality and safety in healthcare has become a field of study and intervention in its own right. Attention to the issue was galvanised by reports on medical error, patient harm and associated poor outcomes (Department of Health 2000; Institute of Medicine 1999), and widely publicised (though disputed) figures that suggested that medical error was among the leading causes of death (see Shojania 2012). Similarly startling and contentious claims have been made regularly ever since. For example, one recent study ranks medical error as the third-highest cause of death in the US, after cancer and heart disease (Makary and Daniel 2016; see Shojania and Dixon-Woods 2017 for critique of the validity of this claim). The World Health Organization (WHO) (2018) suggests that in high-income countries, one patient in ten is harmed as a result of adverse events while receiving hospital care. Over the last two decades, research in several countries, as well as analysis of the wealth of data routinely collected in various national audits, registries and billing systems, had identified areas of high risk, poor reliability and inconsistent outcomes, notably in surgery, anaesthesia and post-operative care. This in turn has given rise to interventions that seek to address these issues, often based on work in other hazardous and safety-critical industries such as civil aviation, and accompanied by often robust, multifaceted efforts at evaluation. While this work has found few if any "magic bullets," it has contributed to an increasing understanding of the complexities of problems of quality and safety in healthcare – and the need for sophisticated interventions to address them (see also Wiig et al., Chapter 13, this volume). We discuss some key mixed-methods studies in the field of patient safety, highlight the contributions to understanding they have offered, and examine some of the challenges that continue to face mixed-methods research and evaluation. In particular, we consider some of the issues that have made realising the promise of mixed-methods research difficult, and the prospects for overcoming these issues.

Our chapter is presented in three main sections. First, we briefly recount the history of research in healthcare, tracing its consequences for dominant assumptions about appropriate methodology and epistemology in the study of interventions to improve safety – and the more recent acknowledgement of the role of mixed methods, particularly in evaluating complex interventions. Next, we trace the rise of concerns about safety, quality and risk in healthcare from the 1980s onward, and discuss examples of increasingly sophisticated mixed-methods evaluation of safety interventions in healthcare that emerged in response. We compare a number of programmes and their evaluations, and note both the insights that have been brought by the incorporation of qualitative methods, and some of the limitations that have emerged. We pick up this theme in the final section, where we argue that although qualitative research has undoubtedly brought analytical advantage, some of its potential may have been overstated – due to the challenges of evaluation in a complex field, epistemological differences between researchers, and conflicting expectations of the endeavour of evaluation itself. We conclude by discussing prospects for evaluation in this field (and in other areas of safety and risk), highlighting the need for more modest ambitions for evaluation of safety programmes and greater attention to the relationship between researchers and practitioners.

RESEARCH AND EVALUATION IN HEALTHCARE

To arguably a greater extent than many other safety-critical industries, the field of healthcare has been dominated for some time by the ideals of evidence-based practice – that is, the notion that routine activities should be guided by strong evidence that interventions are likely to achieve what they intend to (Sackett et al. 1996). Examples of the use of experimental methods in relation to medical treatments from as early as the 16th century have been noted (Oakley 2000), and occasional appeals to the need for evidence can be found from the early 20th century onward, but the origins of healthcare's contemporary evidence-based practice "movement" were in the 1970s and 1980s (e.g. Cochrane 1972). Stemming both from scandals in the 1960s that arose from insufficient trialling of pharmaceutical interventions and from concerns about ensuring the most effective use of scarce healthcare resources (Howick 2011), this movement called for healthcare practice to be informed by high-quality evidence of the effectiveness of interventions deployed. In particular, advocates of evidence-based practice argued that healthcare interventions should be evaluated through fair and objective processes, free of both the prior preferences of researchers and clinicians and the biases of vested interests, such as the pharmaceutical industry, and effectively curated by the clinical–academic community to ensure that the best and most up-to-date evidence was readily available to practitioners. These premises can be seen reflected in the so-called "hierarchy of evidence" (see Figure 11.1),

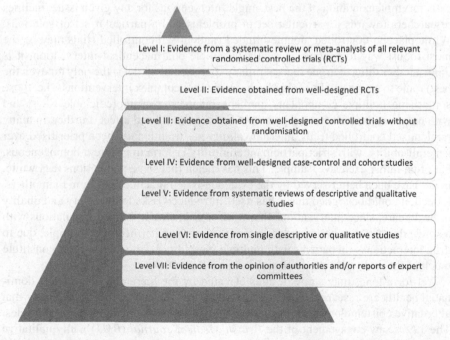

FIGURE 11.1 The hierarchy of evidence. *Source*: Text from Melnyk and Fineout-Overholt (2011: 12).

a model for assessing the validity and reliability of research findings that remains influential in healthcare. This typically places systematic reviews – i.e. integrations of high-quality research evidence, filtered according to explicit inclusion criteria and often incorporating meta-analyses to estimate effect size – at the apex, followed closely by randomised controlled trials – i.e. tests of one intervention against another (or against no intervention), with steps (e.g. randomisation, blinding) taken to eliminate biases arising from differences in sample characteristics, researcher or clinician preferences, and so on.

As a set of aspirations, there is little to object to in the evidence-based practice vision, and as a set of tools for producing a high-quality evidence base to inform practice, it is easy to see its appeal. As an alternative to the kind of judgement-based practice that had predominated in medicine for centuries – and which resulted sometimes in cure, sometimes in harm, often in inefficiency and always in inconsistency – it presents an attractive model of medicine as a science, rather than an art reliant on the skill and attention of the individual practitioner. But the limitations of this vision are also apparent. Originators and critical friends of the evidence-based practice movement alike have highlighted how, for example, an unreflexive application of universal rules indicated by an epidemiologically derived evidence base to individual patients who are all unique can itself produce harm, and fail to incorporate wider considerations – not least patients' own preferences (Greenhalgh et al. 2014; Sackett et al. 1996). But perhaps particularly pertinent to the focus of this volume is the argument that the notion of a universal evidence base, which identifies and calls for implementation of the best single interventions for any given issue, inclines researchers towards a particular set of problems and a particular set of solutions. While placebo-controlled, double-blinded randomised controlled trials may be the most robust way to evaluate a relatively simple pharmaceutical intervention, it is quite another thing to argue that a similar experimental model is the only (or even the best) route to a high-quality evidence base on more complex interventions – i.e. those comprising multiple components, perhaps interacting in unpredictable ways, and with unknown or multifaceted pathways between cause and effect. Further, in many randomised controlled trials, internal validity has traditionally been prioritised over generalisability, with strict participant eligibility criteria to ensure a homogeneous, but often rather exclusive, sample. This has meant that some populations (e.g. white, male) have been better served by the evidence-based practice paradigm than others. Indeed, it could be argued that it has itself introduced risks to the safety and quality of healthcare, since interventions that work well in clearly defined populations with a single disease may be much less effective – or even harmful, for example, due to drug interactions – in patients with multiple morbidities, who increasingly constitute much of the real-world population (Boyd and Kent 2014).

Yet for a long time, study designs elevated by the hierarchy of evidence dominated healthcare research, and it was only towards the end of the 20th century that alternative epistemological frameworks began to get a hearing in biomedical circles. The 1990s saw engagement of the *British Medical Journal* (*BMJ*) with qualitative methods (e.g. Pope and Mays 1993). The reorganisation of healthcare research in the UK in the 2000s, including the inauguration of the National Institute for Health

Research (Department of Health 2006), provided new opportunities for mixed-methods research and evaluation, by providing a platform for large-scale applied health research. At the same time, reflecting realisation of the limitations of the randomised trial model for non-pharmaceutical interventions, the UK Medical Research Council (MRC) introduced a framework for complex interventions which explicitly acknowledged the need for qualitative as well as quantitative methods in their development and evaluation (Campbell et al. 2000). Subsequently, this framework was revised to better acknowledge the role of complexity and the need to tailor evaluation to specific interventions (Craig et al. 2008). Agencies elsewhere, for example in the US (PCORI 2018), have adopted similar sets of standards, and mixed methods are now *de rigueur* in evaluation in healthcare, particularly in relation to multi-component, complex interventions where causal mechanisms, impacts and unintended consequences may vary with context. Alongside this, theory-based evaluation in the mould of Carol Weiss (1995), most prominently realist evaluation (Marchal et al. 2012; Pawson and Tilley 1997), has also become common, and seeks to give explicit attention to understanding causal mechanisms as well as outcomes, with a view to understanding how both might vary through time and space.

Mixed-methods evaluation is thus now widely accepted in health services research in the UK and elsewhere, and even expected by its funders. But as we later explore in more detail, realising the promise of mixed methods – in relation to patient safety and other fields of healthcare research – has not always proven easy, in part because of the legacy of the history outlined above and the persistence of epistemological hierarchies, and in part because of the challenges of integrating rather different modes of understanding in a truly synergistic way. First, however, we turn our attention to the rise of patient safety as a concern in healthcare, and the response to this concern in the form of research, intervention and evaluation.

PATIENT SAFETY, IMPROVEMENT INTERVENTIONS AND MIXED-METHOD EVALUATION

The emergence of patient safety as a concern for clinicians and health services researchers is in some ways bound up with the rise of evidence-based practice précised above (Waring et al. 2016). The development of an increasingly robust (if not always entirely generalisable) evidence base for healthcare interventions was not immediately matched by changes in practice among "rank-and-file" clinicians. The 1990s and 2000s saw the publication of a number of studies that highlighted the extent to which practice lagged behind current evidence (e.g. McGlynn et al. 2003; Schuster et al. 2005) and the frequency of medical errors (e.g. Brennan et al. 1991), and the consequences of these issues in terms of suboptimal outcomes, wasted resources and patient harm. Policy reports from a similar period highlighted the consequences of variation in healthcare practice and deviation from evidence-based standards, and particularly the issue of patient safety problems arising from increasingly complex healthcare systems, suboptimal design and human error, in the UK and the US (Department of Health 2000; Institute of Medicine 1999). These reports contained notable parallels. Both posited that learning from past

problems could have prevented later issues; both drew on insights from human
factors and related fields in acknowledging the limitations of human cognition and
the need for systems-based approaches to improving safety, and both highlighted
the success of other industries in reducing, managing or eliminating safety risks.
The combination of concerns around inconsistent quality of care and failures of
patient safety gave impetus to a range of activities in healthcare policy, practice and
research (see Shojania 2012 for a brief overview). In the UK, this included regulatory interventions such as clinical governance and inspection and audit regimes,
the growth of quality-improvement work driven by local audits of clinical practice,
and the development of large-scale, cross-cutting programmes to address quality
and safety issues, accompanied by similarly ambitious programmes of research
and evaluation.

The rapid rise of patient safety concerns from relative neglect to policy priority
has, however, meant that improvement practice has often exceeded the research base
– in conflict, of course, with the tenets of evidence-based practice discussed above.
Various authors have noted that interventions to reduce risk and improve safety have
emerged in rather a patchwork fashion, with the field characterised in its early stages
by multiple projects springing up, uninformed by existing evidence, with weak bases
in theory, and without high-quality evaluation. "Safety initiatives have been promoted (…) before robust research had demonstrated the effectiveness of these practices" (Wachter 2010: 169; cf. Dixon-Woods and Martin 2016). More sophisticated,
co-ordinated and ambitious attempts to improve safety, recognising the multifaceted
nature of safety problems and the need for well-theorised interventions in response,
have begun to follow. Thus efforts to improve safety based on a single intervention
with a simple theory of change – for example, auditing current practice and feeding back findings to clinicians, premised on the implicit or explicit assumption that
poor practice is due to deficits of knowledge or awareness – have been supplemented
by much more nuanced and programmatic interventions. These interventions have
been more theory-based (drawing, for example, on psychological insights into how
to prompt sustained behaviour change), and have sought to account for influences on
safety at multiple levels, for example, professional norms, patient preferences and
organisational incentives (Shekelle et al. 2011).

In other words, safety interventions in healthcare have quickly moved from
simple to complex. In this way, they have followed the calls of organisations like
the MRC discussed in the previous section (e.g. Craig et al. 2008), and started to
draw on a more sophisticated understanding of the principles of safety science more
broadly, with its emphasis on systems issues, which present complex challenges and
demand complex solutions (Braithwaite 2018). Research and evaluation, however,
have arguably lagged behind. Whether because of funders' or researchers' attach-
ment to the biomedically informed model of evaluation discussed above, or due to an
(understandable) preoccupation with developing an evidence base about *what* works
over understanding *how* it works, much evaluation of complex safety interventions
in healthcare has tended to prioritise (quantitative) demonstration of impact over
(qualitative) understanding of process. Yet understanding how interventions work is
crucial, for the reasons discussed in the previous section. Patient safety interventions

are usually as complex as any, and their reliance on co-ordinated efforts across a variety of actors working at different levels in the system means that consistency of impact can be infuriatingly elusive. Attention to the variety of contextual influences that might mean that a seemingly robust, "proven" intervention that works in one context might not work in another is a fundamental part of developing an actionable, reliable, transferable evidence base.

One example in particular will suffice to demonstrate this point. The 2000s saw a number of studies that showed seemingly impressive impacts from checklist-based safety interventions in healthcare, including in bloodstream infections in intensive care (Pronovost et al. 2006), peri-operative care (De Vries et al. 2010) and surgery (Haynes et al. 2009). These evaluations were published in high-impact journals (the field-leading *New England Journal of Medicine* in these three cases), deploying relatively robust study designs located at respectable tiers of the hierarchy of evidence (controlled or uncontrolled before-and-after studies – Levels III or IV in Figure 11.1). Their approach to change had its roots in an intervention (the checklist) that is widely used in other safety-critical fields such as civil aviation, and to which much success has been attributed (Clay-Williams and Colligan 2015). They thus had some promise in theory – and in practice, results seemed to indicate significant impact and potential to improve patient safety worldwide. In the case of the surgical safety checklist (Haynes et al. 2009), much was made of the universality of the intervention: it was evaluated in a selection of hospitals with divergent safety records, across high-, low- and middle-income countries. Across this diverse and globally representative sample, the evaluation showed a reduction in rates of complication within 30 days of 36 per cent, and a reduction in 30-day mortality rates of 47 per cent (Haynes et al. 2009), both comfortably statistically significant.

Here, then, was the Holy Grail: an intervention that was cheap, relatively easily implemented and proven across a range of contexts rather than being specific to a particular organisation or a narrowly defined patient group. Correspondingly, the WHO, which had produced the guidance on which it was based, endorsed the checklist and encouraged its adoption worldwide, a plea that was enforced by patient safety bodies in many countries, such as the National Patient Safety Agency in the UK.

Yet subsequent experiences, and accompanying research and evaluation, complicated this picture. In particular, extensive qualitative examination of the implementation of the surgical safety checklist and similar tools in multiple contexts has highlighted the challenges of incorporating it into routine work, ensuring consistent use, and – perhaps most importantly of all – securing a culture in which the importance of such activity is embraced. This research reinforces the point that, as the authors of a systematic review put it (Bergs et al. 2015: 781), the checklist and its implementation are "a complex social intervention with an expectation of interaction and cooperation between surgeons, anaesthetists and nurses," not a simple fix with a single, linear, predictable, reproducible causal pathway. Research on other checklist-based interventions has similarly exposed the rich, social mechanisms and contextual conditions that underpin success (Dixon-Woods et al. 2011), and in whose absence such interventions can fall far short of their initial promise (Dixon-Woods et al. 2013).

Findings of this kind will come as no surprise to social scientists. Indeed, they were to some extent anticipated by the investigators of the surgical safety checklist study itself, who found variation in the scale and nature of change, and noted that "the exact mechanism of improvement is less clear and most likely multifactorial" (Haynes et al. 2009: 496–7), including variation in the culture and readiness for change of hospitals and teams. But the desirability of prospective mixed-method evaluation of safety interventions, rather than retrospective analyses that seek to explain what went right in apparently successful trials (or what went wrong when the intervention was incorporated into routine practice), has taken longer to establish, notwithstanding the recommendations of bodies such as the MRC. Where integrated qualitative evaluations have been used prospectively, however, they have offered important insights into both the successes and the failures of efforts to improve healthcare safety.

One notable example, declared "a model for the field" (Pronovost et al. 2011: 341), is the evaluation of the first phase of the Health Foundation–funded Safer Patients Initiative. This large-scale programme involved intervention at multiple levels within acute hospitals to instigate changes that would result in tangible improvements in patient safety, with a principal intended outcome of halving the number of adverse events within two years (Health Foundation 2011). The ambition of the Initiative was matched by its evaluation, which comprised a programme of independent studies, including a mixed-methods controlled before-and-after study that measured change in multiple outcomes, and incorporated an extensive ethnographic examination of the Initiative's realisation in four participating hospitals (Benning et al. 2011). The evaluation's findings, though, were disappointing. They suggested statistically significant improvement relative to control hospitals on only one of several measures of safety climate, and one measure of clinical process (monitoring of vital signs). Other measures were unmoved compared to the controls, or even showed trends favouring the control group. The integrated ethnographic study cast light on some of the reasons for this: the Initiative was greeted with enthusiasm by senior managers in the participating hospitals but failed to penetrate to the level of the ward or operating theatre, where it was seen as an elite preoccupation with limited relevance for day-to-day clinical practice: "somewhere between the blunt end and the sharp end, the model of participative engagement on which [the Initiative] was based had got rather lost" (Benning et al. 2011: 9).

While the findings of the evaluation were disappointing, the evaluation itself is noteworthy, and represents an advance on previous study designs, for at least two reasons. One is the use of contemporaneous controls, and the deployment of a "difference-in-difference" analysis in summatively evaluating effectiveness. Had the Initiative been evaluated using an uncontrolled before-and-after design, it might (probably erroneously) have been declared a success, since there were significant improvements in several safety indicators over time (in both the intervention and control groups) – the so-called "rising tide" effect (Chen et al. 2016). Incorrectly attributing this improvement to the Safer Patients Initiative could plausibly have resulted in the allocation of scarce healthcare resources to an expensive but ineffective intervention. Second, the use of qualitative methods as part of the study design offered explanatory purchase, indicating what went wrong, which aspects of the intervention showed promise and what kinds of modifications might be made in any revised version. Learning of this kind is of course essential to the endeavour of improving safety; null or negative evaluations are not in vain if they provide a resource for improving future improvement efforts. Such insights are likely to come from well-designed qualitative work that provides an explanatory complement to the findings of quantitative evaluation.

Other studies of safety interventions in healthcare have followed this mixed-methods model, combining robust quantitative evaluation that aims for the upper tiers of the hierarchy of evidence with qualitative work that defies its position at the bottom of (some versions of) the hierarchy through its explanatory value. The interventions evaluated have arguably become even more sophisticated, drawing on a range of social scientific insights for their theoretical foundation, and seeking to provide the right kind of impetus, in the right quantity, in the relevant parts of the healthcare system, to instigate improvement (see Table 11.1). But the improvements yielded have often been modest at best – a point to which we return in the next section of the chapter. The most notable examples tend to come from areas such as surgery and intensive care that present high risks to patient safety, but are relatively contained, and thus perhaps more amenable to concerted, focused improvement efforts.

The success of the US-based programme to prevent bloodstream infections in intensive care units briefly mentioned above (Pronovost et al. 2006) led to an effort to replicate this work in the UK. This effort, however, failed to match the spectacular reductions (and in some cases eliminations) achieved in the original programme, demonstrated by a controlled study (Bion et al. 2013) and explained by an integrated ethnography (Dixon-Woods et al. 2013). The ethnography pointed towards the role of a policy and organisational environment that discouraged engagement with the programme, and the absence of the kind of "social movement" and sense of common endeavour that had been instrumental in making a success of the original.

In surgery, one effort to improve safety sought to deploy, in various combinations, the introduction of standard operating procedures, the principles of Lean process improvement and the use of teamwork training based on aviation's Crew Resource Management to pursue improvement in operating theatre practices, based on the theoretically and empirically informed hypothesis that interventions seeking to address culture and systems in tandem would be more likely to succeed than narrower interventions alone. A series of controlled before-and-after studies and an integrative quantitative analysis offered some support to this hypothesis, finding greater impacts in combined-intervention sites (McCulloch et al. 2016). An accompanying qualitative interview-based study sought to explain the differential impact of these approaches, highlighting in particular the significance of intensive expert support in the more successful sites – a finding with important implications for any effort to replicate or roll out the approach (Flynn et al. 2016).

Finally, outside perioperative care, Lawton et al. (2017) and Sheard et al. (2017) presented results from, respectively, the cluster-randomised controlled trial and process evaluation of the Patient Reporting and Action for a Safe Environment (PRASE) intervention. The result of the trial was "negative," with no significant difference

TABLE 11.1

Four Complex Interventions to Improve Quality or Safety in Healthcare, and Their Evaluations

Programme	Setting	Programme Aims and Measures	Programme Methods	Evaluation Approach	Evaluation Findings	Further Information
Safer Patients Initiative	Hospital wide: wards, critical care, perioperative care, medicines management	Improve organisation-wide safety as measured by a variety of indicators, including reductions in mortality and adverse events	Implementation of evidence-based care bundles in key clinical areas, supported by leadership and cultural interventions at the hospital level	Controlled before-and-after study incorporating difference-in-difference analysis, with integrated ethnographic study	Most measures show no significant difference between control and intervention groups; ethnography suggests limited awareness or impact of work at sharp end	Benning et al. (2011)
Matching Michigan	Adult and paediatric intensive care units	Improve practices in intensive care to reduce catheter-associated bloodstream infection rates	Use of technical interventions to ensure consistent use of evidence-based practices known to improve infection control and non-technical interventions to improve systems and culture; introduction of a national catheter-related bloodstream infection reporting system	Non-randomised, stepped study of roll out of intervention, comparing data before and after implementation, with integrated ethnographic study	The rate of improvement in infection control practices did not significantly change after the intervention; ethnography highlights a hostile policy and management environment and a failure to replicate the social movement of the prototype Michigan approach	Bion et al. (2013); Dixon-Woods et al. (2013)

(Continued)

TABLE 11.1 (CONTIUED)
Four Complex Interventions to Improve Quality or Safety in Healthcare, and Their Evaluations

Programme	Setting	Programme Aims and Measures	Programme Methods	Evaluation Approach	Evaluation Findings	Further Information
Safer Delivery of Surgical Services (3S) programme	Surgery	Improve various aspects of surgical care and outcomes, including teamwork, error rates, patient-reported outcomes, mortality, length of stay, readmissions	Use of three interventions, each designed to address a different cause of patient safety issues: teamwork training based on Crew Resource Management; introduction of standard operating procedures; and process improvement based on Lean principles	Controlled evaluations of combinations of the intervention, with meta-analysis; retrospective qualitative interviews with team and participants	Some indication that combined approaches may address some aspects of surgical safety; interviews shed light on the reasons for differential impact and the importance of strong support from the research team that may not be replicable	McCulloch et al. (2016); Flynn et al. (2016)
Patient Reporting and Action for a Safer Environment (PRASE)	Ward-based care	Improve safety of care for hospital inpatients, measured by proportion of patients receiving "harm free care" (Safety Thermometer)	Introduction of action planning cycle: (i) facilitate collection and analysis of patient-reported safety concerns; (ii) feedback to staff, building teams' intelligence about patient experience of safety, areas for improvement, and capacity to improve quality.	Cluster-randomised controlled trial; observation of action planning meetings and qualitative interviews with staff	No significant difference on the primary outcome between control and intervention arms; qualitative study suggests enthusiasm for intervention but differential engagement across wards	Lawton et al. (2017); Sheard et al. (2017)

in the primary outcome between the intervention and control group. The process evaluation, however, demonstrated a wide range of responses to the intervention, and offered rich narratives about its use in participating services. At least according to the qualitative data, then, PRASE had potential. Moreover, the trial relied for its primary outcome measure on a data source – the NHS Safety Thermometer – that the authors themselves acknowledged was a highly flawed way of seeking to gauge the impact of an intervention with much broader improvement ambitions. PRASE sought to empower healthcare staff to focus on improving areas that they saw as a priority, based on the feedback of patients. As Lawton et al. (2017: 629) noted, "this makes it difficult to predict in advance what changes a ward will choose to make and therefore what outcomes it might be appropriate to measure." Researchers' choices about how to measure the impact of a complex, multifaceted intervention are therefore hugely significant, and often limited by the availability of routinely collected data. An ill-chosen outcome measure provides a poor yardstick against which to judge an intervention, and finding an appropriate outcome measure is all the more difficult when evaluating complex safety interventions. Correspondingly, the biomedical tendency to place so much emphasis on the achievement of statistically significant improvement against a single primary outcome measure is all the more problematic – a theme to which we return in the final section.

THE UNFULFILLED PROMISE OF MIXED-METHODS EVALUATIONS?

The increasing methodological sophistication of evaluations of safety interventions in healthcare, then, has in general led to increasing pessimism (or perhaps more accurately, cautious realism) about their impact. As noted above, in itself this is no bad thing. The learning to be derived from rigorous evaluation of unsuccessful interventions is at least as great as the potential learning from success (Peden et al. 2019; Stephens et al. 2018) – though not always as readily published and communicated to the research and practice community (Shojania and Grimshaw 2005). The science of improvement in healthcare is still in its infancy, and the evidence base for the impact and suitability of different forms of intervention in different circumstances is still incomplete. There is much to be gained from the accumulation of knowledge about the what, the how and the when and where, and from the curation of this evidence base in forms that are accessible and useful to researchers and practitioners (Dixon-Woods and Martin 2016; Webb 2011).

On the other hand, a nagging concern for some is that there is something distinctive about intervening in relation to healthcare quality and safety which requires rethinking – perhaps radically – the nature of the evidence base we demand, and the tools we use to produce it. A longstanding debate within the improvement community has focused on whether the traditional methods of health services research are fit for the purpose of evaluating safety interventions, given their complex nature, the iterative way in which they are applied and the challenges of applying techniques such as randomisation and blinding to a dynamic field (Leape et al. 2002; Shojania 2013; Wachter 2010). The studies discussed in the previous section certainly show that, with funding and tenacity, it is *feasible* to apply high-quality quantitative evaluation methods to safety interventions, and to supplement and enrich these with qualitative methods. But the inconsistency of findings, and the disappointment that often follows investment in promising interventions and their evaluation, does raise the question of whether the notion of a "proven intervention" is something of a chimera – at least if we use the word "proven" in the sense traditionally used in health services research. Both the hierarchy of evidence and the established approach to disseminating and implementing evidence in healthcare revolve around the notion of "gold standards" of evidence and practice. In other words, they prize interventions that have survived the most rigorous of evaluation, and are thus seen to be *the (single) best practice*, which should accordingly be applied by all practitioners in all circumstances. Again, one can see how such expectations might reasonably be applied to certain kinds of healthcare intervention, particularly pharmaceutical therapies (though again, there is a danger that this neglects the heterogeneity of patient populations, particularly multimorbid groups, as well as marginalising patient preferences: see, e.g., Greenhalgh et al. 2014). Given the crucial role that contextual variability can play in improvement efforts, however – as well as the difficulties in specifying an intervention so tightly that it can be applied, near-identically, by different practitioners with different skill sets in different circumstances – aspiring to develop and evidence "gold standard" interventions might be a fool's errand.

A more modest – but more pressing and potentially more fruitful – task might be the development of a better understanding of the menu of improvement approaches that is available, and of the likely prerequisites for their effectiveness. Such an evidence base would acknowledge the fallibility of every approach, and make no claim to offer ready-made solutions. It would rather provide as much detailed and helpful guidance as possible on the measures that might be taken to make the approach work – while acknowledging the importance of the skill, determination and individual style given to any intervention by those leading it. The importance of different configurations of contexts and mechanisms that can causally drive different outcomes is, of course, explicitly acknowledged by approaches such as realist evaluation (see, e.g., Randell et al. 2014). Such approaches account for both the range of causal mechanisms that can give rise to outcomes in different circumstances, and the range of desirable endpoints beyond the principal outcomes that traditional biomedical evaluation tends to prize. There is an important balance to be struck, however, between the search for the "right" configuration of context, mechanism and outcome, and the idiographic description a thousand possible combinations, each unique to its circumstance (Marchal et al. 2012). Equally, the reduction of a "successful" intervention in healthcare safety to the achievement of a statistically significant result against a single, narrow measure that cannot account for broader benefits is also unhelpful – and risks the premature disposal of complex interventions to which a simple biomedical model of evaluation cannot do justice.

Part of the challenge for evaluation, then, is to be able to say something more than that "nothing works" (if a truly universal gold standard, according to a particular narrow measure, is the required benchmark) or that "nearly everything works" (in the right, very particular, circumstances). Both the rich description of improvement practices to demonstrate how, when and why, for example, through comparative qualitative case studies, and the evidence provided by robust quantitative evaluation that shows what *can* work, have a role in this. But an equally important task is effective curation of the evidence base, and greater interaction between research and practice communities to assist the task of translating research into practice in a context-sensitive way, that allows practitioners to make their own sense of the evidence base – notwithstanding the variety of challenges involved in such efforts (e.g. Chew et al. 2013; Martin et al. 2011; Ward et al. 2009).

Besides this, debates familiar from other fields where distinct or even conflicting research paradigms have come together continue to preoccupy methodologists in healthcare safety. Methodological purists suggest that the positivist and constructivist roots of epidemiology and qualitative social science, respectively, make them irreconcilable (Doyle et al. 2009). Others take a more pragmatic approach, valuing the complementary insight that qualitative and quantitative understandings can bring in producing local understanding, without concerning themselves with epistemological neatness (see also Kongsvik and Almklov, this volume). Yet these divergent ways of knowing can have very practical ramifications. Rarely, for example, does it prove possible to use qualitative data to exactingly account for divergent outcomes between cases that, by quantitative measures, appear to be "successful" or "unsuccessful." Qualitative insights tend to be more fluid and less certain. But the expectation that qualitative explanations can be neatly mapped onto divergent quantitative results risks giving rise to formulaic, contrived, even positivistic accounts, with a simplicity as beguiling as it is misleading. In this process, the richness and ambiguity of qualitative analysis is reduced to a series of factors: x variables that predict the y variable of the quantitative outcome. Good qualitative analysis should not be reduced to a regression analysis. Similarly, an inappropriately positivistic mindset among qualitative researchers can impede the accumulation of useful knowledge. Mitchell et al. (2017), for example, note a tendency among evaluators to continue to construct checklist-based interventions as simple, stable interventions with predictable intended causal pathways – rather than as something that interacts with its context to produce multiple effects, both intended and unintended, positive and negative. This results, they argue, in the repeated rediscovery of the complex context dependency of the checklist, rather than the steady accumulation of insights that might help improve the intervention.

There are debates, too, within the qualitative healthcare research community about the methods most appropriate to providing explanatory insight into safety interventions. The hierarchy of evidence of the biomedical research tradition has parallels within qualitative research, including the validity of ethnography "in situ" versus reliance on retrospective interview accounts (Silverman 2001), and the utility of insights provided by observational methods that fall short of full ethnographies in the anthropological tradition in their scale and scope (Cupit et al. 2018; Leslie et al. 2014; Waring and Jones 2016). More prosaically, as public finances in many countries remain under tight control and research and evaluation funders demand maximum value from their investments, opportunities for high-quality mixed-methods evaluations may become rarer. Qualitative researchers in the field of healthcare must be vigilant that the biomedical tradition does not result in a return to a situation where their contribution is seen as a desirable, but disposable, add-on.

CONCLUSION

Mixed-methods research and evaluation in healthcare quality and safety offers a useful model for safety research elsewhere, particularly in its increasingly sophisticated efforts to bring together qualitative and quantitative insights. But it also provides some warnings for other researchers, particularly in relation to the need to contain expectations about what mixed-methods approaches can provide, and around the considerations important in deriving the maximum value from both lenses – and reconciling them in a satisfactory way.

We have traced a narrative in healthcare safety research in which an understanding of the need for complex, theory-based interventions has been followed, rather belatedly, by sophistication in evaluation methodology. We have attributed this tardiness in part to the long shadow cast by the biomedical evaluation tradition. Arguably, the translation of the findings of this research base into the kind of useable knowledge that can inform practitioner-led work to improve healthcare safety has been even slower, and here again the tendency to rely on traditional biomedical approaches to knowledge translation may in part be culpable. Sophisticated interventions demand sophisticated evaluation, and sophisticated evaluations demand sophisticated approaches to making findings accessible and relevant. This means overcoming the binary notion that something either works (universally and unequivocally) or does not, and taking seriously the critical, reflexive, intelligent work by practitioners that is needed to make any intervention work. Such approaches to curation and translation remain in their infancy in healthcare.

REFERENCES

Benning, A., Ghaleb, M., Suokas, A., Dixon-Woods, M., et al. (2011) Large scale organisational intervention to improve patient safety in four UK hospitals: mixed method evaluation, *BMJ*. **342**, d195.

Bergs, J., Lambrechts, F., Simons, P., Vlayen, A., et al. (2015) Barriers and facilitators related to the implementation of surgical safety checklists: a systematic review of the qualitative evidence, *BMJ Quality & Safety*. **24**, 12, 776–86.

Bion, J., Richardson, A., Hibbert, P., Beer, J., et al. (2013) "Matching Michigan": a 2-year stepped interventional programme to minimise central venous catheter-blood stream infections in intensive care units in England, *BMJ Quality & Safety*. **22**, 2, 110–23.

Boyd, C. M. and Kent, D. M. (2014) Evidence-based medicine and the hard problem of multimorbidity, *Journal of General Internal Medicine*. **29**, 4, 552–3.

Braithwaite, J. (2018) Changing how we think about healthcare improvement, *BMJ*. **361**, k2014.

Brennan, T. A., Leape, L. L., Laird, N. M., Hebert, L., et al. (1991) Incidence of adverse events and negligence in hospitalized patients: results of the Harvard Medical Practice Study I, *New England Journal of Medicine*. **324**, 6, 370–6.

Campbell, M., Fitzpatrick, R., Haines, A., Kinmonth, A. L., et al. (2000) Framework for design and evaluation of complex interventions to improve health, *BMJ*. **321**, 694–6.

Chen, Y.-F., Hemming, K., Stevens, A. J. and Lilford, R. J. (2016) Secular trends and evaluation of complex interventions: the rising tide phenomenon, *BMJ Quality & Safety*. **25**, 5, 303–10.

Chew, S., Armstrong, N. and Martin, G. (2013) Institutionalising knowledge brokering as a sustainable knowledge translation solution in healthcare: how can it work in practice? *Evidence & Policy.* **9**, 3, 335–51.

Clay-Williams, R. and Colligan, L. (2015) Back to basics: checklists in aviation and healthcare, *BMJ Quality & Safety.* **24**, 7, 428–31.

Cochrane, A. L. (1972) *Effectiveness and efficiency: random reflections on health services.* London: Nuffield Provincial Hospitals Trust.

Craig, P., Dieppe, P., Macintyre, S., Michie, S., et al. (2008) Developing and evaluating complex interventions: the new Medical Research Council guidance, *BMJ.* **337**, a1655.

Cupit, C., Mackintosh, N. and Armstrong, N. (2018) Using ethnography to study improving healthcare: reflections on the "ethnographic" label, *BMJ Quality & Safety.* **27**, 4, 258–60.

De Vries, E. N., Prins, H. A., Crolla, R. M. P. H., den Outer, A. J., et al. (2010) Effect of a comprehensive surgical safety system on patient outcomes, *New England Journal of Medicine.* **363**, 20, 1928–37.

Department of Health (2000) *An organisation with a memory.* London: Stationery Office.

Department of Health (2006) *Best research for best health: a new national research strategy.* London: Department of Health.

Dixon-Woods, M., Bosk, C. L., Aveling, E.-L., Goeschel, C. A., et al. (2011) Explaining Michigan: developing an ex post theory of a patient safety program, *Milbank Quarterly.* **89**, 2, 167–205.

Dixon-Woods, M., Leslie, M., Tarrant, C. and Bion, J. (2013) Explaining matching Michigan: an ethnographic study of a patient safety program, *Implementation Science.* **8**, 70.

Dixon-Woods, M. and Martin, G. P. (2016) Does quality improvement improve quality?, *Future Hospital Journal.* **3**, 3, 191–4.

Doyle, L., Brady, A.-M. and Byrne, G. (2009) An overview of mixed methods research, *Journal of Research in Nursing.* **14**, 2, 175–85.

Flynn, L. C., McCulloch, P. G., Morgan, L. J., Robertson, E. R., et al. (2016) The safer delivery of surgical services program (S3): explaining its differential effectiveness and exploring implications for improving quality in complex systems, *Annals of Surgery.* **264**, 6, 997–1003.

Greenhalgh, T., Howick, J. and Maskrey, N. (2014) Evidence based medicine: a movement in crisis? *BMJ.* **348**, g3725.

Haynes, A. B., Weiser, T. G., Berry, W. R., Lipsitz, S. R., et al. (2009) A surgical safety checklist to reduce morbidity and mortality in a global population, *New England Journal of Medicine.* **360**, 5, 491–9.

Health Foundation (2011) *Learning report: Safer patients initiative.* London: Health Foundation.

Howick, J. H. (2011) *The philosophy of evidence-based medicine.* Oxford: Wiley.

Institute of Medicine (1999) *To err is human: building a safer health system.* Washington, DC: National Academy Press.

Lawton, R., O'Hara, J. K., Sheard, L., Armitage, G., Cocks, K., Buckley, H., Corbacho, B. Reynolds, C., Marsh, C., Moore, S., Watt, I. and Wright, J. et al. (2017) Can Patient Involvement Improve Patient Safety? A Cluster Randomised Control Trial of the Patient Reporting and Action for a Safe Environment (PRASE) Intervention., *BMJ Qual Saf.* **26**, 8, 622–31. https://doi.org/10.1136/bmjqs-2016-005570.

Leape, L. L., Berwick, D. M. and Bates, D. W. (2002) What practices will most improve safety? Evidence-based medicine meets patient safety, *JAMA.* **288**, 4, 501–7.

Leslie, M., Paradis, E., Gropper, M. A., Reeves, S., et al. (2014) Applying ethnography to the study of context in healthcare quality and safety, *BMJ Quality & Safety.* **23**, 2, 99–105.

Makary, M. A. and Daniel, M. (2016) Medical error – the third leading cause of death in the US, *BMJ*. **353**, i2139.

Marchal, B., van Belle, S., van Olmen, J., Hoeree, T., et al. (2012) Is realist evaluation keeping its promise? A review of published empirical studies in the field of health systems research, *Evaluation*. **18**, 2, 192–212.

Martin, G., Currie, G. and Lockett, A. (2011) Prospects for knowledge exchange in health policy and management: institutional and epistemic boundaries, *Journal of Health Services Research & Policy*. **16**, 4, 211–7.

McCulloch, P., Morgan, L., Flynn, L., Rivero-Arias, O., et al. (2016) *Safer delivery of surgical services: a programme of controlled before-and-after intervention studies with pre-planned pooled data analysis*. Programme Grants for Applied Research. Southampton: NIHR Journals Library.

McGlynn, E. A., Asch, S. M., Adams, J., Keesey, J., et al. (2003) The quality of health care delivered to adults in the United States, *New England Journal of Medicine*. **348**, 26, 2635–45.

Melnyk, B. M. and Fineout-Overholt, E. (2011) *Evidence-based practice in nursing and healthcare: a guide to best practice*. Philadelphia, PA: Wolters-Kluwer.

Mitchell, B., Cristancho, S., Nyhof, B. B. and Lingard, L. A. (2017) Mobilising or standing still? A narrative review of Surgical Safety Checklist knowledge as developed in 25 highly cited papers from 2009 to 2016, *BMJ Quality & Safety*. **26**, 10, 837–44.

Oakley, A. (2000) *Experiments in knowing: Gender and method in the social sciences*. Cambridge: Polity Press.

Pawson, R. and Tilley, N. (1997) *Realistic evaluation*. London: SAGE.

PCORI (2018) *PCORI methodology standards*. Washington, DC: Patient-Centered Outcomes Research Institute.

Peden, C. J., Stephens, T., Martin, G., Kahan, B. C., et al. (2019) Effectiveness of a national quality improvement programme to improve survival after emergency abdominal surgery (EPOCH): a stepped-wedge cluster-randomised trial, *The Lancet*. **393**, 2213–21.

Pope, C. and Mays, N. (1993) Opening the black box: An encounter in the corridors of health services research, *BMJ*. **306**, 315–8.

Pronovost, P., Needham, D., Berenholtz, S., Sinopoli, D., et al. (2006) An intervention to decrease catheter-related bloodstream infections in the ICU, *New England Journal of Medicine*. **355**, 26, 2725–32.

Pronovost, P. J., Berenholtz, S. M. and Morlock, L. L. (2011) Is quality of care improving in the UK? *BMJ*. **342**, c6646.

Randell, R., Greenhalgh, J., Hindmarsh, J., Dowding, D., et al. (2014) Integration of robotic surgery into routine practice and impacts on communication, collaboration, and decision making: a realist process evaluation protocol, *Implementation Science*. **9**, 1, 52.

Sackett, D. L., Rosenberg, W. M. C., Gray, J. A. M., Haynes, R. B., et al. (1996) Evidence based medicine: what it is and what it isn't, *BMJ*. **312**, 71–2.

Schuster, M. A., McGlynn, E. A. and Brook, R. H. (2005) How good is the quality of health care in the United States? *Milbank Quarterly*. **83**, 4, 843–95.

Sheard, L., Marsh, C., O'Hara, J., Armitage, G., Wright, J. and Lawton, R. (2017) Exploring how ward staff engage with the implementation of a patient safety intervention: a UK-based qualitative process evaluation, *BMJ Open*. **7**, e014558.

Shekelle, P. G., Pronovost, P. J., Wachter, R. M., Taylor, S. L., et al. (2011) Advancing the science of patient safety, *Annals of Internal Medicine*. **154**, 10, 693–6.

Shojania, K. G. (2012) Deaths due to medical error: Jumbo jets or just small propeller planes?, *BMJ Quality & Safety*. **21**, 9, 709–12.

Shojania, K. G. (2013) Conventional evaluations of improvement interventions: More trials or just more tribulations? *BMJ Quality & Safety*. **22**, 11, 881–4.

Shojania, K. G. and Dixon-Woods, M. (2017) Estimating deaths due to medical error: The ongoing controversy and why it matters, *BMJ Quality & Safety.* **26**, 5, 423–8.

Shojania, K. G. and Grimshaw, J. M. (2005) Evidence-based quality improvement: The state of the science, *Health Affairs.* **24**, 1, 138–50.

Silverman, D. (2001) *Interpreting qualitative data: Methods for analysing talk, text and interaction.* London: SAGE.

Stephens, T. J., Peden, C. J., Pearse, R. M., Shaw, S. E., et al. (2018) Improving care at scale: Process evaluation of a multi-component quality improvement intervention to reduce mortality after emergency abdominal surgery (EPOCH trial), *Implementation Science.* **13**, 1, 142.

Wachter, R. M. (2010) Patient safety at ten: unmistakable progress, troubling gaps, *Health Affairs.* **29**, 1, 165–73.

Ward, V., House, A. and Hamer, S. (2009) Knowledge brokering: The missing link in the evidence to action chain? *Evidence & Policy.* **5**, 267–79.

Waring, J., Allen, D., Braithwaite, J. and Sandall, J. (2016) Healthcare quality and safety: A review of policy, practice and research, *Sociology of Health & Illness.* **38**, 2, 198–215.

Waring, J. and Jones, L. (2016) Maintaining the link between methodology and method in ethnographic health research, *BMJ Quality & Safety.* **25**, 7, 556–7.

Webb, D. (2011) Foreword. In *Evidence: Safer patients initiative phase two.* London: The Health Foundation, pp. iv–vii.

Weiss, C. H. (1995) Nothing as practical as good theory: exploring theory-based evaluation for comprehensive community initiatives for children and families. In Connell, J. P., Kubisch, A. C., Schorr, L. B. and Weiss, C. H. (eds) *New approaches to evaluating community initiatives: concepts, methods, and contexts.* New York: Aspen Institute, pp. 65–92.

World Health Organization (2018) *10 facts on patient safety.* https://www.who.int/features/factfiles/patient_safety/en/. Accessed 26 April 2019.

12 Putting Senior Management in Focus

A Critical Reflection on the Direction of Safety Science Research

Jan Hayes and Sarah Maslen

CONTENTS

INTRODUCTION

For several decades foundational theories in safety science have focused on the extent to which behaviours of top company management determine safety outcomes, due to their influence on all organisational actors. This shift in thinking away from the front line has not been matched by an equivalent move in empirical safety research, which remains focused on the actions of individuals and groups much lower in organisational hierarchies. Studying top management is not without its challenges, but, as we will show, the disconnect between the positioning of current theory and research practice is so distinct that safety science research could be characterised as looking for answers in places that are easy to access, rather than those that are most likely to lead to improvement.

We start by summarising what accident analysis tells us about how senior management practices influence safety outcomes. In contrast, we then review the empirical studies in *Safety Science* in recent years to determine the extent to which this body of work addresses relevant senior management practices. In light of the major research gap we identify, we turn to the literature on elite interviewing to address some approaches applied by social scientists researching other sectors. We also reflect on our own experience researching at this level, propose a new research agenda focused on senior management practices and provide some guidance to assist those who would like to take up this challenge.

A SAFETY SCIENCE RESEARCH AGENDA

SENIOR MANAGERS AND ACCIDENT CAUSATION

The concept of the organisational accident, as popularised by James Reason in the 1990s, continues to inform both safety research and practice. This view of accident causation emphasises the importance of company executives in preventing accidents, given their influence over "organisational factors" (Reason, 1997). Supporting this view, analyses of numerous major disasters (Beamish, 2002; Hayes, 2012; Hayes & Hopkins, 2014; Hopkins, 2008, 2012; Snook, 2000; Vaughan, 1996) have highlighted the key role of senior management in safety performance as they set both formally mandated and implicit, organisational priorities.

Such analyses have also provided insights into problematic management practices. Senior management failings, such as excessive cost cutting, a focus on production without understanding the safety implications and a lack of understanding of the difference between personal and process safety have been consistently highlighted (Hayes, 2012; Hayes & Hopkins, 2014; Hopkins, 2008, 2012; Snook, 2000). It is surely safe to assume that senior managers do not intend to cause disasters, so a common factor in these cases would appear to be a lack of understanding at senior levels of the impact that their decisions can have on safety outcomes i.e. a lack of safety imagination (Pidgeon & O'Leary, 2000). In fact, Quinlan's (2014) analysis of the contributors to ten major mine disasters notes that a common failure in organisations was a top-down management system with the absence of critical feedback loops to ensure that information regarding critical hazards can flow back up to senior management where key decisions are made. Similar observations were made by Hopkins in his analysis of the Moura mine collapse (1999) and the Deepwater Horizon blowout (2012), and by Waring (2015) in his broad review of managerial and other non-technical factors in major accident causation. These studies leave us with a picture of senior managers in poor performing organisations who are isolated from the hazards that their organisation faces and the critical systems in place to control them, although it must be emphasised that this conclusion has been reached largely without studying top managers directly. Rather their practices have been inferred from the resultant outcomes.

SENIOR MANAGERS AND SAFETY CULTURE

Looking beyond the disaster literature, the literature on safety culture also places a strong emphasis on the role of senior executives in driving positive safety outcomes.

Schein's famous definition of culture as "the way we do things around here" stems from priorities and behaviours set from the top. In his words, "[l]eaders create and change cultures, while managers and administrators live within them" (1992, p. 5). Much of the literature focuses on the link between safety and top management. For example, Hale's (2000, p. 12) analysis of what makes for a good "culture of safety" includes "the importance which is given by ... top managers to safety as a goal, alongside and in unavoidable conflict with other organizational goals" as one of eight necessary elements. Similarly, Pidgeon and O'Leary (2000, p. 18) postulate that "senior management commitment to safety" is a factor that both fosters and sustains a good safety culture. Despite this focus on the importance of management actions, there is little research that addresses how excellence in senior management decision-making occurs in practice and so how it can best be fostered. Safety culture implementation programs have attempted to address senior management practice typically by requiring them to develop personal action plans around desired values, such as trust and making safety the top priority, supported by surveys to determine employee perceptions of their behaviour (e.g. Hudson, 2007), although the effectiveness of such actions are rarely investigated. Other organisations have focused on specific executive actions targeting safety, such as management by walking around, despite well-documented limitations (Hopkins, 2012).

EMPIRICAL STUDIES OF SENIOR MANAGEMENT IN THE SAFETY LITERATURE

Similarly, the safety research literature more broadly includes few empirical studies that focus on top managers. The exceptions are notable: drawing on interview data, high reliability researchers emphasise the extent to which management at all levels in high performing organisations are preoccupied with failure and exhibit chronic unease (Fruhen & Flin, 2016; Weick & Sutcliffe, 2001). Other researchers have targeted more structural measures by studying the impact of financial incentives on top management decision-making using interviews, observation and document analysis (Hopkins & Maslen, 2015; Maslen & Hopkins, 2014). In an attempt to look more systematically beyond the disaster literature and investigate our impression that little empirical research with senior management has been done in the normal operations tradition (Bourrier, 2002), we have reviewed 500 abstracts from the journal *Safety Science*. This represents all papers published in the journal in the 20 volumes from July 2016 to June 2018. Eighty-three per cent of papers published in *Safety Science* during this period were empirical. However, our review found few papers that focus on senior management.

We identified 12 papers that provide advice for senior managers about how to improve safety. One group (comprising eight papers) is in the area of safety leadership. Pilbeam, Doherty, Davidson and Denyer (2016) conducted a broad literature review of studies of safety leadership practices. They noted that almost all were aimed at the supervisor / frontline worker dyad and relied on surveys of follower perceptions to draw conclusions. Of the 25 studies they identified, none studied safety leadership by top management. Rather than addressing safety leadership practices of this group directly, this literature tends to provide recommendations to senior management about how best to foster safety leadership at lower levels of the organisation. Similarly, Smith, Eldridge and DeJoy (2016) found in their study of firefighters that transformational leadership has a positive influence on safety climate. The empirical work is based on workers' survey-based assessment of their direct supervisors' behaviour and yet the recommendations of the paper are directed at top management (establish safety as a fundamental value, promote safety behaviours through personal example, etc.). It appears that the manager–subordinate relationship that they have characterised empirically is assumed to cascade upwards in the organisation and so also represent the relationship between those at the top and their immediate subordinates. In a similar way, Zwetsloot, Kines, Ruotsala, Drupsteen and Merivirta (2017) studied factors behind successful implementation of Zero Accident Vision interventions. Their survey of 27 companies found that worker perception of top management support was the most important factor for success and they provide advice to top managers on that basis.

Another small group of papers (four) addresses key performance indicators (KPIs). For example, Gerbec and Kontic (2017) demonstrate how KPIs for business management purposes have been developed for port operations but their paper does not study senior management use of KPIs in practice. In a similar vein, Shirali, Salehi, Savari and Ahmadiangali (2018) develop and model the impact of safety cost indicators on productivity, quality and safety in the steel industry. Again, they propose that top managers (and even stockholders) can use such indicators to predict company performance. These papers try to influence where senior managers direct their attention. They do not study whether this occurs or whether the proposed impacts are delivered in practice.

We found only five papers among the 500 abstracts reviewed that, at least to some extent, address top management safety-related behaviour directly. It is also noteworthy that none of these target major accident risk, but rather all focus on worker safety. Young and Blitvich's (2018) study of two organisations under significant financial pressure included interviews with one general manager and one direct report. Unfortunately, in presenting the results, the researchers do not differentiate between top management and others lower in the management structure. Despite this limitation, it does highlight that productivity pressure impacts organisational focus on worker safety in times of extreme financial pressure and that checks and balances in the system are sometimes removed to save money on the basis that they are unproductive. It is perhaps surprising that a study of such limited scope (based on only eight interviews in two companies overall) was accepted for publication, but the authors acknowledge their rare yet important access to senior managers in these circumstances and the value of even a limited investigation into the potential pitfalls for senior management regarding worker safety in organisations under severe financial pressure.

In their case study regarding implementation of Zero Accident Vision (ZAV) in a Dutch firm, Twaalfhoven and Kortleven (2016) found in interviews that top managers pursue ZAV for both ethical and economic reasons. They also articulate the view that workers' actions are largely a result of the tone and priorities set by top management and yet top managers often take refuge in disciplinary action when programs are seen to fail. Similarly, an interview-based study of four companies that experienced a breakthrough change (i.e. significant improvement) in occupational health and safety (OHS) performance found that key drivers for top management to initiate a strong organisational focus on OHS included concerns about compliance with regulation as well as insurance costs. Some top managers were also driven by ethical considerations, especially following a serious accident in a similar business (Robson et al., 2016). Employer awareness of OHS law in small and medium-sized enterprises in Cyprus was the subject of Isık and Atasoylu's (2017) study using questionnaire-based interviews. They found that awareness of the law and its requirements was extremely low, apparently due to the lack of promotion of OHS law by government including an absence of proactive regulatory inspections. In a similar study, Bonafede et al. (2016, p. 17) examined employers' perceptions and awareness of OHS in Italy. They found that micro and small business managers "perceived OHS as an economic burden much more than an opportunity."

This snapshot of recently published articles in *Safety Science* shows a rich empirical safety research tradition but only 1 per cent of papers published look at the safety practices of top management. We also see that researchers generate advice for top managers by studying those at lower levels, rarely investigating exactly what safety leadership behaviours or performance indicators work in practice for those at the highest level. On the few occasions when senior managers are engaged in empirical studies, our sample suggests that such studies tend to be focused on occupational safety matters. Further, they are undertaken on small- and medium-sized enterprises (SMEs) rather than investigating the potential for major disaster in organisations responsible for complex socio-technical systems. Researchers who claim to have engaged with top levels of management often do not, at least in their reporting of the results, clearly differentiate between actors operating at different organisational levels. Twaalfhoven and Kortleven (2016) even report that some of their surveyed companies refused to disclose the organisational level of survey respondents, bringing into question the extent to which senior managers are represented in the results at all. Despite these clear methodological challenges, issues of access to senior management have not been engaged with in the literature in a sustained way.

Overall, the disaster literature suggests many top management behaviours that may be linked to poor safety outcomes, including lack of awareness of both hazards and the status of systems in place to control them. Despite this, little empirical work has been undertaken either to determine whether such problems are unique to those organisations that have experienced disasters or to investigate how top managers in higher performing organisations may be behaving differently. This is a major research gap that must surely be addressed and so we turn to the broader organisation studies literature to consider the challenges inherent in engaging top management in research.

THE CHALLENGES OF ELITE INTERVIEWING

As the preceding analysis reveals, safety science scholars have almost entirely focused their empirical attentions on frontline workers. This pattern of research focus is not limited to this field. Within disciplines such as sociology and anthropology, researchers mostly study the disadvantaged rather than those in power. The limits of this focus have been highlighted since the middle of the last century. Sociologist C. Wright Mills published a series of empirical studies of elites within the sociological tradition, justified on the basis that power is increasingly concentrated in the hands of a few hundred corporations and their key decision-makers (Mills, 1956, 1957, 1962). Nader (1972), an anthropologist, published an essay on this pressing issue of "studying up" with enduring resonance. She argued that without engaging with those in power, researchers skirt critical social problems and fail to develop "adequate theory and description" (Nader, 1972, p. 290). While studies of elites remain relatively thin on the ground, there are some important exceptions responding to these research calls that span the disciplines of sociology (Dezalay & Garth, 1996; Sklair, 2001), anthropology (Forsythe, 2001; Jackall, 1988; Schwartzman, 1993), human geography (McDowell, 1998; Schoenberger, 1991) and political science (Dexter, 1970; Richards, 1996).

Underscoring the significance of subjects of study, Nader encourages researchers to critically reflect on the consequences of their research. As safety science scholars, we can ask: What have the contributions of empirical research in safety science been? To what extent are we advancing and refining core arguments put forward by high reliability theorists and so on? Are we making safety performance a problem of the workers, by virtue of our research questions and subjects of study? Or are we instead able to get at the role of organisational factors, most influenced by top managers? This is not to say that studying frontline workers is not of value but that in concentrating our studies at this end we are not getting a complete picture. Indeed, we are often missing the critical points in decision-making processes that we are working to understand, if we want to test and further develop the core theories of the field.

There are barriers to researching elite groups that are important to engage with if we are going to effectively broaden our focus. One of the commonly cited barriers to "studying up" is access. Time pressures in corporate environments present a challenge to research participation (see also Bourrier, Chapter 3, this volume). We need to be able to answer the question as to why the elites we are working to recruit would sacrifice their time to participate in the research, particularly given the potentially sensitive nature of the subject matter (Empson, 2018; Laurila, 1997). Common advice is that a topic needs to be of particular elite interest, which may warrant positioning or even designing the research with these interests in mind (Empson, 2018). Researchers also need to persuade potential participants that the research will yield insights that will help them (Coleman, 1996; Greenwood, Whyte, & Harkavy, 1993). Even when an organisation agrees to participate in a given research project, this does not guarantee continuing access and co-operation of individual interviewees and so "access should be considered a continuous negotiating process in which some actors are always in opposition and in which the researcher needs to take the changing and conflicting interests of organizational actors into account" (Laurila, 1997, p. 410).

Access presents several practical challenges that are not entirely solved through research design. Making contact may not be straightforward, given the multiple layers of gatekeepers that may be encountered from executive assistants to other professional staff (Laurila, 1997; Monahan & Fisher, 2015; Thomas, 1993). The time scale for "getting in" may then be considerably longer than other forms of recruitment, and often relies on networks created beyond written approaches (Mikecz, 2012). Researcher reputation can become important. Contacts may be fostered through practitioner conferences. Researchers can also become more visible and, by extension, relevant, through targeted publication in trade journals (Empson, 2018). The funding requirements associated with interviewing a global group also affect the practicalities of doing the data collection post-recruitment (Conti & O'Neil, 2007; Stephens, 2007). Global travel costs more. Researchers will also likely encounter challenges with co-ordinating interviews with multiple participants, making several trips by a researcher to the same location inevitable.

"Studying up" also introduces issues with the presentation of self and power dynamics in the interview. One definition of elite interviewing is not so much centred on the position of power held by the interviewee so much as the power imbalance between the interviewer and interviewee (Mikecz, 2012; Odendahl & Shaw, 2002). The interviewer must work to minimise the appearance of this imbalance. Many scholars who have interviewed elite groups have reflected on the need to adopt specific dress that would be read as appropriate in the interviewee's environment in order to be taken seriously (Alvesson, 2001; Empson, 2018; Grey, Anderson-Gough, & Robson, 1998; Mikecz, 2012; Odendahl & Shaw, 2002). Professionals make split-second judgements about colleagues and clients based on their suit, shoes, watch, mobile devices and so on, and these same judgements extend to the interviewer. Conti and O'Neil (2007) observed that their interview results improved with an upgrade to their wardrobe. Empson (2018, p. 5) explains that such efforts at impression management have little to do with questions of authenticity or the comfort of the interviewer, but instead dress must be guided by "what makes your interviewee feel comfortable, because it is only when they are comfortable that they will relax sufficiently to give you the data you seek."

Sector-specific terminology and narrative structures are also ways that expertise and experience are signalled. As such, adoption of specialised language is also critical, to the point of providing an illusion of a shared background (Alvesson, 2001; Conti & O'Neil, 2007; Laurila, 1997). In some ways, these language demands also reflect the requirements of good interviewing more generally. The interviewer needs to adapt their behaviour and use of language to manage the interaction, including the avoidance of academic jargon (Harvey, 2011). As Empson (2018, p. 6) explains: "If they swear, I swear too. If they treat me 'like a lady,' I try to behave like one. If a woman professional raises issues of 'being a woman in a man's world,' I respond to her signal of sisterly affiliation" (also see McDowell, 1998). The power imbalance inherent in elite interviewing also potentially introduces a need to manage being treated as lesser in the interview, a potentially uncomfortable experience among scholars not practiced at being talked down to (Conti & O'Neil, 2007). Many subjects of elite interviews have also had press training and so are practised at avoiding questions and delivering stock answers, which can limit how deeply the interviewer can delve (Empson, 2018). Interviewers thus need strategies to manage these scenarios.

There are many common themes that emerge in this literature that provide guid-ance on some considerations safety science scholars may take into account if they turn their attentions to researching top management. In the following, we add to these some reflections on our experiences as sociologists interviewing top managers in the oil and gas sector.

REFLECTIONS ON STUDYING TOP MANAGERS IN THE OIL AND GAS SECTOR

Over the last decade, our research program has involved research projects focused on senior management (Hopkins & Maslen, 2015; Maslen & Hopkins, 2014), and a series of studies on senior engineering professionals (Hayes, 2015a, 2015b; Hayes & Maslen, 2015; Maslen, 2014, 2015; Maslen & Hayes, 2014, 2016; Maslen & Ransan-Cooper, 2017). This work involves continuous engagement with the industry beyond research participation to facilitate impact.

One advantage of working in the safety field is that industry leaders generally understand the importance of the subject even though we sometimes define what safety means in different ways, and may have different approaches to maintaining safety. Safety, as a subject, gets us part way in answering the "why bother" question. In this context, in contrast to the viewpoint of the literature described earlier, we mostly do not have a lot of difficulty in getting top management engaged, at least momentarily, in our research plans. However, the details of what research they see as valuable, and so worth participating in, are not always straightforward.

The nature of social science can seem intangible to technical professionals. Requests for interview are sometimes met with offers to send us the relevant pro-cedures, with interview requests turned down on the basis that there is nothing to study. This applies most commonly to requests made to top management to study those at a lower level in their organisation. Recruitment for Hayes' work on opera-tional decision-making (Hayes, 2013) encountered this issue when senior manage-ment gatekeepers in some organisations insisted that the operational personnel made no important decisions, but rather simply followed operating procedures. This and other similar experiences demonstrate that some managers do not understand the difference between what is written in procedures and what happens in practice. Sometimes we need to accept that doors will remain closed for this reason, at least in the short term.

The underlying problem is perhaps that neither top managers nor engineers know how to articulate social science research problems. Despite broad awareness of Swiss cheese (Reason, 1997) and other organisational safety models based on the assumption that individual actions are driven by circumstances, safety problems are still often articulated as a need for behavioural change on the part of some other group of people (operators, contractors, etc.) and the belief that behavioural change is triggered by increased awareness. In the safety context, this often translates into a research scope seeking to understand how to make people more aware, rather than understanding what motivates them to behave in a particular way. We have seen this as the industry starting point in research linked to third-party work around pipelines where engagement with those in the field has shown that the primary problem is contract structures and risk shifting, not lack of hazard awareness (McDermott & Hayes, 2018; McDermott, Henne & Hayes, 2017). In a similar vein, we have received feedback that our research is "just what people say and what they say often isn't right." This person has failed to understand that actions on safety matters, as everything else, are driven by what people believe, not objective truth. In their frame of reference, the research lacks rigour and is, at best, anecdotal.

A practical problem is that companies, and even top managers, operate in the short term and tend to subsume all available resources into fixing immediate problems rather than addressing strategic improvements. This appears to explain why companies sometimes try to use us as consultants, expecting us to give advice regarding incremental improvements on specific transactional problems. On occasions we have come to specific arrangements with participating managers, such as providing an in-house training seminar in exchange for research access. Some managers understand that such a quid pro quo is not quite what it seems. One experience epitomises this: after a week at a company conducting research, interspersed with several pseudo-consultancy meetings, as we were leaving, the contact at the organisation thanked us by saying that we were valuable because they were so busy that they "outsource thinking" (Hopkins & Maslen, 2015). Everyone is under time pressure, and so academics can offer a capacity for sustained attention to problems that are simply too difficult to answer given internal pressures (to say nothing of know-how). As an interesting point of reference, a recent study on research impact suggests that the time to impact for academic research is highly variable but for biomedical and health sciences has been estimated at an average of 17 years (Jones, Castle-Clarke, Manville, Gunashekar & Grant, 2013). This again emphasises the very different time horizons for research as opposed to other management priorities.

Another example of the value of the slower pace of research-based activity has been some "Chatham House Rule" sessions (whereby the researcher is free to draw insights on safety from the sessions themselves, provided individual and organisational anonymity is maintained in later discussion) with senior management teams about the organisational causes of a major accident and the lessons that might be learned by their company.[1] This is beyond the scope of a regular researcher–participant relationship and has the researcher almost acting simultaneously as a workplace therapist (offering time to reflect, as Empsom (2018) says). Ultimately, successful research engagement with top management also often requires researchers to participate in activities beyond data collection and research dissemination. "Chatham House Rule" sessions also require a longer-term commitment beyond the normal relationship, because it takes time to build this trust.

Ethics in human research with professionals are also typically grounded in considerations of confidentiality. For those lower in the organisational hierarchy, the main risk of research participation may be the potential for sanctions from their employer if adverse remarks are reported in a way that identifies the source. When it comes to top managers, confidentiality considerations are different and typically revolve around organisational reputation and/or commercial advantage since they identify much more strongly as owning the identity of the organisation itself. We have dealt with the first issue by careful anonymising of cases, including giving interviewees rights of review of research material to be publicly reported. The second issue returns again to the question of the value of research, but in this case in the context of common good. If a company is not prepared to share with others what they consider to be their best practices, how is any progress to be made? In our experience, some senior managers hesitate over this issue (often driven by the legal department) and some decline to participate for this reason despite our best attempts to persuade.

In terms of the practicalities of data collection once in the interview, previous researchers have described needing to become like the professionals they study. In Odendahl and Shaw's (2002, p. 314) words, the interview should be almost like an "exchange between peers." A lot of the literature focuses on rapport and suggests that the only way to get good data is to have this good rapport, which is mostly founded on a perception that the interviewee and interviewer are not so different in terms of their ways (through language, dress and so on). This accords with our general experience, but it is important to never lose sight of the ultimate goal of engagement. Work to establish rapport, yes, but do not prioritise rapport to the point of avoiding difficult questions or the purpose of the research (Ostrander, 1993).

When it comes to establishing rapport, being junior is one barrier previously identified that has led to other roles such as the "intelligent daughter," which can be a useful role for female researchers seeking to operate in a male-dominated environment which may have sexual overtones. Adopting such a persona puts the interviewee in a teaching role and renders the researcher sexually taboo (Empson, 2018). A version of this role was assumed by one of the authors (Maslen), who was junior and lacked a shared professional background. In some cases, this role has resulted in excellent data, even from hostile participants: they have spelt things out so someone inferior can understand. Asking interviewees to explain fundamentals can sit more easily with this role than when an interviewer is seen as a professional peer.

In summary, maximising research impact requires a long-term view and this is nowhere more apparent than in dealing with senior management. We encourage researchers to work in this space where possible and to take the time to build relationships, even if that means starting by assisting those at the top to address their organisation's short-term needs on a confidential basis (i.e. working as an adviser or consultant, rather than a researcher). Researchers need to be able to cope with multiple roles and to develop clear signals to indicate to industry actors when roles are changing. We have found formal university ethics documentation, such as interview consent forms, particularly useful in this context to flag a change from informal conversation with a colleague, to formal research data collection.

When it comes to research method, we have necessarily focused on elite interviewing since that is both our primary experience and the method of engagement with elites that is most commonly discussed in the organisational literature. Thinking more broadly about research methods, we could aspire to one day conducting workplace observations, not by sitting in a control room but by sitting in a CEO's office or Board room. The insights into safety practices at elite levels and lessons regarding best practice would be significant but such a level of access would be unprecedented and is only in our imagination for now.

CONCLUSION

"Studying up" is a known challenge in social science research with many disciplines grappling with the challenge over decades. In safety science, we have largely failed to address this issue and continued to study frontline personnel, despite a theoretical focus on the actions of senior managers. If we collectively fail to remedy the current disconnect between safety science theory and empirical practice, the discipline risks becoming irrelevant and yet this does not need to be the case. There are lessons to be drawn from broader social science research regarding how to deal with negotiating access, unequal power relationships and unexpected interview dynamics.

With the rise of managerialism and ever more complex systems, those at the top are rarely the "maestros of technology" that Squires' famous study (1986) proposed should be in charge of development of major engineering projects. Varying professional backgrounds across the organisational hierarchy again shine a light on the need to study senior manager professional practice directly. Organisational failures are persistent but still relatively rare. Further research in the high reliability normal operations tradition about common features of the practices of top managers in high performing organisations, and what makes them different from top managers in organisations that fail, is urgently needed.

We are not alone in encouraging such a change in focus of safety research. Other authors (Antonsen, 2009; Dekker & Nyce, 2014) have pointed out that safety research has failed to address issues of power within organisations. Given that top managers are the most powerful organisational actors, our proposed research agenda is consistent with a drive to address power. As we have shown, the safety research community can do well to draw on lessons from organisation studies in other disciplines to most effectively develop methods for addressing a major gap in our corpus of knowledge. Our own experience shows that, despite the inherent challenges, engaging with senior management is worth the bother – for them and for the researchers involved.

ACKNOWLEDGEMENTS

Aspects of this work were funded by the Energy Pipelines Cooperative Research Centre, supported through the Australian Government's Cooperative Research Centres Program. The cash and in-kind support from the Australian Pipeline Industry Association Research and Standards Committee is gratefully acknowledged.

NOTE

1. See https://www.chathamhouse.org/chatham-house-rule

REFERENCES

Alvesson, M. (2001). Knowledge work: Ambiguity, image and identity. *Human Relations*, *54*(7), 863–886. doi:10.1177/0018726701547004
Antonsen, S. (2009). Safety culture and the issue of power. *Safety Science, 47*, 183–191.

Beamish, T. (2002). *Silent Spill, The Organization of an Industrial Crisis*. Cambridge, MA: MIT Press.

Bonafede, M., Corfiati, M., Gagliardi, D., Boccuni, F., Ronchetti, M., Valenti, A., ... Iavicoli, S. (2016). OHS management and employers' perception: Differences by firm size in a large Italian company survey. *Safety Science*, *89*, 11–18. doi:10.1016/j.ssci.2016.05.012

Bourrier, M. (2002). Bridging research and practice: The challenge of "normal operations" studies. *Journal of Contingencies and Crisis Management*, *10*(4), 173–180.

Coleman, S. (1996). Obstacles and opportunities in access to professional work organizations for long-term fieldwork: The case of Japanese laboratories. *Human Organization*, *55*(3), 334–343. doi:10.17730/humo.55.3.138mh1588247263r

Conti, J. A., & O'Neil, M. (2007). Studying power: Qualitative methods and the global elite. *Qualitative Research*, *7*(1), 63–82. doi:10.1177/1468794107071421

Dekker, S. W. A., & Nyce, J. M. (2014). There is safety in power, or power in safety. *Safety Science*, *67*, 44–49.

Dexter, L. (1970). *Elite and Specialized Interviewing*. Evanston, IL: Northwestern University Press.

Dezalay, Y., & Garth, B. G. (1996). *Dealing in Virtue: International Commercial Arbitration and the Construction of a Transnational Legal Order*. Chicago, IL: University of Chicago Press.

Empson, L. (2018). Elite interviewing in professional organizations. *Journal of Professions and Organization*, *5*(1), 58–69. doi:10.1093/jpo/jox010

Forsythe, D. (2001). *Studying Those Who Study Us: An Anthropologist in the World of Artificial Intelligence*. Stanford, CA: Stanford University Press.

Fruhen, L. S., & Flin, R. (2016). "Chronic unease" for safety in senior managers: An interview study of its components, behaviours and consequences. *Journal of Risk Research*, *19*(5), 645–663.

Gerbec, M., & Kontic, B. (2017). Safety related key performance indicators for securing long-term business development – A case study. *Safety Science*, *98*, 77–88.

Greenwood, D. J., Whyte, W. F., & Harkavy, I. (1993). Participatory action research as a process and as a goal. *Human Relations*, *46*(2), 175–192. doi:10.1177/001872679304600203

Grey, C., Anderson-Gough, F., & Robson, K. (1998). *Making Up Accountants: The Organizational and Professional Socialization of Trainee Chartered Accountants*. London: Ashgate.

Hale, A. (2000). Culture's confusions. *Safety Science*, *34*, 1–14.

Harvey, W. S. (2011). Strategies for conducting elite interviews. *Qualitative Research*, *11*(4), 431–441. doi:10.1177/1468794111404329

Hayes, J. (2012). Operator competence and capacity – Lessons from the Montara blowout. *Safety Science*, *50*(3), 563–574.

Hayes, J. (2013). *Operational Decision-Making in High-Hazard Organizations: Drawing a Line in the Sand*. Farnham: Ashgate.

Hayes, J. (2015a). Investigating design office dynamics that support safe design. *Safety Science*, *78*, 25–34.

Hayes, J. (2015b). Taking responsibility for public safety: How engineers seek to minimise disaster incubation in design of hazardous facilities. *Safety Science*, *77*, 48–56.

Hayes, J., & Hopkins, A. (2014). *Nightmare Pipeline Failures: Fantasy Planning, Black Swans and Integrity Management*. Sydney: CCH.

Hayes, J., & Maslen, S. (2015). Knowing stories that matter: Learning for effective safety decision-making. *Journal of Risk Research*, *18*(6), 714–726.

Hopkins, A. (1999). *Managing Major Hazards: The Lessons of the Moura Mine Disaster*. Sydney: Allen & Unwin.

Hopkins, A. (2008). *Failure to Learn: The BP Texas City Refinery disaster*. Sydney: CCH.

Hopkins, A. (2012). *Disastrous Decisions: The Human and Organisational Causes of the Gulf of Mexico Blowout.* Sydney: CCH.

Hopkins, A., & Maslen, S. (2015). *Risky Rewards: How Company Bonuses Affect Safety.* Surrey: Ashgate.

Hudson, P. (2007). Implementing a safety culture in a major multi-national. *Safety Science, 45,* 697–722.

Isık, I. N., & Atasoylu, E. (2017). Occupational safety and health in North Cyprus: Evaluation of risk assessment. *Safety Science, 94,* 17–25.

Jackall, R. (1988). *Moral Mazes: The World of Corporate Managers.* New York: Oxford University Press.

Jones, M. M., Castle-Clarke, S., Manville, C., Gunashekar, S., & Grant, J. (2013). *Assessing Research Impact: An International Review of the Excellence in Innovation for Australia Trial.* Retrieved from Santa Monica, CA: https://www.rand.org/pubs/research_reports /RR278.html

Laurila, J. (1997). Promoting research access and informant rapport in corporate settings: Notes from research on a crisis company. *Scandinavian Journal of Management, 13*(4), 407–418. doi:10.1016/S0956-5221(97)00026-2

Maslen, S. (2014). Learning to prevent disaster: An investigation into methods for building safety knowledge among new engineers to the Australian gas pipeline industry. *Safety Science, 64,* 82–89.

Maslen, S. (2015). Organisational factors for learning in the Australian gas pipeline industry. *Journal of Risk Research, 18*(7), 896–909.

Maslen, S., & Hayes, J. (2014). Experts under the microscope: The Wivenhoe Dam case. *Environment Systems and Decisions, 34*(2), 183–193.

Maslen, S., & Hayes, J. (2016). Preventing black swans: Incident reporting systems as collective knowledge management. *Journal of Risk Research, 19* (10), 1246–1260. doi:10.10 80/13669877.2015.1057204

Maslen, S., & Hopkins, A. (2014). Do incentives work? A qualitative study of managers' motivations in hazardous industries. *Safety Science, 70,* 419–428.

Maslen, S., & Ransan-Cooper, H. (2017). Safety framing and compliance in relation to standards: Experience from the Australian gas pipeline industry. *Safety Science, 94,* 52–60.

McDermott, V., & Hayes, J. (2018). Risk shifting and disorganization in multi-tier contracting chains: The implications for public safety. *Safety Science, 106,* 263–272. doi:10.1016/j. ssci.2016.11.018

McDermott, V., Henne, K., & Hayes, J. (2018). Shifting risk to the frontline: Case studies in different contract working environments. *Journal of Risk Research, 21* (12), 1502–1516.

McDowell, L. (1998). Elites in the city of London: Some methodological considerations. *Environment and Planning A: Economy and Space, 30*(12), 2133–2146. doi:10.1068/ a302133

Mikecz, R. (2012). Interviewing Elites: Addressing methodological issues. *Qualitative Inquiry, 18*(6), 482–493. doi:10.1177/1077800412442818

Mills, C. W. (1956). *White Collar: The American Middle Classes.* New York: Oxford University Press.

Mills, C. W. (1957). *The Power Elite.* New York: Oxford University Press.

Mills, C. W. (1962). *Power, Politics, and People: The Collected Essays of C. Wright Mills.* New York: Ballantine Books.

Monahan, T., & Fisher, J. A. (2015). Strategies for obtaining access to secretive or guarded organizations. *Journal of Contemporary Ethnography, 44*(6), 709–736. doi:10.1177/0891241614549834

Nader, L. (1972). Up the anthropologist: Perspectives gained from studying up. In D. Hymes (Ed.), *Reinventing Anthropology* (pp. 284–311). New York: Pantheon Books.

Odendahl, T., & Shaw, A. (2002). Interviewing Elites. In J. Gubrium & J. Holstein (Eds.), *Handbook of Interview Research: Context and Method* (pp. 299–331). Thousand Oaks, CA: SAGE.

Ostrander, S. A. (1993). "SURELY YOU'RE NOT IN THIS JUST TO BE HELPFUL": Access, Rapport, and Interviews in Three Studies of Elites. *Journal of Contemporary Ethnography, 22*(1), 7–27. doi:10.1177/089124193022001002

Pidgeon, N., & O'Leary, M. (2000). Man-made disasters: Why technology and organizations (sometimes) fail. *Safety Science, 34*, 15–30.

Pilbeam, C., Doherty, N., Davidson, R., & Denyer, D. (2016). Safety leadership practices for organizational safety compliance: Developing a research agenda from a review of the literature. *Safety Science, 86*, 110–121.

Quinlan, M. (2014). *Ten Pathways to Death and Disaster: Learning from Fatal Incidents in Mines and Other High Hazard Workplaces.* Sydney: Federation Press.

Reason, J. (1997). *Managing the Risks of Organizational Accidents.* Aldershot: Ashgate.

Richards, D. (1996). Elite interviewing: Approaches and pitfalls. *Politics, 16*(3), 199–204. doi:10.1111/j.1467-9256.1996.tb00039.x

Robson, L. S., III, B. C. A., Moser, C., Pagell, M., Mansfield, E., Shannon, H. S., ... South, H. (2016). Important factors in common among organizations making large improvement in OHS performance: Results of an exploratory multiple case study. *Safety Science, 86*, 211–227.

Schein, E. (1992). *Organizational Culture and Leadership* (2nd ed.). San Francisco, CA: Jossey-Bass.

Schoenberger, E. (1991). The corporate interview as a research method in economic geography. *The Professional Geographer, 43*(2), 180–189. doi:10.1111/j.0033-0124.1991.00180.x

Schwartzman, H. B. (1993). *Ethnography in Organizations.* Newbury Park, CA: SAGE.

Shirali, G. A., Salehi, V., Savari, R., & Ahmadiangali, K. (2018). Investigating the effectiveness of safety costs on productivity and quality enhancement by means of a quantitative approach. *Safety Science, 103*, 316–322.

Sklair, L. (2001). *The Transnational Capitalist Class.* Oxford: Blackwell.

Smith, T. D., Eldridge, F., & DeJoy, D. M. (2016). Safety-specific transformational and passive leadership influences on firefighter safety climate perceptions and safety behavior outcomes. *Safety Science, 86*, 92–97.

Snook, S. A. (2000). *Friendly Fire: The Accidental Shootdown of US Black Hawks over Northern Iraq.* Princeton, NJ: Princeton University Press.

Squires, A. M. (1986). *The Tender Ship: Governmental Management of Technological Change.* Boston, MA: Springer Science+Business Media, LLC.

Stephens, N. (2007). Collecting data from elites and ultra elites: Telephone and face-to-face interviews with macroeconomists. *Qualitative Research, 7*(2), 203–216. doi:10.1177/1468794107076020

Thomas, R. (1993). Interviewing important people in big companies. *Journal of Contemporary Ethnography, 22*(1), 80–96. doi:10.1177/089124193022001006

Twaalfhoven, S. F. M., & Kortleven, W. J. (2016). The corporate quest for zero accidents: A case study into the response to safety transgressions in the industrial sector. *Safety Science, 86*, 57–68.

Vaughan, D. (1996). *The Challenger Launch Decision: Risky Technology, Culture and Deviance at NASA.* Chicago, IL: University of Chicago Press.

Waring, A. (2015). Managerial and non-technical factors in the development of human-created disasters: A review and research agenda. *Safety Science, 79*, 254–267.

Weick, K. E., & Sutcliffe, K. M. (2001). *Managing the Unexpected: Assuring High Performance in an Age of Complexity.* San Francisco, CA: Jossey-Bass.

Young, S. A., & Blitvich, J. (2018). Safety in hard times – A qualitative analysis of safety concerns in two industrial plants under financial duress. *Safety Science*, *102*, 118–124.

Zwetsloot, G. I. J. M., Kines, P., Ruotsala, R., Drupsteen, L., & Merivirta, M.-L. (2017). The importance of commitment, communication, culture and learning for the implementation of the Zero Accident Vision in 27 companies in Europe. *Safety Science*, *96*, 22–32.

13 Politics, Accident Research and Analysis

The Evolution of Investigation Methods and Practices in Healthcare

Siri Wiig, Jeffrey Braithwaite and Geir Sverre Braut

CONTENTS

INTRODUCTION

The aim of this chapter is to describe and discuss the development and use of methods in healthcare safety investigations of the most severe adverse events. By using the Norwegian healthcare system as a case study, we concentrate on the political and public incentives that have been a driving force to change the existing system, and explore the changes that have emerged in the use of analytical methods and practices in response. We examine how the investigation methods and practices by the supervisory authorities have changed, from pure legal methods of investigation to integration of qualitative methods, in order to understand the development and complex causality of healthcare disasters. We also draw on examples from the safety investigations of high-profile Norwegian cases of the Daniel case and the Benjamin case. The findings are discussed in the context of Turner's Man-Made-Disaster theory.

INCUBATION OF HEALTHCARE DISASTERS

The incubation of internationally well-known healthcare scandals, such as the Mid Staffordshire and the Bristol cases in England (Francis, 2015; Kennedy, 2001), have demonstrated that non-prudent and ineffective healthcare practices may develop and continue over years; this in turn leads to substantial patient harm and a large number of casualties (Braithwaite et al., 2015; Hollnagel et al., 2015; Francis, 2013; Mannion & Davies, 2015; Wiig & Braut, 2018). These healthcare disasters have developed due to lacking or insufficient reactions to warning signals, neglect of outside complaints and whistle-blowing, multiple information-handling difficulties, poor or toxic cultures and failure to comply with regulations and clinical standards. Moreover, there has been a tendency to minimise emergent danger in patient care, neglecting the normalisation of deviance and a cultural acceptance of sub-standard care (Francis, 2015; Kennedy, 2001; Vincent & Amalberti, 2016; Aase & Rosness, 2017), all of which echo Barry Turner's theory about how disasters incubate over time due to human and organisational arrangements and processes in sociotechnical systems (Turner 1976; Pidgeon & O'Leary, 2000).

The Man-Made Disaster theory (Turner, 1976) proposes that accidents and disasters develop over a long chain of events that are undetected due to lack of information flow and lack of understanding of potential risk (see also Gherardi, Foreword and Hopkins, Chapter 1, both this volume). This chain of events develops unnoticed as a result of a cultural failing to detect the warning signals. Leaving aside any philosophical problems about chain logic and working across boundaries (Braithwaite, Hollnagel & Hunte, 2019), the model explains disaster development in six stages. It starts with a "normal" situation, based on a cultural belief that the organisation has the overview of the institution as a whole and is in an acceptable state regarding present and future hazards. Next is the incubation period, characterised by the accumulation of unnoticed events and misperception of signals and decoy phenomena, where action is taken on signals that, in hindsight, diverted attention from the real problems. The incubation period ends when a precipitating event happens, in stage three. A precipitating event is a dramatic event, such as a fire, wrong-site surgery or patient deaths, demonstrating the erroneous cultural beliefs about risk control in the organisation. The onset follows, in stage four, where the direct and unanticipated consequences manifest themselves, such as a number of deaths, an oil spill or construction damage. The onset is closely related to rescue and salvage, stage five, where participants make rapid and ad hoc redefinitions of situations in order to respond. The sixth stage ends with a cultural adjustment, where an inquiry is conducted and cultural beliefs and norms are adjusted to fit the new conceptualisation of the world (Turner, 1976; Rosness et al., 2010).

Analysis of the most severe events in healthcare often reveals that precursors were misunderstood, missed or dismissed by professionals, managers and even by organisations with a mandate to monitor quality and safety (Macrae, 2014; Macrae & Vincent, 2014; Wiig et al., 2018; Wiig & Braut, 2018). This has also been the case in Norway, where some high-profile and severe adverse events (e.g., the Daniel case, the Benjamin case (Helsetilsynet, 2011; 2015)) have initiated a public plea for advancing healthcare safety investigation methods, requiring a repertoire beyond the traditionally applied legal methods. In order to increase understanding of the complex and multifactorial causality in the increasing number of healthcare organisational failures, there has been a call to incorporate principles from qualitative methods, accident theory and user involvement into healthcare safety investigations (NOU, 2015; Wiig et al., 2018; Wiig & Braut, 2018). There has been political and public pressure to conduct safety investigations in healthcare, not only in new ways but also by new bodies with learning purposes only (NOU, 2015; Wiig et al., 2018; Wiig & Braut, 2018; Wiig & Macrae, 2018). In this chapter we analyse this development.

THE NEED FOR METHODS INNOVATION IN HEALTHCARE INVESTIGATIONS

The rationale for methods innovation is supported by recent research and academic debate in the healthcare and safety science communities (Vincent et al., 2017; Hulme et al., 2019; Wiig & Lindøe, 2009). A systematic review of systems thinking in accident analysis (Hulme et al., 2019) showed that despite arguing for systems thinking, the investigations mainly identified contributing factors as mono-causal, from the sharp end of the sociotechnical system. This was most likely a result of the information and data available to support the analysis, authors argued, and concluded that researchers in general should search for opportunities to develop novel approaches that incorporate key system thinking principles in accident causation. In healthcare, Vincent et al. (2017) have also argued in favour of major changes in safety investigations. The time frame should widen and cover the entire patient journey, and collaboration with patients and families is desirable to identify and prioritise safety issues. Fewer and deeper analyses are suggested, as healthcare is characterised by such a high number of events. Investigating all events with sufficient depth is too resource-demanding. Other suggestions emphasise investigation into both success and failure in addition to exploring influencing factors at different intervals, as patients may deteriorate due to poor care at different points and stages in the care pathway over time (Vincent et al., 2017; Vincent & Amalberti, 2016). Safety recommendations also need to target wider interventions than the traditional people-focused methods and, to a much larger degree, deal with the fundamental underlying safety problems (Vincent et al., 2017).

CONTEXT OF THE NORWEGIAN INVESTIGATORY SYSTEM OF THE MOST SEVERE ADVERSE EVENT

The Norwegian Board of Health Supervision (NBHS) is the national regulatory body for health and care services in Norway, established as an autonomous governmental body by law. It has been responsible for government investigation of the most severe adverse events in healthcare for several decades. From the mid-1980s, there has been a substantial rise in the number of cases investigated. The NBHS's role was strengthened when the Act on patients' rights was enforced in 2001, giving patients a right to get their case evaluated by the NBHS (Act, 1999a). The scope then is to assess if there is a breach of any legal requirements relating to provision of healthcare, not least the requirement to practice according to sound professional practice (as related to health personnel and service provision by institutions) (Act, 1999b; Act, 1999c; Act, 2011; Braut, 2011: 139).

Most of the cases are investigated at a regional level by the offices of the county governors. In 2003, the formerly independent office of the chief county medical officer was merged with the county governor's office. The cases where formal reactions towards health personnel are involved are now transferred to the central office of NBHS for final decisions. The cases where purely system deficiencies are considered will mostly be completed at regional level, according to procedures laid down by the central office of NBHS.

NBHS is an inspection body with sanctioning power against healthcare professionals and organisations. In a sample of the most severe adverse events, the NBHS conducts on-site visits to collect information for the regulatory follow-up. This type of inspection activity was implemented as a pilot project in 2010 and established by law in 2012. An inspection report is published when the regulatory investigation is completed. Still, the majority of the most severe cases are handled by the county governors, at the regional level.

In 2017, the Accident Investigation Board for the Health and Care Norway (AIBHCN) services was established by law (Act 2017). This is an independent and third-party investigatory body that undertakes regular learning-focused and system-wide investigations into serious patient safety issues across the healthcare system, similar to the Healthcare Safety Investigation Branch in England (Wiig & Macrae, 2018). The AIBHCN organisation of 20 investigators has been in operation since May 2019. No investigation report had yet been issued at the time of the data collection and analysis.

METHOD

This chapter presents a case study (Yin, 2014) of the development of the Norwegian healthcare safety investigation regime of the most severe adverse events. It is based on a qualitative thematic content analysis (Pope et al., 2006) of national policy documents such as reports to Parliament, white papers, public hearing documents and selected high-profile investigation reports from the NBHS. Case reports from the Daniel and the Benjamin case are used as data in our study, in addition to the NBHS's annual reports from the unit conducting the inspection activities (these are related to the most severe adverse events).

The analysis of policy documents includes a Norwegian Official Report (NOU) from a law commission (Arianson-commission) with a mandate to "Follow-up of serious adverse events and suspected legal violations in the health and care services" (NOU, 2015: p.14), including an assessment of positives and negatives, associated with an independent investigation board for the health and care services. Moreover, we include an analysis of a Parliament Act, with its preparatory documents on "The National Investigation Board for the Health and Care Services" (Act, 2017; Prop. 68 L, 2016–2017). Finally, four reports to the Parliament on patient safety and quality are included in the data material (Meld. St. 10 (2012-2013); Meld. St. 11 (2014-2015); Meld. St. 12 (2015-2016), Meld. St. 13 (2016-2017).

The first and third authors read the materials. They then analysed the content, producing the Findings section, developing themes and commentary centred on Norwegian healthcare safety investigation authority development, including political and public engagement. Then, two cases were extracted and synthesised from the materials. The data were assessed for trends in Norwegian accident investigation, and an analysis of investigation methods and competence was documented. All the data, including clinical data concerning the two specific case reports, are publicly available and downloaded from the Internet. No ethical approvals or informed consent were required for the study.

FINDINGS[1]

POLITICAL AND PUBLIC ENGAGEMENT TO CHANGE ANALYTICAL METHODS

The findings reveal an increasing attention from the public, including patients, relatives and next of kin organised groups, and politicians who aim to emphasise · patient and system perspectives in accident investigation. Since 2014, the political engagement has been supported by the Minister of Health's presentation of an annual patient safety and quality report to the Parliament. Key topics in these reports are user involvement, a culture and system for openness, and the role of leadership in quality and safety improvement (Meld. St. 10 (2012-2013); Meld. St. 11 (2014-2015); Meld. St. 12 (2015-2016), Meld. St. 13 (2016-2017).

These reports followed NHBS annual reports and reports from other relevant bodies such as the National Registration System of Adverse Events, the National Surveillance System of Hospital Infections and the Norwegian System of Patient

Injury Compensation. In addition, the Norwegian healthcare system has improved the instruments for gaining oversight of types and scope of adverse events, by, for example, establishing the opportunity for patient and next of kin to report their cases to the AIBHCN. The AIBHCN can choose which case to take on, and do not have to investigate all cases.

The Arianson-commission, nominated by the Cabinet elaborating NOU 2015, was divided in the conclusion relating to the law proposal of establishing the AIBHCN.[2] The majority faction opposed the proposal of an independent safety investigatory body, claiming that the resources should instead be used for strengthening activities related to safety improvement and investigation activities at an enterprise level, and on a local level. However, this majority still acknowledged that the findings from investigations, by an independent commission, would probably produce unique findings that differ from internal investigations and regulatory inquiries, thus providing new knowledge to healthcare providers and decision-makers.

The minority faction claimed that establishing an independent investigation board on a national level was a community responsibility that was aimed at involving patients and their relatives. According to this faction, the work of such a board could produce knowledge that other bodies (such as police and supervisory institutions) are not in a position to obtain, due to methodological constraints and sanctioning powers. This would stimulate new learning and practice change. The argument was that an independent investigation board can take a broader approach and a wider scope than traditional investigation systems. It can go deeper into the causal chains and multifactorial contributions to an accident, applying a more extensive set of methods and adopting a longer time frame than is usually accomplished by administrative and judicial bodies (NOU, 2015).

REGULATORY ANALYSIS OF ADVERSE EVENT

As of 2010, investigation reports from the NBHS show a gradual change in the analytical approach from an individual perspective to a system perspective, increasingly emphasising the learning aspects of the event investigations. The methodological approach in data collection also developed from being mainly oriented towards written information exchange (e.g. complaints, medical records, written statements from healthcare personnel and healthcare organisations, log data from medical technical equipment) to include methods also focusing on qualitative interviews with involved healthcare personnel and managers. In addition, steps were taken to include data collection from patients and next of kin, thereby incorporating their perspective into the investigation.

The Daniel Case

The Daniel case, in which a three-year-old child died after a routine tonsillectomy in 2009 (Helsetilsynet, 2015a, b), provides more rounded insights to this chapter's findings. The regulatory follow-up and police investigation were initially closed in 2010, but new information presented in the news media led to the regulatory case being reopened in 2014 (Wiig et al., 2018). Upon review, the Daniel case was characterised by the failure of the initial regulatory response and hospital follow-up. It was claimed that cover-up actions were performed in the aftermath, but also that there had been serious trouble with workplace culture and power struggles within the hospital organisation for years prior to the death of Daniel. During the reopening of the case, the NBHS and its subordinate officer (County Governor's office) were heavily criticised for using unduly narrow investigatory methods and for incompetence related to the investigation, as medical doctors were only provided with a limited orientation towards organisational and work environment factors (Aftenposten, 2015). The NBHS subsequently took advantage of new data collection methods to understand the case based on a broader mix of information from multiple stakeholders. Stakeholders that usually would not be involved in a legal investigation process were invited, adding new perspectives and fresh information to the case. The NBHS supplemented their data collection by way of face-to-face meetings with medical specialists, next of kin and family members, and conducted in-depth interviews with the involved personnel and managers. Helsetilsynet (NBHS) stated: "We have invited several of those who provided input to meet us and elaborate their points of view, to ensure we have a correct understanding of their view" (2015b:14).

Based on the additional information, the NBHS (Helsetilsynet, 2015b) changed parts of their conclusion in the regulatory follow-up and developed safety recommendation for system-wide learning, with an aim to prevent similar adverse tonsillectomy events. Prior to this safety investigation, the NBHS had never issued safety recommendations. The Daniel investigation also identified several preconditions for the severe adverse event, such as a negative psychosocial working environment, professional collaboration climate, hierarchies, technical equipment, misperception and miscommunications. In addition, the management did not meet the legal requirements of managing adverse events and neglected to perform a sound follow-up with the next of kin.

The Benjamin Case

The Benjamin case concerned a young man who had undergone knee surgery due to injury (NOU, 2015). During the night after surgery, he suffered great pain and was given several doses of an analgesic pharmaceutical (Ketorax). The following morning, Benjamin was found deceased. The relatives claimed that the hospital had not observed the patient properly. In the first instance, the supervisory authority did not initiate further investigations, but after critical questions from the parents, both the police and the supervisory authorities investigated the case further. This led to critical remarks being made towards the hospital by the supervisory authorities. The police decided to fine the hospital for not working in accordance with the legal requirements to provide sound professional care and practice. However, the parents claimed that several questions remained unanswered, even after the case ended. Among them were questions at the intersection of different legislations (working environment and health service provision). Questions related to the individual actions of involved personnel also seemed to be difficult to raise by the ordinary administrative and judicial systems, as they were not able to gain evidence in a way that avoided self-incrimination of the personnel involved, referring to the interpretation of article 6, number 1 in the European convention on human rights by the European Court of Human Rights (Council of Europe, 2019).

Amendments in Regulatory Safety Investigation Methods

Public debate regarding the two particular cases described above, combined with long-lasting intra-organisational discussions and developmental projects based on other cases, have led to amended investigation procedures and methods in the NBHS, with increasing communication now deemed essential with the patient and family throughout the investigation process. This is becoming increasingly evident when studying the procedures for investigations, and when comparing investigation reports from the last few years with reports from former periods. In addition, the NBHS has allocated funding for research and development projects at the regional level to develop methods and recommendations for user involvement in regulatory investigations (Wiig et al., 2019a, b; Fylkesmannen i Sør-Trøndelag, 2017; Helsetilsynet, 2019). This has resulted in an improved quality of legal investigation, due to additional information provided by the family of the affected patient (including after death in the most severe cases). Involving the family in face-to-face dialogue has extended data collection methods, contributed to a more holistic picture of the event and contributed information on human and organisational factors. These factors include identifying new actors and their role in the event, new information about the patient journey and a greater appreciation of preconditions prior to the active failures causing harm, disability or death (Wiig et al., 2019a, b).

IMPLEMENTING SYSTEM CHANGE TO ENHANCE SYSTEMIC LEARNING

Widening the Methods Repertoire

The document analysis demonstrated that since 2010 and the Benjamin case, strong arguments have been mobilised in favour of developing an independent investigation board for the most severe events in the Norwegian healthcare system. The Norwegian Official Commission (NOU, 2015) (Arianson-commission) developed a proposal for organising the new independent investigation board, including a new legal act for this body. The new act relating to "The National Investigation Board for the Health and Care Services" was approved by Parliament, on June 2, 2017, despite the public hearing showing major resistance to this proposal on the part of healthcare organisations. Notwithstanding the opposition, the Norwegian Minister of Health and Care Services and the majority of the Parliamentary members argued for, and voted in favour of, establishing the independent investigation board as a prerequisite for in-depth investigation solely for the purposes of learning (Prop. 68 L, 2016-2017).

One of the key arguments for establishing the new independent investigation board was the need for a broad investigatory approach to understanding the complex casual factors in healthcare disasters that goes beyond the current methods of regulatory follow-up, with a focus on deviation from regulation and poor professional practice. Moreover, the results focused on how diversity in investigative methods has the potential to generate richer information. Dialogue-based data collection and the application of appropriate accident theories can move us beyond a traditional linear cause–effect conceptualisation of how healthcare failures happen. Central to such an approach is the need for a system-wide perspective addressing multiple stakeholders' (e.g., regulators, hospital, general practitioner, nursing home and healthcare professionals) roles in the adverse event. This helps strengthen understanding of the creation of the contextual conditions under which healthcare professionals operate. A new, potentially important scrutiny of the role played by regulatory bodies, national policy makers, ministries and top managers is subsequently created, and the ability to observe patterns in accidents and adverse events over time and space is new to healthcare safety investigations in Norway. The principle of a no-blame investigation is identified as crucial in order to bring forward safety-critical information that should form the basis for system learning and improvement. Paramount to the new approach is to combine data collection methods with a safe space for patients, family members and healthcare personnel directly involved to share information. This in turn will allow for an improved analytical approach and will integrate all directly affected parties into subsequent processes (Prop. 68 L, 2016-2017).

Widening Investigatory Competence

The role of multi-disciplinary teams and competent investigation teams (e.g., have the ability to address risk management, have effective organisation and leadership, employ accident investigation methods and meet the legal requirements and working environment) are key enablers of improved investigatory practice. In addition, it is suggested that including staff from different healthcare professions is a prerequisite to operationalising the independent investigation board (Prop. 68 L, 2016-2017). This more diverse, richer approach differs significantly from the basic team competence and composition required in legal regulatory investigation, which is traditionally staffed with medical doctors and legal professionals only. The multi-disciplinary teams of the AIBHCN are equipped according to the law, with rights enabling the team to contact all involved healthcare personnel. Involved healthcare personnel cannot be sanctioned in any way by their employer, police or regulatory body, for sharing information with the AIBHCN. The AIBHCN also has access to all information sources from the regulatory bodies and the police who may have relevance for their independent investigation. The AIBHCN is autonomous in decisions about scope, depth, methodological approach and time frame in an investigation (Prop. 68 L, 2016-2017).

DISCUSSION

The rationale and principles underlying the development of adverse event investigation and analysis in Norway have strong links to Turner's theory (Turner, 1976, 1994; Turner & Pidgeon, 1997) and highlight how an understanding of complex adverse events relies on information from multiple sources (e.g., patients, next of kin, healthcare professionals, managers and inspectorates) at different system levels over time. These findings have a range of implications in light of Turner's work on the development and analysis of disasters in complex organisations.

Conceptualising Disaster Incubation in Healthcare

Our case study demonstrates that the investigation methods and analysis in the Norwegian healthcare system have developed to take into account complex causality of accidents and multiple contributing factors (such as working environment, culture, team-work, technical equipment and procedures). Central questions in this analytical approach concern if and how these factors constitute underlying preconditions for the incubation period of the events under investigation (Turner, 1976; Turner, 1994; Turner & Pidgeon, 1997). The need for investigation methods that are able to uncover preconditions and contributing factors beyond the initial triggers of disaster, combined with a need for improved learning from adverse events, have resulted in amended regulatory investigation procedures and the establishment of an independent, no-blame investigation board, similar to the model adopted in England in 2017 (Macrae & Vincent, 2017; Wiig & Macrae, 2018; Macrae 2019). In both Norway and England, these developments occurred because of the persistent efforts of the next of kin who had terrible experiences with the healthcare system (Macrae and Vincent, 2014; Wiig & Braut, 2018; NOU, 2015). They had seen healthcare scandals emerging over time, while the healthcare system's internal stakeholders appeared to be treating such situations as normal. Associated to this, there also appeared to be a prolonged lack of progress in implementing preventive measures, despite warnings from both concerned families and whistle-blowing from healthcare professionals. According to Turner's Man-Made Disaster model, this represents a disruption in cultural beliefs and norms about patient hazards and the actual state of the system (Turner's stage 1) (Turner, 1976). By analysing the healthcare system through the lens of Turner's Man-Made Disaster model, we can discern that changes in the Norwegian healthcare system over time emerged from an incubation period over years (Turner's stage 2), across which multiple healthcare events took place (including those illustrated in our case studies, discussed above). In addition, change resulted from a cultural collapse with a precipitating event of a high-profile newspaper story about the fundamental organisational failure behind the Daniel case (Turner's stage 3). The onset of this cultural collapse (Turner's stage 4) was followed by reopening the regulatory investigation, as the NBHS appeared to have no other option but to restart their investigation of a case they had actually closed five years earlier (Wiig et al., 2018). The Daniel case stands out in the Norwegian system. The regulatory body's immediate response and salvage (Turner's stage 5) included integration of broader investigation scope (additional parties invited), new data collection methods and ad hoc adjustments of regulatory investigation practice (meetings, interviews, document analysis over time). The political and public pressure led the healthcare system towards cultural readjustment (Turner's stage 6) where: (1) the regulator began issuing recommendations for improvement based on their legal safety investigations and (2) they established the new independent investigation board solely for the purpose of learning. Both readjustments correspond to a new world view of how a healthcare system should work, aligned with principles of learning and accountability, and one focused on patient and family involvement and improvement.

Promoting Requisite Imagination and Qualitative Methods in Safety Investigations

Our analysis highlights the need to develop skills in qualitative methods among investigators and establish multi-disciplinary investigation teams. There is a need for a diverse range of methodological approaches, scientific perspectives and investigatory skills in order to understand how and why accidents happen, both in research and practice (Wiig & Macrae, 2018; Aase & Braithwaite, 2017; Braithwaite et al., 2018). The developments described here are in line with Westrum's work on requisite imagination (Westrum, 2004; Rosness et al., 2010). Westrum's theory complements Turner's work, by investigating cultures of information flow in organisation. Westrum focuses on the importance of diversity and making use of different perspectives to identify and react to problems. No matter the position, rank or professional role of the participants involved, they and their organisations should take advantage of all information available. Multi-disciplinary and diverse teams are better placed to help understand, through multiple prisms, the situations encountered and the extent to which organisational problems can be avoided. We argue that requisite imagination is fundamental in safety investigations, to understand how and why organisations fail. Requisite imagination is relevant for both regulatory investigations and independent learning-focused investigations, as healthcare systems are so complex that a narrow medical or legal perspective is an insufficient basis for properly understanding the emergence of adverse events (Schaefer & Wiig, 2017).

Our results argue in favour of a varied methods repertoire and a broad set of perspectives in accident investigation. This equates with Martin and Turner's (1986:p155) position on how the contribution of grounded theory in organisational research can provide important components for the researcher's toolkit – offering ways of making sense of, and improving, organisational behaviours (see also Le Coze, this volume). By taking advantage of a range of data-gathering models such as qualitative interviews, focus groups, observations and face-to-face data collection methods, in addition to traditional written information exchange, the investigators' toolkits are far more comprehensive (Wiig et al., 2019a, b). This can feed into a richer understanding of how and why adverse events happen – and preventive measures can then be formulated (see also Kongsvik and Almklov, this volume). It remains an important, and only partly solved, question of how to become aware of those minor signals often preceding major calamities in the healthcare system. It is not clear whether investigation boards or reporting systems are able to detect malfunctioning service systems sufficiently early in an incubation period to take corrective action (Pidgeon & O'Leary, 2000).

There is still considerable underused potential in harnessing the variety of qualitative methods available to current safety investigations. As Rapport and colleagues (2019) pointed out, qualitative methods are expanding in scope and are gaining recognition internationally for their ability to illuminate processes of healthcare delivery. Qualitative methods supplement or complement causal explanations and offer a more nuanced understanding of care processes and patient experiences (see Martin, O'Hara and Waring, this volume). In line with Turner's model, we suggest safety investigations can usefully explore the potential benefits of qualitative methods beyond individual interviews, document analysis and meetings. There are benefits to be secured in triangulating methods (interviews, focus groups, observation), data sources (photo, videos, narratives, transcripts, medical records, documents) and analysts (healthcare professionals, patients, investigators) (Rapport et al., 2019; Patton, 2002).

In healthcare, major advances lie in utilising patient representatives as co-researchers. This is becoming more common, and investigation teams could experiment with this core idea by inviting patients as co-investigators similar to early experiments which have tested this with success in internal hospital investigations (Kok et al., 2018) and in regulatory investigations of adverse events (Adams et al., 2013). Patients and family input adds new information to any investigation. Methods harnessing focus groups, interviews and more active participation from potential informants, in workshops, or through nominal group techniques, for example, could also add new types of information and lead to more reflection on the range of problems that organisations have struggled with over time.

Ethnography and rapid ethnography represent another powerful yet underexploited approach in healthcare (Dixon-Woods, 2003; Cupit et al., 2018). Ethnographic research is well suited to identifying conditions of risk in organisations that involves human and organisational dynamics (see also Kuran, Chapter 7, this volume). A potential way forward would be to use ethnographic approaches for gaining insight into the everyday understanding staff and managers have of their organisation and to thereby search for latent conditions that may trigger healthcare failure (Dixon-Woods, 2003; Cupit et al., 2018; Braithwaite et al., 2015; Rapport et al., 2019). We acknowledge ethnography may be a challenging and time-consuming approach when investigating accidents. Nevertheless, investigators could consider a more proactive use of it to identify particularly interesting areas and topics for future investigation. Creativity in methods innovation should be encouraged in both regulatory and blame-free investigatory work.

POLICY MAKERS' PUSH FOR MANAGERS AS TARGETS FOR SYSTEM IMPROVEMENT

Policy documents and investigation practice have developed in a direction focusing increasingly on the role of culture, openness and in particular the managers' role in controlling risks. This can be related to a need to prevent "sloppy management" as a cause of disaster, in the way Turner critically conceptualised this in his work (Turner, 1994). Sloppy management may result in information being misunderstood or not communicated. Healthcare managers need to ask questions of themselves and those in their organisations. Do they have a realistic view of their organisation and do they allow staff to bring forward alternative views and interpretations of their operations (Turner, 1994)?

Going further, it is noteworthy that the public discourse on how to deal with adverse events is all-too-often ambiguous. On the one hand, it is frequently claimed that the main objective with investigation is learning for the future. On the other hand the question about accountability for existing systems soon comes up, thus raising the spectre of who is responsible and, often, who is to blame (Reason, 1997; Quick, 2017). Finding the balance of the individual and system perspectives is likely to be forever debated in healthcare investigations. In the Daniel and the Benjamin cases, for example, professional accountability and systems accounts were intermingled. That said, the acknowledgement that healthcare is a complex adaptive system and that humans operate within that complexity with degraded information that is variably disjunctive, and under conditions of bounded rationality (Turner and Pidgeon, 1997), and most often with good intentions but sometimes poor outcomes, is now much more commonly understood. This understanding has pushed the Norwegian system in the direction of conceptualising healthcare from a systems perspective, emphasising the role of everyone involved, but with a specific emphasis on managers to develop sound working conditions and to act diligently and with the appropriate levels of competence, within a sound safety culture.

This systems thinking approach is echoed in the major 2018 report, *Crossing the Global Quality Chasm: Improving the Quality of Health Care Globally* (National Academics of Sciences, Engineering, and Medicine, 2018) and is in line with Turner's ideas of the complex interaction between humans and organisations which are mutually constituted in disasters. The concept of the variable disjunction of information is central in Turner's theory and is, in our view, becoming increasingly more relevant for healthcare provision. When services are becoming more and more specialised and fragmented, with, for example, multiple handovers within and across care levels (Aase et al., 2017), the capacity for a large number of parties to have a mutual understanding of a patient safety problem is degraded. Differing conceptualisations of a problem can lead to adverse events developing because healthcare professionals have different pieces of information about what is happening, diverse perspectives on events and distinctive interpretations of even what the problem is – implying that it is increasingly possible for unanticipated events with unintended consequences to emerge (Pidgeon & O'Leary, 2000).

CONCLUSION

In this chapter we have analysed the development of analytical approaches in healthcare safety investigation, and the associated methods and practices, in light of Turner's work on disaster incubation. Healthcare disasters develop in different ways to industrial disasters, but the theoretical model provided by Turner has demonstrated explanatory power for healthcare systems and provides useful concepts, ideas and levels that explain change. Based on our analysis we suggest that healthcare investigators should nurture ideas of requisite imagination within their teams, and test new models of team composition, including involving patient and family representatives as co-investigators. We suggest experimenting with and exploiting a range of qualitative methods, such as ethnography and focus groups, to gain rich data and insight into organisational dynamics and work practices in complex contexts characterised by increasing specialisation of health professional roles and clinical work.

In the years to come we will hopefully develop new knowledge and better models and theories by which to understand the causal chains and complexities behind adverse events in healthcare. The challenges of how to use this knowledge for improvement purposes remain. At this point, when the investigation of an event is finalised, the different investigatory systems should optimally see it as an obligation to work together rather than operate as separate bodies, if the goal is to provide synergies, varied perspectives and richer learning across their investigations. This is likely to build greater levels of integrity and credibility into investigatory activities. Synthesising learning potential using a variety of methods across diverse investigation approaches and perspectives would most likely add to our potential to strengthen systems learning and improvement in many cases.

ACKNOWLEDGEMENTS

Authors would like to thank research assistant Meagan Warwick, Centre for Healthcare Resilience and Implementation Science, Australian Institute of Health Innovation, Macquarie University, Australia, for helpful comments and editorial support in finalising the manuscript.

NOTES

1. A preliminary report on this project was presented at the Working On Safety conference, Prague 2017, cfr. Wiig & Braut, (2018).
2. Siri Wiig and Geir Sverre Braut were both members of this commission. Wiig was part of the minority fraction, Braut of the majority fraction.

REFERENCES

Aase, K. & Braithwaite, J. (2017). What is the role of theory in research on patient safety and quality? Is there a Nordic perspective to theorizing? In: Aase, K. & Schibevaag, L. (eds.), *Researching patient safety and quality in health care: a Nordic perspective*. Boca Raton, FL: CRC Press, Taylor & Francis Group, pp. 57–74.

Aase, K. & Rosness, R. (2017). Organisatoriske ulykker og resiliente organisasjoner i helsetjenesten – ulike perspektiver. In: Aase, K. (Red), *Pasientsikkerhet – teori og praksis (Patient safety – theory and practice)*. Oslo, NO: Universitetsforlaget, pp. 27–48.

Aase, K., Waring, J. & Schibevaag, L. (eds.) (2017). *Researching quality and safety in care transitions: international perspectives*. Cham, CH: Palgrave MacMillan.

Act (1999a). Act of 2 July 1999, No. 63 relating to Patients' Rights (the Patient's Rights Act).

Act (1999b). Act of 2 July 1999, No. 64 relating to Health Personnel etc.

Act (1999c). Act of 2nd July 1999 relating to specialised health care.

Act (2011). Act of 24th June 2011 relating to municipal health and care services.

Act (2017). Act of 6th June 2017 relating to The National investigation board for the health and care services.

Adams, S.A., van de Bovenkamp, H. & Robben, P. (2013). Including citizens in institutional reviews: Expectations from the Dutch Healthcare Inspectorate. *Health Expectations* 18(5):1463–73. doi: 10.1111/hex.12126.

Aftenposten (2015). Norwegian hospitals can learn a lot from the Daniel case. http://www .aftenposten.no/nyheter/iriks/Professor---Mye-a-lare-av-Daniel-saken-for-norske -sykehus-8064285.html

Braithwaite, J., Churruca, K., Long, J.C., Ellis, L.A. & Herkes, J. (2018). When complexity science meets implementation science: A theoretical and empirical analysis of systems change. *BMC Medicine* 16:63. doi: 10.1186/s12916-018-1057-z.

Braithwaite, J., Hollnagel, E. & Hunte, G.S. (eds.) (2019) *Resilient health care volume 5: Working across boundaries.* Boca Raton, FL: Routledge. ISBN: 9780367224592.

Braithwaite, J., Wears, R.L. & Hollnagel, E. (2015). Resilient health care: Turning patient safety on its head. *International Journal of Quality in Health Care* 27(5):418–20. doi: 10.1093/intqhc/mzv063.

Braut, G.S. (2011). The requirement to practice in accordance with sound professional standards. In: Molven, O. & Ferkis, J. (eds.), *Healthcare, welfare and law. Health legislation as a mirror of the Norwegian welfare state.* Oslo, NO: Gyldendal Akademisk, pp. 139–149.

Council of Europe (2019). *Guide on article 6 of the European convention on human rights.* Strasbourg: Council of Europe & European Court of Human Rights. https://www.echr.coe.int/Documents/Guide_Art_6_criminal_ENG.pdf

Cupit, C., Mackintosh, N. & Armstrong, N. (2018). Using ethnography to study improving healthcare: Reflections on the "ethnographic" label. *BMJ Quality and Safety* 27(4):258–60. doi: 10.1136/bmjqs-2017-007599.

Dixon-Woods, M. (2003). What can ethnography do for quality and safety in health care? *BMJ Quality and Safety* 12:326–7. doi: 10.1136/qhc.12.5.326.

Francis, R. (2015). Freedom to speak up. London: TSO. http://webarchive.nationalarchives.gov.uk/20150218150343/https://freedomtospeakup.http://org.uk/wp-content/uploads/2014/07/F2SU_Executive-summary.pdf

Francis, R. (2013). *Report of the Mid Staffordshire NHS Foundation Trust Public Inquiry Executive summary.* London: Stationery Office.

Fylkesmannen i Sør-Trøndelag (2017). *Styrket involvering av pasienter, tjenestemottakere og pårørende i tilsyn.* Rapport. Fylkesmannen i Sør-Trøndelag, Trondheim.

Helsetilsynet (2011). System failure at Vestre Viken Hospital. https://www.helsetilsynet.no/historisk-arkiv/nyheter-og-pressemeldinger/nyheter-2004-2012/Systemsvikt-ved-Vestre-Viken-Ringerike-sykehus/

Helsetilsynet (2015). Overview page of the Daniel Case. https://www.helsetilsynet.no/tilsyn/tilsynssaker/daniel-saken-samleside/

Helsetilsynet (2015a). 4.6.2015. Investigation of adverse event – draft report sent to public hearing. https://www.helsetilsynet.no/no/Tilsyn/Tilsynssaker/Utkast-til-rapport-i-tilsynssak-dodsfall-etter-postoperative-komplikasjoner-etter-tonsillektomi/

Helsetilsynet (2015b). 18.11.2015. Final investigation report. https://www.helsetilsynet.no/upload/tilsyn/varsel_enhet/Danielsaken-endelig-rapport-nov-2015.pdf

Helsetilsynet (2019). Saman om betre tilsyn. Tilrådingar om brukarinvolvering i tilsyn. Rapport fra Helsetilsynet 2/2019. (Improving supervision together – Recommendation for user involvemen in supervision). https://www.helsetilsynet.no/publikasjoner/rapport-fra-helsetilsynet/2019/saman-om-betre-tilsyn-tilradingar-om-brukerinvolvering-i-tilsyn/

Hollnagel, E., Wears, R.L. & Braithwaite, J. (2015). *From safety-I to safety-II: A white paper.* University of Southern Denmark, University of Florida, USA, and Macquarie University, Australia.

Hulme, A, Stanton, N.A., Walker, G., Waterson, P. & Salmon, P. (2019). What do applications of systems thinking accident analysis methods tell us about accident causation? A systematic review of applications between 1990 and 2018. *Safety Science*, 117(August 2019):164–183.

Kennedy, I. (2001). *Learning from Bristol: The report of the public inquiry into children's heart surgery at the Bristol Royal Infirmary 1984–1995.* London: Stationary Office, Crown Copyright. ISBN: 0101536321.

Kok, J., Leistikov, I. & Bal, R. (2018). Patient and family engagement in incident investigations: Exploring hospital manager and incident investigators' experiences and challenges. *Journal of Health Services Research & Policy* 23(4):252–261. doi: 10.1177/1355819618788586.

Macrae, C. (2014). Early warnings, weak signals, and learning from healthcare disasters. *BMJ Quality and Safety* 23:440–445. doi: 10.1136/bmjqs-2013-002685.

Macrae, C. (2019). Investigating for improvement? Five strategies to ensure national patient safety investigations improve patient safety. *Journal of the Royal Society of Medicine* 22:1–5. doi:10.1177/0141076819848114.

Macrae, C. & Vincent, C. (2014). Learning from failure: The need for independent safety investigations in healthcare. *Journal of the Royal Society of Medicine* 107(11):439–43. doi: 10.1177/0141076814555939.

Macrae, C. & Vincent, C. (2017). A new national safety investigator for healthcare: The road ahead. *The Royal Society for Medicine* 110(3): 90–2. doi: 10.1177/0141076817694577.

Mannion, R. & Davies, H.T.O. (2015). Cultures of silence and cultures of voice: the role of whistleblowing in healthcare organizations. *International Journal of Health Policy Management* 4(8):503–5. doi: 10.15171/IJHPM.2015.120.

Martin, P.Y. & Turner, B. (1986). Grounded theory and organizational research. *Journal of Applied Behavioral Science* 22(2):141–57. doi: 10.1177/002188638602200207.

Meld. St. 10 (2012–2013). God kvalitet – trygge tjenester. Kvalitet og pasientsikkerhet i helse og omsorgstjenesten. *Det kongelige helse- og omsorgsdepartement,* Oslo.

Meld.St. 11 (2014–2015). Kvalitet og pasientsikkerhet. *Det kongelige helse- og omsorgsdepartement,* Oslo.

Meld.St. 12 (2015–2016). Kvalitet og pasientsikkerhet. *Det kongelige helse- og omsorgsdepartement,* Oslo.

Meld.St. 13 (2016–2017). Kvalitet og pasientsikkerhet. *Det kongelige helse- og omsorgsdepartement,* Oslo.

National Academies of Sciences, Engineering, and Medicine (2018). *Crossing the Global Quality Chasm: Improving Health Care Worldwide.* Washington, DC: National Academies Press. doi: 10.17226/25152.

NOU (2015:11). Med åpne kort. *DSS,* Oslo. https://www.regjeringen.no/contentassets/daa ed86b64c04f79a2790e87d8bb4576/no/pdfs/nou201520150011000dddpdfs.pdf

Patton, M. (2002). *Qualitative research and evaluation methods.* 3rd ed. Thousand Oaks, CA: SAGE.

Pidgeon, N. & O'Leary. (2000). Man-made disasters: why technology and organizations (sometimes) fail. *Safety Science* 34(1–3):15–30. doi: 10.1016/S0925-7535(00)00004-7.

Pope, C., Ziebland, S. & Mays, N. (2006). Analysing qualitative data. In: Pope, C. & Mays, N. (eds.), *Qualitative research in health care.* Oxford: Blackwell Publishing, pp. 63–81.

Prop.68 L (2016–2017). Proposisjon til Stortinget (forslag til lov). *Lov om Statens undersøkelseskommisjon for helse og omsorgstjenesten. Det kongelige helse og omsorgsdepartementet.* https://www.regjeringen.no/no/dokumenter/prop.-68-l-20162017/id2544823/sec2

Quick, O. (2017). *Regulating patient safety: The end of professional dominance?* Cambridge: Cambridge University Press.

Rapport, F., Hogden, A., Faris, M., Bierbaum, M., Clay-Williams, R., Long, J., Shih, P. & Braithwaite, J. (2019). *Qualitative research in healthcare, modern methods, clear translation – a white paper.* Sydney, NSW: Australian Institute of Health Innovation, Macquarie University.

Reason, J. (1997). *Managing the risks of organizational accidents.* Abingdon: Ashgate Publishing.

Rosness, R., Grøtan, T.O., Guttormsen, G., Herrera, I., Steiro, T., Størseth, F., Tinmannsvik, R.K. & Wærø, I. (2010). *Organizational accidents and resilient organizations. Six perspectives. Revision 2.* Trondheim: Sintef Technology and Society. ISBN: 978-82-14-05056-1.

Schaefer, C. & Wiig, S. (2017). Strategy and practise of external inspection in healthcare services – a Norwegian comparative case study. *BMC Safety in Health* 3:3. doi: 10.1186/s40886-017-0054-9.

Turner, B. (1994). Causes of disaster: Sloppy management. *British Journal of Management* 5:215–9. doi: 10.1111/j.1467-8551.1994.tb00172.x.

Turner, B. & Pidgeon, N. (1997). *Man-made disasters.* Oxford: Butterworth-Heinemann.

Turner, B.A. (1976). The organization and interorganizational development of disasters. *Administrative Science Quarterly* 21(3):378–97. doi: 10.2307/2391850.

Vincent, C. & Amalberti, R. (2016). *Safer healthcare.* London: Springer Open.

Vincent, C., Carthey, J., Macrae, C. & Amalberti, A. (2017). Safety analysis over time: Seven major changes to adverse event investigation. *Implementation Science* 12:151. doi: 10.1186/s13012-017-0695-4.

Westrum, R. (2004). A typology of organisational cultures. *BMJ Quality & Safety in Health Care* 13(Suppl II): ii22–7. doi: 10.1136/qshc.2003.009522.

Wiig, S., Bourrier, M., Aase, K. & Røise, O. (2018). Transparency in health care: Disclosing adverse events to the public. In: Bourrier, M. & Bieder, C. (eds.), *Risk communication for the future.* Cham: Springer Open, pp. 111–25.

Wiig, S. & Braut, G.S. (2018). Developments in analysis of severe adverse events in healthcare – policy and practice in Norway. In: Bernatik, A., Locurkova, L. & Jørgensen, K. (eds.), *Prevention of accidents at work.* London: CRC Press, pp. 39–45.

Wiig, S., Haraldseid-Driftland, C., Tvete Zachrisen, R., Hannisdal, E. & Schibevaag, L. (2019a). Next of kin involvement in regulatory investigation of adverse events that caused patient death: a process evaluation (part I – The next of kin's perspective). *Journal of Patient Safety.* Published ahead of print. doi: 10.1097/PTS.0000000000000630.

Wiig, S. & Lindøe, P.H. (2009). Patient safety in the interface between hospital and risk regulator. *Journal of Risk Research,* 12(3–4):411–27. doi: 10.1080/13669870902952879.

Wiig, S. & Macrae, C. (2018). Introducing national healthcare safety investigation bodies. *British Journal of Surgery,* 105:1710–2. doi: 10.1002/bjs.11033.

Wiig, S., Schibevaag, L. Tvete Zachrisen, R., Anderson, J.A., Hannisdal, E. & Haraldseid-Driftland, C. (2019b). Next of kin involvement in regulatory investigation of adverse events that caused patient death: A process evaluation (part II – the inspectors' perspective). *Journal of Patient Safety.* Published ahead of print. doi: 10.1097/PTS.0000000000000634.

Yin, R. (2014). *Case study research: Design and methods.* 5th ed. Thousand Oaks, CA: SAGE.

14 Exploring the Grey Area of Cyber Security in Industrial Control Systems (ICS)

A Qualitative Research Matter

Alberto Zanutto

CONTENTS

INTRODUCTION

For many years, companies have considered production, and the security of production sites, as separate domains. Production sites have historically needed to be secured against intruders with, for example, several kinds of fences for business secrecy. Technological changes and the development of new methods of remote control have forever changed the vision of organisations as productive spaces that need to be protected locally. Studies in this field have shown the conspicuous increase in the risks to which organisations are subject through the adoption of new technological paradigms, which can distabilise the equilibrium of organisations. As Macrae (2014) and Pettersen (2016) state, these changes can generate problems of "understanding" within organisations, due to the expansion of the required knowledge domains associated with the difficult work of re-articulating organisational systems and practices. At the same time, in recent decades, the context of security practices has changed. These practices can no longer be represented as an intervention only on "local" organisational complexities. These practices now confront the risks generated by complexity that emerges in the link between local and global security threats and working practices must continually include the possibility of cyber-attacks at a global level. Today, security practices are no longer separated from production practices in many industrial sites. Industries and manufacturing organisations must deal with an increasing number of attacks through globalised network communications that may affect each workplace and each work task. For example, in December 2015 in Ukraine, a cyber-attack targeting part of the national power grid caused large financial losses and also placed the public in danger. This event, and many others, shows how vulnerable local Industrial Control System (ICS) infrastructures have now become a site of global security threats.

Broader questions of security have progressively expanded and now touch every aspect of our daily lives. For this reason, acting on these domains of knowledge has become particularly challenging, and deeply integrated into other domains of practice. Bieder and Gould (2020: 6) remind us that "despite the number of years where safety and security have coexisted as approaches, there seems to be limited research on how safety and security are managed in practice at all levels." By focusing here specifically on security, this chapter considers a small but particularly interesting set of challenges in the industrial sector: all industrial environments and technologies have a long history of development, and these historical infrastructures and security practices are now part of the context within which a new era of security practices must be generated to respond to industrial transformations such as the advent of the "Internet of Things," where huge numbers of devices are interconnected and controlled remotely through ICS.

This specific field of observation allows us to highlight the blurred contours of the knowledge domains and the practices implemented by organisations. When an ICS is under attack, any protection or response is the result of complex collaboration between a variety of professionals and infrastructures using a variety of technological devices and systems. Professionals and technologies work closely together within organisations, often requiring external support from other expert professionals. This needs special attention to be paid to the heterogeneous sociotechnical practices and to actors' accounts. The emergence of new forms of security practices creates a "grey area" where many things happen: a space where technology and people, official representations and peculiarities of workplace practices are combined in a complex environment. How can these environments be studied? How can the sociotechnical complexity of these environments be revealed? Following Turner's (1983) approach, we emphasise how industrial organisations based on sociotechnical infrastructures for security are a particularly relevant research setting for a qualitative research programme in this field. Exploration of this domain can aid understanding of how security practices work in complex organisations and how practices confront uncertainty, hidden procedures and complex collective and cultural environments (Dourish and Anderson 2006; Dourish et al. 2004; Zanutto et al. 2017).

The chapter focuses on this new organisational task, suggesting how qualitative research practices can help in describing the ways in which a wide range of organisational actors are building their knowledge in the cyber security domain. The rest of this chapter offers an exploration of how technical infrastructures are socially shaped and maintained inside organisations and across security professionals, to reveal how cyber insecurity is inherent to everyday experiences within a wide range of organisations.

HANDLING CYBER THREATS: FOCUSING ON THE ORGANISATIONAL GREY AREAS

The work of Barry Turner (1976) helps to affirm that when a disaster happens, it is very often the result of disjunctive information managed by different and separate organisations. Individuals and groups operating in different knowledge layers can come to develop a variety of different interpretations of the same problem (Turner 1976) (see also Gherardi, Foreword and Hopkins, Chapter 1, both this volume). Disasters have shown the importance of technological failures but, as suggested by Weick, "technology is less of a problem than is the way people are organized" (1998: 74). One of the main issues concerning cyber security in ICS is the combination of the secrecy of data as well as limits on the disclosure of the work conducted by security teams. This secrecy can be amplified by the organisational context of modern industrial workplaces, where technical staff and operators often work alone, dealing with both everyday tasks and design complexities (Luff et al. 2000). This can create a "shadow space" where workers are faced with their tasks alone and in a hidden way. Star and Strauss' work (1999) highlighted this "grey area" to show that when workers are represented as non-persons – that is, when the background work and workers have been erased from their settings, or when everyday discourses about work are framed in abstract terms that use "indicators" and other indirect performance indicator – we are dealing with the "invisibility" of work. We should remember that the technological work done by software engineers, automation engineers, software designers, computer programmers, computer systems analysts and many others is in any case more or less invisible to the wider public, and very often to a company's wider workforce as well. Indeed, through the course of these studies many informants confirmed their experience that the work of cyber security is invisible not only outside companies but also inside. The work is used by managers, vendors, insurance companies and institutional players, but the nature of the underlying work can remain invisible to them. This feeds the idea that organisations often refer to their industrial plants and business goals, forgetting how the sensitive information is actually managed and handled with complex ICS. Often the management's position is to primarily pay special attention to the technological side of production (Busby & Bennet 2008).

As stated by Gherardi and colleagues (2017), these places where organisations permit actors to organise themselves is a "grey zone" (they call it "shadow," but with the same meaning) located in a "space and time of ambiguous definition and of mutable relations, as in the tidemark between the sea and the sand, or in the liminal space of the threshold" (2017: 8). This metaphor helps us consider security practices as an organisational issue that are deeply connected to "the entanglement of transparency and secrecy." Authors like Turner (1983) and Law (2012) have explored how this area of uncertainty plays a major role in the security domain: any action that aims to reduce uncertainty through imposing mandatory constraints or protocols can actually increase risk by leading to a restricted and narrow view of the associated security tasks. Law argues instead that "it will be important to tolerate ambiguity and non-coherence in those plural knowledges," and states that "no doubt our own non-coherences are more or less productive too as we tinker our way through our projects and practices" (Law 2012: 9). The aim in this chapter is to explore this tide-mark space "between the sea and the sand," where industrial cyber security takes place, paying close attention to how people act in these situations and how organisa-tions help (or do not help) them in coping with cyber security in these grey areas.

One way of analysing people and their often-hidden complex practices dealing with the uncertainty of cyber threats is to go through the practices of dealing with alerts, clues, and any anomalous signals coming from sociotechnical infrastructures. How could we explore these grey zones? Turner's advice (1983) was mainly aimed at considering workplace practices and other organisational practices related to safety.

> When studying a practice, we focus on how practitioners determine the nature of a situation, how they select effective and efficient techniques (materials, implements, bodily postures, methods, etc.), determine deviations from what has been assumed in the rule formulations, deal with routine problems, recover from breakdowns, etc. (Schmidt 2014: 9)

The peculiarity of security practices is that people are required to be able to deal with unknown quantities: the nature of attackers, their motivation, their technical means, their *modus operandi*. The public discourse about this domain is a reminder that the field is full of statistics regarding attackers or incident reports that are either incomplete or lacking. Official reporting, like ICS-CRT in the US in 2016, accounted for just 700 threats to ICS across 250,000 manufacturing firms active in 2016.[1] But only general details are shared by the companies about the threats they have faced. Security engineers are asked not to disclose details of attacks so as not to disclose sensitive information. That is one reason why debates in this field are mainly focused on the technologies to be adopted by organisations: by definition, technology is the first concern of companies in regard to cyber threats. Within the industrial sec-tor, however, security technologies play an ambiguous role. Digital infrastructures arranged to fight the threats are considered to be highly confidential, and the details of these infrastructures need to be concealed from everybody both inside and out-side the organisation. This is one reason why, socially sensitive research of security practices can provide important contributions to our understanding of such a closed and complex field.

RESEARCH DESIGN IN AN UNDEFINED FIELD OF PRACTICES

Sociologists have a mandate to develop fieldwork by adopting any previous explor-atory qualitative approach that best allows us to learn from events, practices and theories. Focusing on several contributions, starting with the fundamental ones by Turner (1978) and Perrow (2011), the research design has to pay particular attention to different representations provided by managers, professionals, technicians, maintenance engineers and technology vendors. Alongside these representations there are public narratives about industrial security developed by the media, institutional actors, service providers and vendors, and hackers. Engaging with this diverse range of actors requires methodological adjustment, as explained below, in order to approach the different informants with different search strategies.

As Caelli, Ray and Mill (2003) describe, a qualitative approach can generate a risk of confusion and superficiality in research work. However, the flexibility of the tools and the careful work on the collected accounts allow the development of a multifaceted approach in the comparisons of the research object. The work of cyber security is understandable only to colleagues who share work practices and who are rarely called to externally account for their work. For this reason, it is useful to structure a multi-layer research design that allows for the reconfiguration of data collection strategies according to the different informants.

In this scenario a special space has been occupied by the description of critical events, whether they are the most well known or simply those resulting from one's own work. This too is a legacy of Turner (1978) and his invitation to explore every detail that can be retrieved through narratives, descriptions and specialist reports but also by individual rescue volunteers or workers who happened to be on the scene of disasters. In the ambit of security, this can involve large providers of security services and single professionals that enterprises call after critical events occur. These stories are often contextualised in a deliberately imprecise way because they constitute the specific testimony of the professionals. However, after the data collection and analysis we can explore the questions Garfinkel suggested about ethnographic work: "what did we do? What did we learn? How can we teach it?" (1996: 9).

The initial aim was to conduct fieldwork following an explorative ethnographic approach, lasting over the two-year course of the project. The initial purpose of this fieldwork was to open three multi-sited ethnographic stages in three large UK companies to observe directly some organisational activities at the workplaces and understand how social construction of risk in ICS has a direct impact on everyday practices. Unfortunately, after almost one year spent defining an agreement, this approach was abandoned due to the challenges of negotiating access (see also Bourrier, Chapter 3, this volume; and Hayes and Maslen, Chapter 12, this volume). The management wished to avoid access to their plants and ultimately decided that security is not an area to be explored by external scholars. We adjusted our approach to the topic by contacting people that had their main core business in security: technicians, sellers, scholars, researchers in fields connected to cyber security and so on. As our starting point was to explore how people in companies construct their representations regarding cyber security on ICS and how these representations generate work practices themselves, we decided to move towards a discursive analysis (Buchanan and Bryman 2009). This approach necessitated the introduction of oneself as a legitimate researcher in confidential meeting contexts of security experts. These spaces always foresee a multiplicity of accesses to information (reports, official documents, research, etc.) and a continuous compensation between official and informal speeches (see also Kuran, Chapter 7, this volume). But in the end, they are spaces in which to shape stories and support narratives and a socially identified way of "doing security" among people who belong to the same community. This requires using an ethnographic approach so that the cyber risk theme is seen "through participants' eyes" (Hine 2000). And at the same time, as suggested by Boellstorff and colleagues (2012: 67) the ethnographic approach allows us to "seek to understand shared practices, meanings, and social contexts, and the interrelations among them."

The research presented here, which lasted two years, began with a collection of contextual accounts used to describe local or general situations related to cyber security in industries. We then selected a wide range of key informants to collect a plurality of opinions on cyber risk in the industrial sector (Hilgartner 1992; Star and Strauss 1999). The interviewees' selection followed a theoretical sampling approach, which privileged heterogeneity of testimonials, drawing on speakers met at conferences, start-up companies working on security, university research programmes, research centres specialising in security and personal contacts to provide liaisons with hackers and security consultants (Caelli et al. 2003). We included three scholars working in Italy and the UK to gain a clearer idea of the main research programmes and scientific production concerning cyber security in ICS in these two countries. Their competencies were related to organisational studies and technological development of ICS devices. Hence, we interviewed them about their fieldwork to obtain further accounts about cyber security in the ICS domain. In total 25 interviews were conducted in the UK and Italy, which included contributions from various security professionals, scholars, IT experts, IT security managers, security consultants, insurance brokers, white-hat hackers and vendors (Table 14.1).

The interviews were organised as contextual interviews conducted in the informants' workplaces (Fairclough 2003) and lasted one hour on average. They were all recorded and transcribed, apart from five, for which we only have notes, due

TABLE 14.1
Interviewees and Their Competencies

Role	No.	Job Descriptions
Scholars	3	Sociologist of risk and safety analysis; research coordinator of a European project on security management; software engineering researcher working on security in critical infrastructure
Managers	2	Engineer in selling and controlling appliances; software engineer and consultant in ICS cyber security
Engineers	10	Control engineers; maintenance engineers; electrical engineers; software engineers; network engineer
IT or security managers and experts	7	IT security managers; security analyst experts; senior software engineers; penetration test engineers; social engineering experts
Hackers	2	White-hat hackers in the market acting as security event analysts
Vendors	1	Systems for off-shore equipment

to a specific request from the interviewees. During the research phase, numerous documents, verbal accounts, shared stories from companies and web reports about industrial cyber security were collected. Several speeches from more than ten national and international conferences regarding cyber security (in the UK and Italy) were recorded. An additional field activity consisted of a focus group with the Institute of Engineering and Technology UK (IET), held with ten participants in London. Analysis started simultaneously with data collection and these progressed from initial data coding to a reflexive, constructive grounded approach that focused the analysis on identifying concepts, detecting emerging issues, and labelling and commenting on every possible converging point within the data (Glaser & Strauss 1967; Seale 2004; King and Brooks 2015). In a third phase, the most salient categories were selected for detailed examination and conceptualisation, and the corresponding data excerpts were further analysed through a template analysis to identify critical descriptions, testimonies and excerpts. In this way it was possible to create a circularity in the analysis by creating a continuous comparison between field material and explored categories. This approach made it possible to identify the salient categories and to continuously compare them with the extracts that could best identify the concepts that emerged from the analysis (Maarten 1996; Holstein and Gubrium 2013).

A QUALITATIVE START: CHASING GOOD STORIES

The most prominent insight from documents and speeches delivered at security conferences and informal talks with actors involved in ICS mostly have one common representation: "soon it will be your turn to be hacked!" Their accounts call for urgent action to be taken to protect the ICS in order to prompt people to be mindful of their risky "nature", and the possibility that some attacks will affect all companies sooner or later. For example, the most notorious ICS attacks like Stuxnet, German steel-mill, Ukraine blackout, etc., covered by several analysis (Langner 2011, Lin et al. 2017, Liang et al. 2017), are cited to generate uncertainty in the audience. Another issue is the practicality and reality of these cases: they have become the "real" benchmark for everybody in this sector since 2008. Cases like "Stuxnet" have become the first "brick" of every "social representation" in this field.

> I think once you get people into that mental thought of "okay, I understand the risks and I personally don't want to be the one who has just introduced 'Stuxnet' onto the network" or any of the other popular ones that, you know, it's almost like buzzword bingo at conferences. Once people understand that these exist, because if I ring one of them up now and say, "do you know what Stuxnet is?" They'll say "no, I've no idea" or "yeah, it's something to do with some Iranian thing, isn't it?" They won't really understand what happened. (Power grid maintenance engineer, UK #1)

One signal that these accounts were unspecific is that the same rhetoric was also evoked when Small and Medium Enterprises (SME) were involved. For these companies the cyber security issue is often at an immature stage and Stuxnet, a hacking strategy that targeted an Iranian nuclear plant, cannot represent a good proxy for them. Consequently, in many discourses these threats were performed "as real," although the meaning to the participants could be vague.

These kinds of stories show how the ICS territory can be poorly defined. Above all, these stories emphasise the social nature of the definition of the domain of cyber security in the industrial field. As suggested by O'Reilly (2015: 13), collecting practice stories helps to

> understand the making of the social world as ongoing processes, both shaped by and shaping general patterns, arrangements, rules, norms, and other structures. Ethnography that pays attention to both wider structures and to the thoughts and feelings of agents, within the context of action, is thus an ideal methodology.

In this first phase of the project, we considered stories concerning the variety of attacks. It was clear that the cyber security world, and the nature of these threats, can still remain complex and ambiguous to a wide range of participants in the industrial and manufacturing sectors. Many respondents recounted various kinds of attacks to highlight the fact that the industrial sector is at an immature stage. One automation engineer stated that from his experience in selling control systems to oil companies, if he added some security devices in a call for tender for an ICS technology, the extra costs could induce the buyer to select another vendor. Other interviewees stated that the topic is considered important only by IT technicians and security managers. Some actions have been taken by companies only because of laws or other external requirements imposed by authorities.

Stories collected in different contexts attracted our attention to a field of investigation with many different possibilities of access to information: how personnel produce discourses and good stories about the relevance of the security topic within companies. As Marcus (1995) suggests, the complexity of the object under investigation leads to the adoption of a plurality of investigation models. Stories are particularly well suited to transfer generic information between companies and security meeting contexts. These stories also allowed us to collect micro-events that describe the typology of actions carried out in organisations, even if these cannot be observed directly. For example, a technician can informally say that if asked to go and fix something which has occurred within a production plant that was highlighted by an alert, the people in charge of it tend to act only in consideration of the technical aspect. They are in charge of fixing things to save production. However, without having any sensitivity with regard to cyber risk, employees end up working in a grey area, often invisible to management, which can prevent potential attacks being recognised if, by chance, a serious security attack lies behind some mundane technical problem.

> It is briefly mentioned that if you see anything strange, if you suspect something, then you have to let your manager know, report it to the 24-hour control centre. They monitor the infrastructure for things like vibration, and then they can listen to the conversations if they need to, they have cameras and so on, which they can pull up to see what's going on. So that is triggered, but regarding myself, personally, I have never

encountered any individual event. Not really, I've never seen any cyber-attack person-ally but as I say we don't really monitor for cyber-attacks, so if there was one happen-ing, I wouldn't know about it anyway. I like to have procedures and documents that I can refer to, and to be honest, I've always been left wanting, because it is very poor, that no, in essence, no, we don't have anything or any procedure you know, that we can refer to, regarding that. (Maintenance oil engineer, IT #3)

Raising issues and constructing a discourse about security here is totally left to the respondent's sensitivity, and the respondent, in a loosely structured situation, is expected to fix things and employ his experience as a maintenance engineer. Of course, sometimes malicious attacks do happen, and generic attacks may be very dif-ficult to analyse, particularly because the countermeasures are driven by the knowl-edge domain representation.

The virus is treated as a mechanical failure of one part, which is the interface, or the breakdown of a controller, but in this sense, whoever has designed the emergency systems has been able to make them really secure because a technical breakdown is something we want to exclude from the start. I don't expect to face an emergency with a system if later I leave the system unprotected. The safety system needs to be isolated; very often, it's disconnected from the external net and that is connected only to the safety controllers' net. The only thing that is left besides the safety system is the mechanical control. Therefore, if the safety system fails because it's being attacked, even if it's difficult, only the electro mechanic remains. The last block is a threshold that will be overcome by the tank that needs to explode with the gas inside. If all the antecedent alarms fail, there is anyhow a threshold that will be identified by a sensor that opens an airway valve. There aren't planned things. (IT Security engineer, UK #5)

This excerpt indicates the sorts of things that we may be missing in these contexts. Analysing the accounts of participants in the industrial and manufacturing sector, it became clear that securing vulnerabilities is often viewed as an individual task for each of the people involved: each has to tackle the problem or define a posi-tion related to themselves and their work. Beyond specialist security staff, the com-munity may not have yet developed a collective representation of risk that includes attack scenarios.

"TOO LATE!": THE SOCIAL CONSTRUCTION OF THE DANGER IN ICS

Having developed a qualitative sensitivity to the challenges of managing security threats encountered by participants in this sector, in this next section accounts are provided to describe how ICS and the work to protect companies from cyber-attacks take place in organisations. Through their accounts, security staff could address their target audience in many different ways (Berti 2017). When working with organisational members, a common objective of security staff is asking people to face this world of unknown hackers in a way that assumes they already know what to do.

Moreover, security problems are quite banal. We would only need to put a patch and carry out the update that Microsoft already offers for free. We usually don't do it not just because of a matter of carelessness from the operators' side but it's also about the desire of leaving the infrastructure criticism unvaried without having to touch it so that perhaps it would become unmanageable unless something happens to put the system at risk. I also think that there is a responsibility problem because it is the operator who needs to make a decision. (Maintenance engineer, IT #9)

Cyber security experts describe a grey area where operators are able to leave technical systems untouched and software not updated, simply because updating is not a mandatory requirement of the operator's job. Companies may also sometimes not consider it mandatory either, and can be sensitive to the economic and technical risks of updating and upgrading systems, where the unpredictable effects of an accidental stoppage of a machine can subsequently lead to economic loss. The grey area therefore revolves around understanding and balancing these different risks inherent to these machineries and the ICS, and production and protection becomes a space that must be managed specifically using the available knowledge to reduce damage of different forms. It is interesting to observe that accounts – and those who assume the responsibility for critical infrastructures – usually refer to many organisationally complex issues such as legacy systems and unpredictable accidents. Rhetorically speaking, stories link problems with solutions by illustrating how people have the know-how to work in the grey zone, by articulating organisational practices to control a particular set of problems. In other words, in this context the main collective assumption is adopting the "best choice" metaphor: company managers must decide which approach and depiction is better for their organisation, particularly in cases where guidelines and protocols set by public agencies at national or international levels can conflict with local organisational culture and practices. There is an open, grey area that organisations repeatedly seek to resolve in different way by addressing the question, What is considered to be dangerous in this context?

Yes, but I'm convinced that if you are always looking for possible threats, it is easier to anticipate anomalies during everyday life. You have to think at the disasters that you can avoid, as much as possible. It is quite common for safety, but not really implemented in the case of cyber threats. If at a certain time a great amount of traffic is observed on the firewall, today it is not considered an anomaly by most companies. In fact, they are waiting for a real and concrete disaster. Just after, you can project something. Something that can be done, but it will be late. (IT Security engineer, UK #7)

The wide range of possible anomalies in organisations cannot be simply fixed, and so many professionals are continually engaged in trying to find remedies for this open set of threats. Things can be done in order to, for instance, detect network traffic, but again the question of "how much" traffic is dangerous for a company in terms of how its firewall must be managed is clearly a "qualitative" definition. However, experts are used to saying that when something happens to the company, it is invariably "too late."

GOING THROUGH A GOOD KEY STORY: PLC MANAGEMENT

One of the primary objectives of qualitative research is to link the particular to the general. Each story is never an isolated, local story about a specific workplace. Each story is always a way of "doing" security, of saying how you want to do it and defining the fields of action and degrees of freedom of the actors involved in their specific workplaces. One of the most critical features within the technical environment of a firm, specifically in a production site, are PLCs (Programmable Logic Controllers). These machines are very "simple" and do their job in a very task-oriented way. As stated by several engineers interviewed, PLCs have been used in firms since the 1980s and since then they have not developed much in terms of security, for many (reasonable, practical) reasons. Firstly, in essence they were machines designed to work in local settings and for specific automated tasks. Secondly, engineers were responsible for groups of restricted staff who trusted each other, as well as being responsible for the machines and developing heterogeneous networks that were oriented to production in crafted, local ways.

Due to these constraints, the best solution in the past was to put (digital) "security fences" around these machines to prevent attacks from intruders. From the 1990s, however, as the need increased to be able to check production in real time and to do so remotely, PLCs have become particularly vulnerable to attacks – both as specific machines to hack, and also through the workstation connected to them and to the Internet. One more aspect that makes this technology an easy target is that only seven brands produce 80 per cent of the total PLCs used in industries and, between them, Siemens owns over 30 per cent of the market. Since they are usually controlled by a Microsoft Windows workstation it is clear why these systems must be protected. In looking at organisations' practices in using these technologies, we can observe that operators still have a special relationship with these machines, and consequently highlight the need to have "direct access" to them.

> Let's say that the operator always wants to have access to the PLC without having supplementary problems. In this regard, Siemens has released newer PLC that allow authentication and message cryptography … Thus, they provide basic security. It isn't being used by the operator because he is the one who always wants to have direct access. (IT Security engineer, UK #9)

In a production plant, operators and technicians want to have direct "control" of the machines. However, many complexities are emerging. First, these machines survive for a long time, and despite their original and "local" duty, their work must now be checked remotely, as well as being connected to the Internet, even though their design was not originally conceived with this in mind. At the same time, upgrading them to fix bugs or confront the problem of insufficient protection is not an easy task as the machines age.

> So, he/she sends back the choice that needs to update the [PLC] system by someone else who in turn isn't capable of understanding how important this is. Only at that point it's easier to say "let's leave everything as it is, rather that throwing ourselves into

updates that could cause economic discomfort, if not worse." (IT Security engineer, UK #4)

Ultimately, these machines are connected to workstations, which in turn must be updated as well. Accounts portray these machines as stable and person-dependent. Security is considered to be an option that staff are apparently able to manage at a local level. Another complexity emerging from stories collected through the interviews concerns the fact that industries and technicians working with these "simple" machines say that they should never be altered or touched, since they have been functioning well for a long period of time.

> Note that in many industrial systems many offensive experiments are forbidden. And many actions provided by the offensive security practice, also a simple port scanning, in an industrial device, could be potentially dangerous. A PLC could go mad if you launch a simple procedure like a port scanning. It depends on one hand by legacy system, from the other side it depends on the machine's interactions which you don't completely know. It is again often totally unknown by the operator of cyber security. (…) So you are often embarrassed. You don't know how the industrial system works exactly, how it could react to your tests and you are not allowed to make penetration tests nor applied specific procedures if never documented in this site … What can I do? Often an error in these tests could produce many problems for production, or worse, it could endanger people as well. It is too dangerous, so you can't do any test you like. (IT Security engineer, IT #5)

When IT security staff arrive in order to carry out technical assessments regarding possible attacks, and are asked to perform some kind of technical testing which requires simple actions such as port scanning or switching off the PLC, the crew working with the machine, as stated by a security consultant, become "crazy with fear." These machines are used in almost all industrial sites and in all critical infrastructures, thus in a nuclear plant, for example, or in a power grid distribution unit, conducting PLC testing is not simply an "ordinary" action. The engineers do not seem to express any faith in "real life" testing. Organisational accounts show that production plants have many such grey areas where the lack of knowledge is problematic.

> Typically, customers don't know what is on theirs networks, really. They don't know the number of PLCs; they don't know what IT components are existing. They have tables and spreadsheets of what they think they have on the system, but when you look at what is really going on, only about 20% is covered, i.e. of what they believe, or regarding the initial spreadsheet we were given. What we do also is to provide a kind of asset – I don't like the term asset – business guys like to call it asset, but we create a network map of all the network. (IT Security engineer, UK #5)

Practical organisational stories collected through interviews show that large organisations with "stable businesses," such as power plant production and distribution, and any critical infrastructures which are obliged to add new systems to old ones, report facing considerable challenges. Ultimately, no one person has an exhaustive and complete knowledge of how a system works, and discourses recount the notion that only external experts are able to carry out complete, comprehensive work on these technological networks. This is also true in terms of PLC maintenance. The complexities inherent in the industrial plants require such a thorough knowledge that no one person is able to manage a situation easily. When some parts, for example, were found to be faulty, a maintenance crew simply substituted parts detected as broken, and that was the end of it.

> The same thing applies to the PLC: if you have a PLC card with sixteen entry and exit breaks, you get another card and you replace it, even if you don't fully understand how it was broken. (IT Security engineer, UK #7)

CONCLUSION: A METAPHOR TO HELP UNDERSTAND THE "GREY ZONE"

The research about cyber security in the industrial and manufacturing sector, as we have seen, is particularly complex due to the specific constraints, sensitivities and challenges of the field. There are limits to the ready diffusion of information and the professionals who dedicate themselves to this area are constantly moving in an unstable, invisible field, a grey area defined by much research (Pettersen 2016; Turner 1983). This emphasises why qualitative research, which seeks to closely look at work practices and the stories and representations that surround them, must yield to this uncommon complexity and must be able to draw on a plurality of sources. The methodological tools of the work, as illustrated in this chapter, are the observations, interviews, frequenting of environments related to the promotion of security, the collection of stories on security practices and the rhetorical recommendations on the risks posed by cyber treats. Throughout this chapter we have shown how various materials have been collected through a very complex methodological "tinkering." We focused on the practice stories told by security workers to define some peculiarities of the semantic fields of security work, and among these we selected an object that appears in the technological articulations of industrial security: the PLC. This work is similar to that of anthropologists; with their techniques, the professionals of "cultures" can help us in recognizing the types of contexts and practices that these spaces support (see also Kuran, Chapter 7, this volume).

A last step that we can explore here, and that discusses material that is in some ways always "fragmented," is the use of explanatory, typological frameworks that emphasise feelings of content and form. To this end, a good framework may be offered by the metaphor of being "in limbo" – being in neither one place or another – as a way of understanding the data presented and the challenges faced by the emergence of cyber security practices in industrial settings. As suggested by Gareth Morgan (1997), metaphors can be considered a particularly simplified and useful symbolic space for organisational analysis. The concepts that underlie the origins of the idea of "limbo" include *neglect, oblivion and confinement* and help to explore the idea of a grey space in security practices. The security practices recounted here were *neglected* in terms of the limited attention paid to events and alerts when producing a new vision of security issues, and the indications of neglect towards protocols and the everyday practices that could have a bearing on security. The idea of *oblivion* may help explore the organisational practices concerning people's work performance, specifically when they are unable to imagine exactly what type of threat can potentially exist within various environments. Some practices lapse into oblivion because they are connected to the technologies that are not totally mapped or understood within organisations, so that often maintenance staff change parts without having any clear understanding of the original cause of the damage. The idea of *confinement* recalls the idea of a place where many things can happen in a hidden, trapped space without any particular understanding of the impact on organisations. Local habits and bad practices are part of the confinement that pertains to this grey area. The confinement is also connected to practices using technologies in production lines. Exploring and resolving this "limbo" may be a way, following the metaphor, of defining how organisations can change, by deciding to live in the grey zone where complex and non-linear collective decisions should be taken. But how much can a metaphor help, and why pursue it in this context? As we have seen, data and situations are very fragmented within the ICS and the metaphor tool may allow an emphasis on some conceptual dimensions and organisational processes that are otherwise hidden in strategies and practices of cyber threat management. Practice stories, reread in light of new metaphors, should be considered in conjunction with the contexts from which they come, allowing a re-reading of cultural differences that challenges established stereotypes and typologies. An informed empirical study, according to the logic of the qualitative analysis of practices, will always be situated in a specific time and place. In this case the suggestion here for exploring a *limbo strategy* is to open our minds to methodological synthesis and explore the convergence of many organisational actions, perspectives and sources of data. The aim is to switch on all the lights so that we can illuminate the grey area, calling for a substantive strategy that endeavours to understand more profoundly what is, in essence, a poorly explored area.

NOTE

1. https://www.us-cert.gov/sites/default/files/Annual_Reports/FY2016_Industrial _Control_Systems_Assessment_Summary_Report_S508C.pdf (retrieved May 15, 2020).

REFERENCES

Berti, M. (2017). *Elgar Introduction to Organizational Discourse Analysis*. Edward Elgar Publishing.
Bieder, C., & Gould, K. P. (Eds.). (2020). *The Coupling of Safety and Security: Exploring Interrelations in Theory and Practice*. New York: Springer International Publishing.
Boellstorff, T., Nardi, B., Pearce, C., & Taylor, T. L. (2012). *Ethnography and Virtual Worlds: A Handbook of Method*. Princeton University Press.
Buchanan, D., & Bryman, A. (2009). *The Sage Handbook of Organizational Research Methods*. London: SAGE Publications Ltd.

Busby, J. S., & Alcock, R. E. (2008). Risk and Organizational Networks: Making Sense of Failure in the Division of Labour. *Risk Management*, 10(4), 235–256.

Caelli, K., Ray, L., & Mill, J. (2003). 'Clear as Mud': Toward Greater Clarity in Generic Qualitative Research. *International Journal of Qualitative Methods*, 2(2), 1–13.

Dourish, P., & Anderson, K. (2006). Collective Information Practice: Exploring Privacy and Security as Social and Cultural Phenomena. *Human–Computer Interaction*, 21(3), 319–342.

Dourish, P., Grinter, E., Delgado de la Flor, J., & Joseph, M. (2004). Security in the Wild: User Strategies for Managing Security as an Everyday, Practical Problem. *Personal and Ubiquitous Computing*, 8(6), 391–401.

Fairclough, N. (2003). *Analysing Discourse: Textual Analysis for Social Research*. London: Routledge.

Garfinkel, H. (1996). Ethnomethodology's Program. *Social Psychology Quarterly*, 59(1), 5–21.

Gherardi, S., Jensen, K., & Nerland, M. (2017). Shadow Organizing: A Metaphor to Explore Organizing as Intra-Relating. *Qualitative Research in Organizations and Management: An International Journal*, 12(1), 2–17. doi:10.1108/QROM-06-2016-1385

Gherardi, S., & Nicolini, D. (2000). The Organizational Learning of Safety in Communities of Practice. *Journal of Management Inquiry*, 9(1), 7–18.

Glaser, B. G., & Strauss, A. L. (1967). *The Discovery of Grounded Theory: Strategies for Qualitative Research*. Chicago, IL: Aldine Publishing Company.

Hilgartner, S. (1992). The Social Construction of Risk Objects: Or, How to Pry Open Networks of Risk. In: *Organizations, Uncertainties, and Risk*, eds. James F. Short and Lee Clarke Boulder, CO: Westview Press, pp. 39–53.

Hine, C. (2000). *Virtual Ethnography*. SAGE Publications Ltd. https://www.doi.org/10.4135/9780857020277

Holstein, J. A., & Gubrium, J. F. (2013). *Handbook of Constructionist Research*. New York: Guilford Press.

King, N., & Brooks, J. M. (2015). *Template Analysis for Business and Management Students*. London: SAGE Publications.

Knowles, W., Prince, D., Hutchison, D., Disso, J. F. P., & Jones, K. (2015). A Survey of Cyber Security Management in Industrial Control Systems. *International Journal of Critical Infrastructure Protection*, 9(C), 52–80.

Langner, R. (2011). Stuxnet: Dissecting a Cyberwarfare Weapon. *IEEE Security & Privacy*, 9(3), 49–51.

Law, J. (2012). Reality Failures. In: *Agency without Actors? New Approaches to Collective Action*, eds. J.-H. Passoth, B. Peuker, & M. Schillmeier (no. 58; issue 58). Milton Park, Abingdon-on-Thames, Oxfordshire, England, UK: Routledge, pp. 146–160.

Liang, G., Weller, S. R., Zhao, J., Luo, F., & Dong, Z. Y. (2017). The 2015 Ukraine Blackout: Implications for False Data Injection Attacks. *IEEE Transactions on Power Systems*, 32(4), 3317–3318.

Lin, C., Wu, S., & Lee, M. (2017). Cyber Attack and Defense on Industry Control Systems. In: *2017 IEEE Conference on Dependable and Secure Computing*, Taipei Taiwan, 524–526.

Luff, P., Hindmarsh, J., & Heath, C. (2000). *Workplace Studies: Recovering Work Practice and Informing System Design*. Cambridge: Cambridge University Press.

Maarten, S. (1996). *Selective Ethnographic Analysis - Qualitative Modeling for Work Place Ethnography*. Westchester Ave. White Plains, NY: NYNEX Science & Technology Collaboration and Practice.

Macrae, C. (2014). *Close Calls: Managing Risk and Resilience in Airline Flight Safety*. London: Palgrave Macmillan.

Marcus, G. E. (1995). Ethnography in/of the World System: The Emergence of Multi-Sited Ethnography. *Annual Review of Anthropology*, 24, 95–117.

Martin, R. A. (2001). Managing Vulnerabilities in Networked Systems. *Computer*, 34(11), 32–38.

Morgan, G. (1997). *Images of Organization by Gareth Morgan (26 February 1997) Paperback*. Thousand Oaks: SAGE Publications, Inc.

O'Reilly, K. (2015). Ethnography: Telling Practice Stories. In: *Emerging Trends in the Social and Behavioral Sciences*. American Cancer Society, 1–14. https://doi.org/10.1002/9781118900772.etrds0120.

Perrow, G. (2011). *Normal Accidents: Living with High Risk Technologies*(Updated edition with a New afterword and a new postscript by the author edizione). Princeton University Press.

Pettersen, K. A. (2016). Understanding Uncertainty: Thinking Through in Relation to High-Risk Technologies. In: *Routledge Handbook of Risk Studies*. Abingdon: Routledge, pp. 39–48.

Schmidt, K. (2014). The Concept of "Practice": What's the Point? In: *Proceeding of the 11th International Conference on the Design-COOP2014*, eds. C. Rossitto, L. Ciolfi, D. Martin, and B. Conein. Nice, France, 2014 (11th Conference on the design of cooperative systems conference COOP14).

Seale, C. (2004). *Qualitative research practice*. SAGE.

Seale, C., Gobo, G., Gubrium, J. F., & Silverman, D. (eds.). (2004). *Qualitative Research Practice* (1 edition). Newbury Park, California: SAGE Publications Ltd.

Star, S. L., & Strauss, A. (1999). Layers of Silence, Arenas of Voice: The Ecology of Visible and Invisible Work. *Computer Supported Cooperative Work (CSCW)* 8(1–2), 9–30.

Turner, B. A. (1976). The Organizational and Interorganizational Development of Disasters. *Administrative Science Quarterly*, 21(3), 378–97.

Turner, B. A. (1978). *Man-Made Disasters*. London: Wykeham.

Turner, B. A. (1983). Making Sense of Unseemly Behavior in Organizations. *International Studies of Management & Organization*, 13(3), 164–181.

Weick, K. E. (1998). Foresights of Failure: An Appreciation of Barry Turner. *Journal of Contingencies and Crisis Management*, 6(2), 72–75.

Zanutto, A., Rashid, A., Busby, J. S., Frey, S., & Follis, K. (2017). The Shadow Warriors: In the No Man's Land between Industrial Control Systems and Enterprise IT Systems. In: *Thirteenth Symposium on Usable Privacy and Security, SOUPS*, July 12–14, 2017, Santa Clara, CA.

Afterword: Connoisseurship, the Sociological Imagination and Turner's Qualitative Method

Nick Pidgeon

> If we distrust the idea that social science knowledge is something special, which can only be developed according to particular scientific canons, and think of it instead as an activity which extends common sense knowledge, this has implications for the way in which social science is carried out and for the way in which methodological issues are passed on. It becomes more important for research activities to be regarded as akin to craft skills, and to devise ways of transmitting and improving such skill so that "bad practice" or poor skill levels are not perpetuated. This requires the research practitioner to be a "reflective practitioner," using reflection to understand and to improve on existing practice. It also requires greater attention to be paid to the modes of knowing which are appropriate to social research, with particular emphasis upon the possibility that researchers need to develop a particular form of "connoisseurship," appraising the act of knowing in order to participate in skilful knowing as well as in coherent analysis. (B.A. Turner, 1990, p1)

As several of the contributors to this volume kindly acknowledge, I had the very great fortune to spend a significant part of my early academic career working alongside Barry Turner, together with our great friends and colleagues David Blockley and Brian Toft. How I came to read and appreciate the first edition of *Man-Made Disasters* in April 1980, its transformative impact upon my thinking and the subsequent history of our productive collaboration throughout the mid-1980s and beyond has been documented elsewhere (Pidgeon, 2019). Reflecting upon the various contributions in the current volume, it seems both timely and fitting to return to Turner's pioneering thinking on qualitative method just over 40 years on. And to do so while observing the world beset by a very real collective disaster, a pandemic foreseen in many national risk assessments and yet in its global scope and impacts not entirely prepared for. Over the spring and summer of 2020 I have repeatedly found myself asking the question, "what would Barry have said here?", as I did so many times also when preparing the second edition of *Man-Made Disasters* in the year after his untimely death (Turner and Pidgeon, 1997). I am sure that had he lived he would have had some very clear views about the sociological and organisational origins of our current predicament.

TURNER'S FORESIGHT – THE IMPORTANCE OF WORK AND FIT IN GROUNDED THEORY

In offering this afterword I would first congratulate the editors and individual authors for bringing together a vibrant collection of chapters on a topic, qualitative methodological innovation in safety research, that often gets overlooked in the field. The very label "safety science" means that it is inevitably hard to gain acceptance for "softer" qualitative methods within the quantification-dominated science and engineering disciplines which form the core of risk and safety studies. As commented by Gherardi and Turner (1987), adopting a metaphor from the ironic title of a book on masculinity by Bruce Feirstein – "real men don't collect soft data!"

Fortunately, times have changed somewhat, even in the safety sciences. If we place the often hotly contested debates about fundamental epistemologies and paradigms aside for one moment, one can argue here that there is no intrinsic incompatibility between the qualitative and quantitative social sciences if numbers and words simply serve as different *modes of representation* for constructing our understanding of material phenomena in the world (Henwood and Pidgeon, 1992). This was Turner's pragmatic philosophy, too, and the detailed methodology that he pioneered in the safety domain – perceptively labelled by Hopkins (Chapter 2, this volume) as "desktop ethnography" – still holds a key relevance for contemporary disasters researchers in ways which extend beyond the scope of Turner's original work.

Turner himself was modest about the methodological origins of *Man-Made Disasters*, making no reference whatsoever in the first 1978 edition to the debt he owed to Glaser and Strauss (1967), and his use of grounded theory to progress his PhD analysis (Turner, 1976), which subsequently and very rapidly became the published book. He explained afterwards that as the book had emerged so very easily from the thesis he wished its conclusions to stand on their own – to speak to practitioners and other academics in the field as a substantive account, unencumbered with unnecessary methodological details or protocol. In that respect Turner always viewed social science method and methodologies as a set of craft tools, in part passed on in tacit form from researcher to researcher (cf Polanyi, 1958; Ravetz, 1971), but always secondary to the main task of attaining a substantive empirical or theoretical goal rather than as an end to be pursued in and of themselves.

In reading the draft chapters for the current book I could not resist a smile when I came across the well-known hierarchy of evidence (Figure 11.1, Martin, O'Hara & Waring, Chapter 11, this volume), still promoted vociferously as the only foundation for drawing secure scientific conclusions and "objective" knowledge by so many within the medical and natural sciences communities. The irony here is that the Man-Made Disasters analysis would, by this standard, be viewed as a "very weak" form of evidence indeed – based as it was upon expert committee reports (Level VII) and case/qualitative studies (Level VI). This in turn begs the important question: how might an account which could be deemed so "methodologically challenged" by widely accepted criteria be in the same breath so conceptually powerful and long-lived? Of course, making reference to a now longstanding set of arguments, one can posit alternative indicators of trustworthiness and quality to fully appraise the strength or otherwise of the qualitative social sciences (Lincoln and Guba, 1985). In Turner's case, it suffices to say that it is a feature of any carefully constructed grounded theory conducted across suitably sampled comparative cases (also Antonsen and Haavik, Chapter 5, this volume), that it suggests its own relevance, fit and work. Disaster Incubation Theory, as it later became known, was immediately recognisable by those who first encountered it in the late 1970s as having both wide empirical scope and deep novel theoretical value, and as a model that had the power to transform our fundamental understanding of the way events unfolded in the world (Pidgeon, 2019).

As this and other volumes attest (Le Coze, 2019a), Turner's analysis has survived the many challenges of time well, serving both as a counter to the narrow and universalising nature of much mainstream quantitative thinking that still dominates risk assessment and safety sciences, but also as an encouragement to other qualitative accident researchers in their pursuit of contextually sensitive modes of knowledge production. This may also explain a feature of Turner's analysis that only becomes clear in hindsight, and when reading some of the current chapters – that it was *so well done* that it correctly anticipated developments and conclusions that other more prominent accident researchers would subsequently lay claim to (system complexity and uncertainty, how multiple failures undermine layered safety defences, the cultural blinkers that organisations adopt), even when those who followed claimed only superficial or no knowledge of Turner's original writings. As Hopkins correctly notes (Chapter 2, this volume), *Man-Made Disasters* did not so much give rise to the tradition of organisational safety science as it served to mark the beginning of a whole genre of thinking. And if not always a strict "precursor" to more recent developments in the field of safety sciences, Turner's qualitative account of his fully saturated and well-grounded analysis has time and time again proven foresighted in both its details and scope of application. This observation alone should encourage readers and authors of this volume to consider where qualitative research can and should add further deep insight and value to the contemporary safety sciences field.

SOCIOLOGICAL IMAGINATION AND THE "CRAFT" OF DOING WELL-GROUNDED QUALITATIVE SAFETY RESEARCH

Turner favoured an intellectual strategy akin to bricolage (also Bourrier, Chapter 3, this volume) when approaching everyday life, his version of the sociological method, and the academic theories that he drew upon. Accordingly, when I first joined the research project that he led with David Blockley investigating accidents in the UK construction industry, his first recommendation was that I read C. Wright Mills' (1959) classic account of the Sociological Imagination – and with it the idea explored in its Appendix that the researcher should hone "connoisseurship and craft skills" for making sense of and interpreting what would inevitably prove messy real-world qualitative data (Turner, 1988). Such a stance is well suited, perhaps even requisite, for many qualitative forms of social sciences inquiry (Bailyn, 1977). On that basis Turner developed in his paper published in *Quality and Quantity* (Turner,

1981a) a nice template for a workshop format that I myself have used to pass on grounded theory analytic practices. Put simply, an experienced data analyst can lead intensive small group workshops to collaborate in segmenting, coding, defining concepts, writing memos and assembling an analysis from a small segment of trial data. Within this process, Turner was always clear on the role of prior sociological theory, and with it sensitising concepts (Glaser and Strauss, 1967), to the development of any well-grounded qualitative analysis. As Le Coze (Chapter 4, this volume) describes, rather than pure induction of concepts (a logical impossibility anyway) Turner emphasised in his informal discussions about method the concept of abduction as developed by the philosopher Charles Sanders Pierce, as a description of the doing of grounded theory. In practice this becomes a process of cognitive "flip-flop" between data and its conceptualisation (see Bulmer, 1979; Pidgeon and Henwood, 2004; also Bailyn, 1977). He was also clear that not all those who were new to qualitative research could achieve the necessary cognitive flexibility to live with the uncertainties that this messy process of analysing unstructured qualitative data inevitably brings (Turner, 1981a). It would need significant application and effort on the part of the analyst, allied with cognitive flexibility and theoretical insight. Although we cannot know for sure, in his meticulously researched piece Le Coze (Chapter 4, this volume) argues that Turner would have engaged or endorsed enthusiastically the subsequent debates about "constructivist" or "new" grounded theory had he lived longer. In one of his few notes on postmodernism written to himself, following a visit to an especially lively conference on the topic hosted by the British Sociological Association at Lancaster University, he concluded with the comment:

> there's not much agreement about what it [postmodernism] is, but it's slap bang in the middle of symbolism, cultural analysis and many other forms of inquiry ... and it doesn't look as if it is going to go away. I look forward to discussions with those more expert than I who may be able to extend understanding of the postmodern problematic. (Tuner, 1987, p 6–7)

This somewhat ambivalent quote ably sums up the way that I recall Turner expressing a hesitant personal position regarding the rise of postmodernist thinking within philosophy and sociology as a whole.

MAJOR SOCIOTECHNICAL RISKS AND DECISIONS UNDER IGNORANCE: THE VIEW FROM SCIENCE AND TECHNOLOGY STUDIES

As chapters in this volume document, Turner is widely credited with working and publishing across three very different domains, engaging completely separate academic networks and gathering insights as he went: critical management studies and organisational symbolism, major sociotechnical failures, and of course qualitative research methods and the practical application of grounded theory for sociology. It was only in subsequent appreciations, by both Turner (1995) and others who knew him (this volume), that the close interdependencies between these different areas of intellectual activity become clear to us. We would not have *Man-Made Disasters* – an account of technological–organisational–cultural practices under fundamental uncertainty – without both the close craft of grounded theory and the theoretical orientations provided by organisational symbolism. The emphasis upon the symbolic and cultural in organisational life brings into focus "the world as envisaged" rather than "the world as is" as a core analytic concern in the model (cf also Rae, Webber and Dekker, Chapter 8, this volume). Culture in effect provides the lenses through which we both observe and, importantly, sometimes overlook emerging properties of the world – hence culture serving as "a way of seeing as well as a way of not seeing" (Turner and Pidgeon, 1997, p 49).

Although he never published in its associated journals, as Gherardi (Foreword, this volume) correctly points out, much of Turner's thinking about disasters and organisations could today also be counted as located within the (then fledgling) academic discipline of Science and Technology Studies (STS). Indeed, Turner's work is largely neglected in much of the STS field of today.[1] I do myself recall him recommending that I read Thomas P. Hughes' classic STS book *Networks of Power* on the history of electricity systems, as an illustration of the ways in which apparently technical systems were also sociotechnical networks of social arrangements. He also invited key STS colleagues to present seminars in Exeter and London on topics spanning the philosophy and sociology of risk and safety, including Brian Wynne, Jerry Ravetz, Silvio Funtowicz, Arie Rip and David Collingridge. In particular, Turner believed that the now influential work of Collingridge (1980) on decision-making under ignorance held a number of obvious connections with the fundamental position on uncertainty and risk that he had adopted in *Man-Made Disasters*. One of the defining characteristics of a disaster (rather than a simple accident) is the generation of surprise when a significant disjunction between "the world as envisaged" and "the world as is" comes to light. Collingridge had likewise studied cases of major project failure, in his case in the development of new complex technological systems. He argued that in the early stages of formulating complex projects (in planning, project scheduling, or in engineering design) some uncertainties are very hard to anticipate; hence the possibility of error always exists. Collingridge's own response was to orient design around options which would be corrigible: that is, ones which were capable of being readily monitored for mistakes and corrected cheaply and which allowed time for correction were errors to be discovered and were flexible and well behaved such that future options were kept open (Turner 1981b). For example, by designing multiple small-scale units that give opportunities for learning rather than single mega-projects for which decisions would become locked in before their full consequences could be appreciated. Turner also believed that preventing failures required an understanding of how forms of structural uncertainty and ignorance arise within organisations and design teams (also Blockley, 1980, 1992) rather than to simply focus upon conventional probability and "risk" of failure. From this perspective Turner and Blockley were among the very first in the UK to argue, alongside Collingridge (1980), Funtowicz and Ravetz (1992) and Wynne (1992), that the tools and axioms of conventional quantitative risk assessment needed to be carefully scrutinised (Turner, 1981b). In particular, they need to be supplemented by an understanding of a more qualitative appraisal of uncertainties where fundamental ignorance arises, something which is now a more mainstream view both within safety sciences and STS theorising (see e.g. Pidgeon, 1998; Stirling, 2010).

TIME, ORDER AND REPRESENTATION

One aspect of Turner's methodological contribution appears not so well represented in this volume: his attention to time and the cultural patterning of events in the disaster incubation period (also Gherardi, this volume). The concept of time is central to the idea of an incubation period, and he represented this in diagrammatic form, in a technique that clearly predates Rasmussen's Accimap (see Waterson, et al. 2017). This technique (Toft and Turner, 1987) takes a schematic form in which the reader can readily discern events observed, understood or assumed prior to the disaster ("the world as envisaged") overlain by events hidden or only partially understood in advance for their significance in the accident chain ("the world as is"). It was used by Turner to analyse and communicate his core case studies (Cambrian Colliery, Hixon Crossing, Summerland), and in other contexts it has been used to characterise construction failures (Pidgeon, Bockley & Turner, 1986, 1988) and military intelligence failure (Stech, 1979). We also realised subsequently that such diagrams can be used to map the impacts of specific accident inquiry recommendations onto the pre-accident sequence, to understand how particular recommendations were designed to prevent different branches or interactions within the incubation period from unfolding in the future. Critically, they also show how the diagram, named subsequently as the Schematic Report Analysis Diagram (SRAD), can be used to identify where an accident inquiry has *failed* to make a recommendation which might otherwise have prevented a part of the accident chain from occurring (Toft and Turner, 1987; Turner and Toft, 1988; Toft and Reynolds, 1994). In this way one might also begin to diagnose the power and political processes that inevitably shape most formal inquiries into accidents (Pidgeon, 1998). In the research project that I worked on alongside Blockley and Turner, we had also proposed to develop a data-base of SRADs, as a means of discerning incubating failure patterns across different case studies (Pidgeon et al, 1990). Although that idea probably retains some merit today in the age of big data algorithms, the machine-learning tools of the 1980s were not sufficient for the difficult task of learning across such complex qualitative cases.

Interestingly, the SRAD did come to the notice of the wider qualitative methods community. One of the often-overlooked aspects of good craft in qualitative data analysis is finding the right means for the succinct communication of core findings to audiences – indeed crafting such a public representation forms the final and critical analytic move of good qualitative data analysis. The SRAD always serves to succinctly summarise a complete disaster case study but in terms also clearly anchored in disaster incubation theory (energy, misinformation, organisational boundaries and blinkers, drift over time, etc.). Accordingly, an example of Turner's schematic diagram was solicited to appear in the 2nd edition of the highly influential book *Qualitative Data Analysis* by Miles and Huberman (1984), as one of their paradigm examples of how different forms of schematic diagram can be used to represent qualitative research outcomes. It's curious that in comparison to much else of Turner's methodological work this core technique seems to have gained little subsequent significance, perhaps inevitable given the move away from concepts such as linear causes or "conditions" in the turn to text and discourse that had begun to preoccupy mainstream and qualitative sociology during the 1980s and beyond.

SAFETY II AND SAFETY III: THE MOVE FROM HUMAN FACTORS TO THE ETHNOGRAPHIC METHOD

Turner moved from Exeter to Middlesex University in the late 1980s. In his subsequent collaborations with risk researchers in London one can discern the emerging outlines of the debates about Safety I and Safety II (see e.g. Hood et al., 1992; Turner, 1994; Pidgeon, 1998). Was it better to anticipate a class of broadly "known" failures through conducting risk assessments and specifying better operating procedures (Safety I), or in the face of ultimately unpredictable hazards, was it better to foster locally embedded resilience and flexibility by learning from successful operations (Safety II)? While safe operations have always needed a degree of both, striking the correct balance for any particular circumstance remains the safety practitioner's goal. The notion that key trade-offs might exist between anticipation and resilience had already been addressed by the political scientist Aaron Wildavsky in his book *Searching for Safety* (Wildavsky, 1988; Hood et al, 1992), although Safety II stresses less a trade-off than the continual search for excellence in *both* anticipation and resilience, alongside the negotiation of any trade-offs that might occur between the two.

This debate has a clear methodological dimension, and one which led me recently to consideration of the idea of Safety III (see Pidgeon, 2019). Hopkins (2014) argues that Safety II is simply the high reliability organisation in another guise, or perhaps a sub-set of what high reliability organisations purport to do. Le Coze (2019b) by contrast disagrees, highlighting the different disciplinary origins of the two – specifically, human factors and individualised cognitive science underpinning resilience engineering, as opposed to the more socially distributed view from sociology, social psychology and politics evident in high reliability organisations. As most readers of this volume will appreciate, such distinctions matter – whatever we might think about the desirability of achieving a requisite synthesis – since disciplines socialise their members into particular ways of understanding how the world works and with this how to study it. In my own experience of working on interdisciplinary risk research in many contexts, it is clear that mathematicians, sociologists, psychologists and engineers, by the very nature of a profession, think with different concepts (probability, power, human error, reliability), address different phenomena (people, society, systems) and sometimes defend their boundaries vigorously in debates about epistemologies, method and what counts as "legitimate" knowledge. Not only does that bring with it the longstanding problem of correctly communicating between diverse disciplines where terms differ, or where the same term means somewhat different things, but at times it brings heated border disputes over whose approach is to be considered the most appropriate or the "best" way forward.

I have argued elsewhere (Pidgeon 2019), following the important thinking of Le Coze (2019b), that we may now in reality be facing three distinctive perspectives on systems safety and risk: namely Safety I (conventional risk analyses and anticipatory standards), Safety II (human factors and resilience engineering) and finally Safety III (organisational culture and high reliability organisations). As Turner's own fusion of ethnographic method and cultural symbolism, and the many valuable contributions to this current volume attest (e.g. Le Coze, Chapter 4, Kuran, Chapter 7, Hayes and Maslen, Chapter 12, all this volume), fine-grained interpretive social sciences thinking and method have considerable value if applied systematically to the problem of studying how social systems underlie disasters and safety. By highlighting a close inspection of the organisational and management origins of safety performance, whether through "desktop" or more conventional forms of field-based ethnography, Safety III emphasises the *social processes* of interpretation and networking, negotiation between teams, and collective sensemaking (e.g. Weick and Sutcliffe 2001, Macrae 2014). Thus, the radical methodological step in Turner's work was in approaching safety from the perspective of (primarily micro-sociological) organisational culture, with its attendant detailed qualitative emphasis through the ethnographic and interpretive method. The work of *Man-Made Disasters* – and that of key researchers who followed this lead (Vaughan, Hopkins) – foresaw this radical step change in methodological practice.

GLOBALISATION AND SAFETY IV: NEW PROBLEMS, NEW CHALLENGES?

In advancing the metaphors of Safety I, II, and III, we should not forget that risks and safety practice do not stand in isolation from the wider trends that are occurring within and across societies. In particular, since the late-1970s neo-liberal thinking has driven a pattern of globalisation in many countries: characterised by extensive deregulation, privatisation of state assets, outsourcing of responsibility for standards and audits, and a retreat by the state from the direct provision of many social goods formerly taken for granted by citizens such as welfare, health care, employment, water and energy, and financial protection. One would be highly surprised to learn that safety had been completely insulated from such intense global pressures (also Dekker, 2020). In an entirely separate development, contemporary sociologists turned in the mid-1980s to the topic of globalised risk following the Chernobyl accident and the increasing neo-liberal trends emerging in the world economy. Their concern was to describe the changing conditions of late modernity alongside that of globalisation and the place of risk within it. Both Beck (1992) and Giddens (1990) sought to describe the increasingly complex and interconnected nature of technological, financial and environmental risks, alongside novel emerging technologies and systemic risks. In such a "risk society" individualised responsibility for securing safety from harm sits alongside an erosion of trust in institutions and the unwinding of the social certainties that traditional identities (such as being an employee, trade unionist, home owner, even safety engineer) had always brought us (Henwood and

Pidgeon, 2014). Beck himself did not entirely foresee that the first genuine empirical test of this thesis would come in the US and Britain, amidst the near-systemic collapse in 2008 of global banking systems because of the global proliferation of opaque, and ultimately worthless, financialisation instruments.

Le Coze (2017) raises this important question of globalisation, and in so doing points to what I believe to be a genuine deficit in our contemporary thinking about safety. Risk society is founded in an analysis from mainstream sociology, and the work of Beck, Giddens and others tends to lend its focus solely to social critique, and through this to ways of characterising the conditions of contemporary society. This points to a gap for the safety sciences, as one requiring a Safety IV (macro-global oriented) approach (Pidgeon, 2019). What is clear for contemporary accident research and practice is that the trends embedded within globalisation may well be bringing with them unanticipated, unwelcome forms of societal and organisational "drift" (Pettersen-Gould and Fjæran, 2019), ones that may serve to attenuate our perceptions of risk while simultaneously undercutting the established norms and processes for dealing with hazards. In the worlds of Safety II and III the hope was always that an organisation's striving for excellence would ultimately be to the benefit of both profitability and safety performance. In the globalised, deregulated, financialised and digitised world of Safety IV that assumption may no longer hold true.

Since the 1980s we have made significant and steady progress in our understanding of organisational safety and risks (Safety I, III and III) but Safety IV may well set the toughest challenges yet for the coming generation of researchers and practitioners. In his own glance back at former times, Le Coze (2017) observes perceptively that the safety fingerprint of globalisation may now be unfolding in an uncanny re-run of the disasters which were so common in the 1980s, and which served as the main stimulus for the first phase of sustained organisational safety research. The more recent case of the crashes and grounding of the Boeing 737-Max8 aircraft is also an example of the ways in which global competitiveness between different regions and organisations, in this case leading to a highly complex and ultimately flawed aircraft design, has precipitated major loss of life and near catastrophic economic consequences even for a company the size of Boeing. The global scale of risk is also a feature of our current pandemic predicament – as a virus first identified in a city in China has exploited the organised properties of our complex interconnected transportation and social systems to spread rapidly around the world with major health and economic impacts. Note here that much of this could be summarised in terms of Turner's original idea of an organised "anti-task": that is, a process which subverts the existing normal properties of organised life to unintentionally amplify and disperse risk. It seems unlikely then that research and safety practice should ignore such global developments. The world of Safety IV clearly presents an entirely new set of methodological challenges: to understand how global financial flows are affecting safety investments; the nature of complex common mode failures and globalised "anti-tasks"; the ways in which global digital systems lead to information difficulties; personnel flows and connectivity, as well as organisational fragmentation along transboundary lines; and the novel logistics and supra-national operations that underlie current modes of organised industrial design and production.

And what Sage advice might Barry have given to an early career safety researcher contemplating their own use of qualitative research methods right now? I am sure he would have commended the contributions in this volume and their very obvious methodological and theoretical pluralism. He would almost certainly have also recommended a strategy of methodological "bricolage" backed up with a sizeable dose of carefully crafted interpretive sensemaking. That advice should, I hope, serve to set a radical research agenda for the new generation of qualitative safety researchers, as well as a light to illuminate the way for practitioners in both the present day and into the next 40 years of safety sciences research.

NOTE

1. I was forcibly reminded of this fact recently when academics presenting at the principal Science and Technology Studies conference 4S/EASST held in 2016, in a session devoted to a sociological analysis of the events following the Töhoku tsunami and subsequent Fukushima nuclear accident, announced their firm conclusion that Fukushima was not a random happening or Act of God but clearly a "Man-Made Disaster." They were certainly correct in that evaluation but were also at the same time entirely unaware of Turner's earlier works.

REFERENCES

Bailyn, L. (1977) Research as a cognitive process: Implications for data analysis. *Quality and Quantity*, 11, 97–117.

Beck, U. (1992) *Risk Society. Towards a New Modernity*. London: SAGE.

Blockley, D. I. (1992) *Engineering Safety*. London: McGraw-Hill.

Blockley, D. I. (1980) *The Nature of Structural Design and Safety*. Chichester: Ellis-Horwood.

Bulmer, M. (1979) Concepts in the analysis of qualitative data. In M. Bulmer (ed) *Sociological Research Methods*. London: Macmillan, pp. 241–262.

Collingridge, D. (1980) *The Social Control of Technology*. Milton-Keynes: Open University Press.

Dekker, S. W. A. (2020) Safety after neoliberalism. *Safety Science*, 125, 104630.

Funtowicz, S. O., and Ravetz, J. R. (1992) Three types of risk assessment and the emergence of post-normal science. In S. Krimsky and D. Golding (eds) *Social Theories of Risk*. Westport: Praeger, pp. 251–274.

Gherardi, S., and Turner, B. A. (1987) Real men don't collect soft data. Quaderno 13, Departimento di Political Sociale, Università di Trento.

Giddens, A. (1990) *Modernity and Self-Identity*. Cambridge: CUP.

Glaser, B. G., and Strauss, A. (1967) *The Discovery of Grounded Theory*. Chicago: Aldine.

Henwood, K. L., and Pidgeon, N. F. (1992) Qualitative research and psychological theorising. *British Journal of Psychology*, 83, 97–111.

Henwood, K. L., and Pidgeon, N. F. (2014) *Risk and Identity Futures*. Foresight Future Identities Programme. London: Government Office of Science.

Hood, C., Jones, D., Pidgeon, N. F., Turner, B., and Gibson, R. (1992) Risk management. Ch 6 of F. Warner (ed) *Risk Analysis, Perception and Management: Report of a Royal Society Study Group*. London: Royal Society, pp. 135–192.

Hopkins, A. (2014) Issues in safety science. *Safety Science*, 67, 6–14.

Le Coze, J.-C. (2017) Globalization and high-risk systems. *Policy and Practice in Health and Safety*, 15, 57–81.

Le Coze, J.-C. (ed) (2019a) *Safety Sciences Research: Evolution, Challenges and New Directions.* London: CRC Press.

Le Coze, J.-C. (2019b) Vive la diversité ! High reliability organisations (HRO) and resilience engineering (RE). *Safety Science*, 117, 469–478.

Lincoln, Y., and Guba, E. G. (1985) *Naturalistic Inquiry.* Beverly Hills: SAGE.

Macrae, C. (2014) *Close Calls: Managing Risk and Resilience in Airline Flight Safety.* London: Palgrave MacMillan.

Miles, M. B., and Hubermann, A. M. (1984) *Qualitative Data Analysis: A Methods Sourcebook*, 2nd Ed. London: SAGE.

Mills, C. W. (1959) *The Sociological Imagination.* New York: Oxford University Press.

Pettersen-Gould, K., and Fjæran, L. (2019) Drift and the social attenuation of risk. In J.-C. Le Coze (ed) *Safety Sciences Research: Evolution, Challenges and New Directions.* London: CRC Press, pp. 119–132.

Pidgeon, N. F. (1998) Safety culture: Key theoretical issues. *Work and Stress*, 12(3), 202–216.

Pidgeon, N. F. (2019) Observing the English weather – A personal journey from Safety I to IV. In J.-C. Le Coze (ed) *Safety Sciences Research: Evolution, Challenges and New Directions.* London: CRC Press, pp. 269–289.

Pidgeon, N. F., Blockley, D. I., and Turner, B. A. (1986) Design practice and snow loading: Lessons from a roof collapse. *Structural Engineers*, 64(A), 67–71.

Pidgeon, N. F., Blockley, D. I., and Turner, B. A. (1988) Site investigations: Lessons from a late discovery of hazardous waste. *Structural Engineers*, 66(19), 311–315.

Pidgeon, N. F., Turner, B. A., Blockley, D. I., & Toft, B. (1991) Corporate Safety Culture: Improving the Management Contribution to System Reliability. In: *Reliability '91*, (Ed. Matthews, R. H.). Elsevier, 682–690.

Pidgeon, N. F., and Henwood, K. L. (2004) Grounded theory. In M. Hardy and A. Bryman (eds) *Handbook of Data Analysis.* London: SAGE, pp. 625–648.

Pidgeon, N. F., Stone, J., Blockley, D. I., and Turner, B. A. (1990) Management of safety through lessons from case histories. In M. H. Walter and R. F. Cox (eds) *Safety and Reliability in the 90s.* London: Elsevier Applied Science, pp. 201–216.

Polanyi, M. (1958) *Personal Knowledge.* London: Routledge.

Ravetz, J. R. (1971) *Scientific Knowledge and Its Social Problems.* Oxford: Clarendon Press.

Stech, F. (1979) *Political and Military Intention Estimation.* Bethesda: MATHTECH, Inc.

Stirling, A. (2010) Keep it complex. *Nature*, 468, 1029–1031.

Toft, B., and Reynolds, S. (1994) *Learning from Disasters: A Management Approach.* Oxford: Butterworth-Heinemann.

Toft, B., and Turner, B. A. (1987) The schematic report analysis diagram: A simple aid to learning from large scale failures. *International CIS Journal*, 1, 12–23.

Turner, B. A (1976) *The Failure of Foresight.* PhD Thesis, University of Exeter.

Turner, B. A. (1981b) Some questions about the use of concepts of risk and probability in connection with decisions about hazardous events. Working paper, University of Exeter, September 1981.

Turner, B. A. (1981a) Some practical aspects of qualitative data analysis: One way of organising the cognitive processes associated with the generation of grounded theory. *Quality and Quantity*, 15, 225–247.

Turner, B. A. (1987) Postmodernism and organisational culture. Unpublished manuscript. Exeter University.

Turner, B. A. (1988) Connoisseurship and the study of organizational cultures. In A. Bryman (ed) *Doing Research in Organizations.* London: Routledge, pp. 108–122.

Turner, B. A. (1990) If research methods are commonsense, can they be taught? Paper presented at *International Sociological Association Conference*, Madrid, July.

Turner, B. A. (1994) Flexibility and improvisation in emergency response. *Disaster Management*, 6(2), 84–90.

Turner, B. A. (1995) A personal trajectory through organization studies. *Research in the Sociology of Organizations*, 13, 275–301.

Turner, B. A., and Pidgeon, N. F. (1997) *Man-Made Disasters*, 2nd ed. Oxford: Butterworth-Heinemann.

Turner, B. A., and Toft, B. (1988) Organizational learning from disasters. In H. B. F Gow and R. W. Kay (eds) *Emergency Planning for Industrial Hazards*. London: Elsevier, 297–313.

Waterson, P., Jenkins, D. P., Salmon, P. M., and Underwood, P. (2017) 'Remixing Rasmussen': The evolution of AcciMaps within systemic accident analysis. *Applied Ergonomics*, 59, 483–503.

Weick, K. E., and Sutcliffe, K. M. (2001) *Managing the Unexpected: Assuring High Performance in an Age of Complexity*. San Francisco: Jossey-Bass.

Wildavsky, A. (1988) *Searching for Safety*. New Brunswick: Transaction Books.

Wynne, B. (1992) Risk and social learning: Reification to engagement. In S. Krimsky, and D. Golding (eds) *Social Theories of Risk*. Westport: Praeger, pp. 275–300.

Bibliography

Aase, K., & Braithwaite, J. (2017). What is the role of theory in research on patient safety and quality? Is there a Nordic perspective to theorizing? In: Aase, K., & Schibevaag, L. (eds.), *Researching Patient Safety and Quality in Health Care: A Nordic Perspective.* Boca Raton, FL: CRC Press, Taylor & Francis Group, pp. 57–74.

Aase, K., & Rosness, R. (2017). Organisatoriske ulykker og resiliente organisasjoner i helsetjenesten – ulike perspektiver. In: Aase, K. (Red.), *Pasientsikkerhet – teori og praksis (Patient Safety – Theory and Practice).* Oslo, NO: Universitetsforlaget, pp. 27–48.

Aase, K., Waring, J., & Schibevaag, L. (eds.). (2017). *Researching Quality and Safety in Care Transitions: International Perspectives.* Cham: Palgrave Macmillan.

Abdelhamid, T., & Everett, J. (2000). Identifying root causes of construction accidents. *Journal of Construction Engineering and Management* 126(1), 52–60. doi: 10.1061/(ASCE)0733-9364(2000)126:1(52)

Abelès, M. (2011). *Des anthropologues àl'OMC: scènes de la gouvernance mondiale.* Paris: CNRS èditions.

Act. (1999a). Act of 2 July 1999, no. 63 relating to patients' rights (the Patient's Rights Act).

Act. (1999b). Act of 2 July 1999, no. 64 relating to health personnel etc.

Act. (1999c). Act of 2nd July 1999 relating to specialised health care.

Act. (2011). Act of 24th June 2011 relating to municipal health and care services.

Act. (2017). Act of 6th June 2017 relating to the national investigation board for the health and care services.

Adams, S. A., van de Bovenkamp, H., & Robben, P. (2013). Including citizens in institutional reviews: Expectations from the Dutch Healthcare Inspectorate. *Health Expectations* 18(5), 1463–1473. doi: 10.1111/hex.12126

Aftenposten. (2015). Norwegian hospitals can learn a lot from the Daniel case. http://www.aftenposten.no/nyheter/iriks/Professor---Mye-a-lare-av-Daniel-saken-for-norske-sykehus-8064285.html

Albrechtsen, E., & D. Besnard (eds.). (2013). *Oil and Gas, Technology and Humans: Assessing the Human Factors of Technological Change.* Farnham: Ashgate.

Almklov, P. G. (2008). Deltakelse og refleksjon. Praksis og tekst. [Participation and reflection. Practice and text.] *Norsk antropologisk tidsskrift* 19(01), 38–52.

Almklov, P. G., & Antonsen, S. (2010). The commoditization of societal safety. *Journal of Contingencies and Crisis Management* 18(3), 132–144.

Almklov, P. G., & Antonsen, S. (2014). Making work invisible: New public management and operational work in critical infrastructure sectors. *Public Administration* 92(2), 477–492.

Almklov, P. G., & Hepsø, V. (2011). Between and beyond data: How analogue field experience informs the interpretation of remote data sources in petroleum reservoir geology. *Social Studies of Science* 41(4), 539–561.

Almklov, P. G., Rosness, R., & Størkersen, K. (2014). When safety science meets the practitioners: Does safety science contribute to marginalization of practical knowledge? *Safety Science* 67, 25–36.

Alruqi, W. M., Hallowell, M. R., & Techera, U. (2018). Safety climate dimensions and their relationship to construction safety performance: A meta-analytic review. *Safety Science* 109, 165–173.

Alvesson, M. (2001). Knowledge work: Ambiguity, image and identity. *Human Relations* 54(7), 863–886. doi: 10.1177/0018726701547004

American Petroleum Institute. (2015). *Pipeline Safety Management Systems: Recommended Practice 1173*. https://www.api.org/~/media/files/publications/whats%20new/1173_e1%20pa.pdf

Andersen, S. S. (2005). *Case-Studier og Generalisering* (In Norwegian. Case Studies and Generalization). Bergen: Fagbokforlaget.

Antonsen, S. (2009). Safety culture and the issue of power. *Safety Science* 47, 183–191.

Antonsen, S. (2009). Safety culture assessment: A mission impossible? *Journal of Contingencies and Crisis Management* 17(4), 242–254.

Antonsen, S. (2009). *Safety Culture: Theory, Method and Improvement*. Farnham: Ashgate.

Antonsen, S., Ramstad, L. S., & Kongsvik, T. (2007). Unlocking the organization: Action research as a means of improving organizational safety. *Safety Science Monitor* 11(1), Article 4.

Ashby, W. R. (1958). Requisite variety and its implications for the control of complex systems. *Cybernetica* 1(2), 83–99.

Australian Transport Safety Bureau. (2011). A systematic review of the effectiveness of safety management systems. https://www.atsb.gov.au/media/4053559/xr2011002_final.pdf

Aven, T. (2014). What is safety science? *Safety Science* 67, 15–20. doi: 10.1016/j.ssci.2013.07.026

Aven, T. (2019). *The Science of Risk Analysis: Foundation and Practice*. New York: Routledge.

Bailyn, L. (1977). Research as a cognitive process: Implications for data analysis. *Quality and Quantity* 11, 97–117.

Barad, K. (2007). *Meeting the Universe Halfway: Quantum Physics and the Entanglement of Matter and Meaning*. Duke University Press.

Barad, K. (2011). Erasers and erasures: Pinch's unfortunate 'uncertainty principle'. *Social Studies of Science* 41(3), 443–454.

Barley, S. R. (1986). Technology as an occasion for structuring: Evidence from observations of CT scanners and the social order of radiology departments. *Administrative Science Quarterly* 31(1), 78–108.

Barth, F. (1994). *Manifestasjon og Prosess*. Oslo: Universitetsforlaget.

Bateson, G. (1972). *Steps to an Ecology of Mind*. New York: Ballantine.

Beals, F., Kidman, J., & Funaki, H. (2020). Insider and outsider research: Negotiating self at the edge of the emic/etic divide. *Qualitative Inquiry* 26(6), 593–601. doi: 10.1177/1077800419843950

Beamish, T. (2002). *Silent Spill, The Organization of an Industrial Crisis*. Cambridge, MA: MIT Press.

Beck, U. (1992). *Risk Society: Towards a New Modernity*. London: SAGE.

Becker, H. S. (2014). *What about Mozart? What about Murder? Reasoning from Cases*. Chicago, IL: University of Chicago Press.

Becker, H. S. (2017). *Evidence*. Chicago, IL: University of Chicago Press.

Bell, C., & Roberts, H. (eds.). (1984). *Social Researching: Politics, Problems, Practice*. London: Routledge & Kegan Paul.

Benning, A., Ghaleb, M., Suokas, A., Dixon-Woods, M., et al. (2011). Large scale organisational intervention to improve patient safety in four UK hospitals: Mixed method evaluation. *BMJ* 342, d195.

Bergs, J., Lambrechts, F., Simons, P., Vlayen, A., et al. (2015). Barriers and facilitators related to the implementation of surgical safety checklists: A systematic review of the qualitative evidence. *BMJ Quality & Safety* 24(12), 776–786.

Bergstrom, J. (2020). The discursive effects of safety science. In: J.-C. Le Coze (ed.), *Safety Science Research: Evolution, Challenges and New Directions*. Boca Raton, FL: CRC Press, pp. 173–186.

Bernard, H. R. (2006). *Research Methods in Anthropology* (5th ed.). Lanham, MD: AltaMira Press.

Berti, M. (2017). *Elgar Introduction to Organizational Discourse Analysis*. Edward Elgar Publishing.

Bieder, C., & Bourrier, M. (2013). *Trapping Safety into Rules: How Desirable or Avoidable is Proceduralization?* Farnham: Ashgate & CRC Press.

Bieder, C., & Gould, K. P. (2020). Exploring the interrelations between safety and security: Research and management challenges. In: Bieder, C., & Gould, K. P. (eds.), *The Coupling of Safety and Security*. Cham: Springer Open, pp. 105–113.

Bieder, C., & Gould, K. P. (eds.). (2020). *The Coupling of Safety and Security: Exploring Interrelations in Theory and Practice*. New York: Springer International Publishing.

Bion, J., Richardson, A., Hibbert, P., Beer, J., et al. (2013). "Matching Michigan": A 2-year stepped interventional programme to minimise central venous catheter-blood stream infections in intensive care units in England. *BMJ Quality & Safety* 22(2), 110–123.

Blockley, D. I. (1980). *The Nature of Structural Design and Safety*. Chichester: Ellis-Horwood.

Blockley, D. I. (1992). *Engineering Safety*. London: McGraw-Hill.

Bloor, D. (1976). The strong programme in the sociology of knowledge. *Knowledge and Social Imagery* 2, 3–23.

Bloor, D. (1999). Anti-latour. *Studies in History and Philosophy of Science Part A* 30(1), 81–112.

Blumer, H. (1969). *Symbolic Interactionism. Perspective and Method*. Los Angeles, CA: University California Press.

Bochatay, N. (2018). *Continuities in Health Care Work: Group Processes and Training at Two Academic Medical Centers in Switzerland and California* (Doctoral dissertation, University of Geneva).

Boellstorff, T., Nardi, B., Pearce, C., & Taylor, T. L. (2012). *Ethnography and Virtual Worlds: A Handbook of Method*. Princeton University Press.

Boje, D. M. (1995). Stories of the storytelling organization: A postmodern analysis of Disney as "Tamara-Land." *Academy of Management Journal* 38(4), 997–1035. doi: 10.2307/256618

Bonafede, M., Corfiati, M., Gagliardi, D., Boccuni, F., Ronchetti, M., Valenti, A., Iavicoli, S. (2016). OHS management and employers' perception: Differences by firm size in a large Italian company survey. *Safety Science* 89, 11–18. doi: 10.1016/j.ssci.2016.05.012

Born, G. (2011). *Uncertain Vision: Birt, Dyke and the Reinvention of the BBC*. London: Random House.

Borys, D. (2009). Exploring risk-awareness as a cultural approach to safety: Exposing the gap between work as imagined and work as actually performed. *Safety Science Monitor* 13, 1–11.

Bosk, C. (2001). Irony, ethnography, and informed consent. In: Hoffmaster, B. (ed.), *Bioethics in Social Context*. Philadelphia, PA: Temple University Press, pp. 199–220.

Bosk, C., & De Vries, R. (2008). Bureaucracies of mass deception: Institutional review boards and the ethics of ethnographic research. In: C. Bosk (dir.), *What Would You Do? Juggling Bioethics and Ethnography*. Chicago, IL: University of Chicago Press, pp. 87–99.

Bourrier, M. (1996). Organizing maintenance work at two American nuclear power plants. *Journal of Contingencies and Crisis Management* 4(2), 104–112.

Bourrier, M. (1998). Elements for designing a self-correcting organisation: Examples from nuclear plants. In: Hale, A., & Baram, M. (eds.), *Safety Management: The Challenge of Change*. Oxford: Pergamon, pp. 133–146.

Bourrier, M. (1999). *Le nucléaire à l'épreuve de l'organisation*. Paris: Presses Universitaires de France.

Bourrier, M. (2002). Bridging research and practice: The challenge of "normal operations" studies. *Journal of Contingencies and Crisis Management* 10(4), 173–180.

Bourrier, M. (2011). The legacy of the high reliability organization project. *Journal of Contingencies and Crisis Management* 19(1), 9–13.

Boyd, C. M., & Kent, D. M. (2014). Evidence-based medicine and the hard problem of multimorbidity. *Journal of General Internal Medicine* 29(4), 552–553.

Braithwaite, J. (2018). Changing how we think about healthcare improvement. *BMJ* 361, k2014.

Braithwaite, J., Churruca, K., Long, J. C., Ellis, L. A., & Herkes, J. (2018). When complexity science meets implementation science: A theoretical and empirical analysis of systems change. *BMC Medicine* 16, 63. doi: 10.1186/s12916-018-1057-z

Braithwaite, J., Hollnagel, E., & Hunte, G. S. (eds.). (2019). *Resilient Health Care Volume 5: Working across Boundaries*. Boca Raton, FL: Routledge.

Braithwaite, J., Wears, R. L., & Hollnagel, E. (2015). Resilient health care: Turning patient safety on its head. *International Journal of Quality in Health Care* 27(5), 418–420. doi: 10.1093/intqhc/mzv063

Bråten, M., Hovi, B., Jensen, R. S., Leiren, M. D., & Skollerud, K. H. (2013). *Arbeidsforhold i Vegsektoren. Et Forprosjekt*.

Braut, G. S. (2011). The requirement to practice in accordance with sound professional standards. In: Molven, O., & Ferkis, J. (eds.), *Healthcare, Welfare and Law. Health Legislation as a Mirror of the Norwegian Welfare State*. Oslo, NO: Gyldendal Akademisk, pp. 139–149.

Brennan, T. A., Leape, L. L., Laird, N. M., Hebert, L., et al. (1991). Incidence of adverse events and negligence in hospitalized patients: Results of the Harvard Medical Practice Study I. *New England Journal of Medicine* 324(6), 370–376.

Broadhead, R. S., & Rist, R. C. (1976). Gatekeepers and the social control of social research. *Social Problems* 23(3), 325–336.

Bryman, A. (ed.). (1988). *Doing Research in Organizations*. New York: Routledge.

Buchanan, D., & Bryman, A. (2009). *The Sage Handbook of Organizational Research Methods*. London: SAGE Publications Ltd.

Bulmer, M. (1979). Concepts in the analysis of qualitative data. In: M. Bulmer (ed.), *Sociological Research Methods*. London: Macmillan, pp. 241–262.

Burawoy, M. (1998). The extended case method. *Sociological Theory* 6(1), 4–33.

Burawoy, M. (2003). Manufacturing the global. *Ethnography* 2(2), 147–159.

Burawoy, M. (2004). For public sociology. *American Review of Sociology* 70(1), 4–28.

Burawoy, M. (2009). Challenges for a global sociology. *Contexts* 8(4), 36–41.

Busby, J. S., & Alcock, R. E. (2008). Risk and organizational networks: Making sense of failure in the division of labour. *Risk Management* 10(4), 235–256.

Byrne, D., & Callaghan, G. (2013). *Complexity Theory and the Social Sciences: The State of the Art*. London: Routledge.

Caelli, K., Ray, L., & Mill, J. (2003). 'Clear as mud': Toward greater clarity in generic qualitative research. *International Journal of Qualitative Methods* 2(2), 1–13.

CAIB (Columbia Accident Investigation Board). (2003). *Report*, Vol. 1. Washington, DC: NASA.

Callon, M. (1986). Some elements of a sociology of translation: Domestication of the scallops and the fishermen of St Brieuc Bay. In: J. Law (ed.), *Power, Action and Belief: A New Sociology of Knowledge*. London: Routledge, pp. 196–223.

Callon, M., & Latour, B. (1981). Unscrewing the big Leviathan: How actors macro-structure reality and how sociologists help them to do so. In: Knorr-Cetina, K. D., & Cicourel, A. V. (eds.), *Advances in Social Theory and Methodology: Toward an Integration of Micro-and Macro-Sociologies*. Boston, MA: Routledge and Kegan Paul, pp. 277–303.

Callon, M., & Law, J. (1982). On interests and their transformation: Enrolment and counter-enrolment. *Social Studies of Science* 12(4), 615–625.

Campbell, M., Fitzpatrick, R., Haines, A., Kinmonth, A. L., et al. (2000). Framework for design and evaluation of complex interventions to improve health. *BMJ* 321, 694–696.

Carson, W. G. (1982). *The Other Price of Britain's Oil*. Oxford: Martin Robertson.

Cefkin, M. (ed.). (2010). *Ethnography and the Corporate Encounter: Reflections on Research in and of Corporations*. New York: Berghahn Books.

Charmaz, K. (2000). Constructivist and objectivist grounded theory. In: N. K. Denzin, & Y. Lincoln (eds.), *Handbook of Qualitative Research* (2nd ed.). Thousand Oaks, CA (pp. 509–535).

Chen, Y.-F., Hemming, K., Stevens, A. J., & Lilford, R. J. (2016). Secular trends and evaluation of complex interventions: The rising tide phenomenon. *BMJ Quality & Safety* 25(5), 303–310.

Chew, S., Armstrong, N., & Martin, G. (2013). Institutionalising knowledge brokering as a sustainable knowledge translation solution in healthcare: How can it work in practice? *Evidence & Policy* 9(3), 335–351.

Cifkin, M. (ed.). (2010). *Ethnography and the Corporate Encounter, Reflections on Research in and of Corporations*. New York and Oxford: Berghahn Books.

Clarke, A. (2003). Situational analyses: Grounded theory after the postmodern turn. *Symbolic Interaction* 26(4), 553–576.

Clarke, A. (2009). From grounded theory to situational analysis: What's new? Why? How? In: J. M. Morse, P. N. Stern, J. M. Cordin, K. C. Charmaz, B. Bowers, & A. E. Clarke (eds.), *Developing Grounded Theory: The Second Generation*. Walnut Creek, CA: Left Coast Press, Inc., 194–233.

Clarke, L. (2001). *Mission Improbable: Using Fantasy Documents to Tame Disaster*. Chicago, IL: University of Chicago Press, 1999. Paperback 2001.

Clarke, L., & Short, J. F. (1993). Social organization and risk: Some current controversies. *Annual Review of Sociology* 19(1), 375–399.

Clay-Williams, R., & Colligan, L. (2015). Back to basics: Checklists in aviation and healthcare. *BMJ Quality & Safety* 24(7), 428–431.

Cochrane, A. L. (1972). *Effectiveness and Efficiency: Random Reflections on Health Services*. London: Nuffield Provincial Hospitals Trust.

Coghland, D., & Brannick, T. (2010). *Doing Action Research in Your Own Organization* (3rd ed.). London: SAGE.

Coleman, S. (1996). Obstacles and opportunities in access to professional work organizations for long-term fieldwork: The case of Japanese laboratories. *Human Organization* 55(3), 334–343. doi: 10.17730/humo.55.3.l38mh1588247263r

Collingridge, D. (1980). *The Social Control of Technology*. Milton-Keynes: Open University Press.

Comfort, L. K., Boin, A., & Demchak, C. C. (eds.). (2010). *Designing Resilience. Preparing for Extreme Events*. Pittsburgh, PA: University of Pittsburgh Press.

Conti, J. A., & O'Neil, M. (2007). Studying power: Qualitative methods and the global elite. *Qualitative Research* 7(1), 63–82. doi: 10.1177/1468794107071421

Cooperrider, D. L. (2018). *The Appreciative Inquiry Handbook* (2nd ed.). Berrett-Koehler.

Corbin, J. M., & Strauss, A. (1990). Grounded theory research: Procedures, canons, and evaluative criteria. *Qualitative Sociology* 13(1), 3–21. https://doi.org/10.1007/BF00988593

Council of Europe. (2019). *Guide on Article 6 of the European Convention on Human Rights*. Strasbourg: Council of Europe & European Court of Human Rights. https://www.echr.coe.int/Documents/Guide_Art_6_criminal_ENG.pdf

Craig, P., Dieppe, P., Macintyre, S., Michie, S., et al. (2008). Developing and evaluating complex interventions: The new Medical Research Council guidance. *BMJ* 337, a1655.

Crompton, R., & Jones, G. (1988). Researching white collar organizations: Why sociologists should not stop doing case studies. In: A. Bryman (ed.), *Doing Research in Organizations*. London, NY: Routledge, pp. 68–81.

Cunliffe, A. L. (2011). Crafting qualitative research: Morgan and Smircich 30 years on. *Organizational Research Methods* 14(4), 647–673. doi: 10.1177/1094428110373658

Cupit, C., Mackintosh, N., & Armstrong, N. (2018). Using ethnography to study improving healthcare: Reflections on the "ethnographic" label. *BMJ Quality and Safety* 27(4), 258–260. doi: 10.1136/bmjqs-2017-007599

Danner, C., & Schulman, P. (2018). Rethinking risk assessment for public utility safety regulation. *Risk Analysis* 39(5) (November), 1044–1059.

Davis, M. S. (1971). That's interesting!: Towards a phenomenology of sociology and a sociology of phenomenology. *Philosophy of the Social Sciences* 1(2), 309–344. doi: 10.1177/004839317100100211

Davoudian, K., Wu, J. S., & Apostolakis, G. (1994). Incorporating organizational factors into risk assessment through the analysis of work processes. *Reliability Engineering and System Safety* 45(1–2), 85–105.

De Vries, E. N., Prins, H. A., Crolla, R. M. P. H., den Outer, A. J., et al. (2010). Effect of a comprehensive surgical safety system on patient outcomes. *New England Journal of Medicine* 363(20), 1928–1937.

Deeb, H. N:, & Marcus, G. E. (2011). In the green room: An experiment in ethnographic method at the WTO. *PoLAR: Political and Legal Anthropology Review* 34(1), 51–76.

Dekker, S. (2004). *Ten Questions about Human Error: A New View of Human Factors and System Safety* (1st ed.). CRC Press.

Dekker, S. (2012). Resilience engineering: Chronicling the emergence of confused consensus. In: *Resilience Engineering: Concepts and Precepts*. Ashgate, pp. 77–92. https://research-repository.griffith.edu.au/handle/10072/63581

Dekker, S. (2014). Deferring to expertise versus the prima donna syndrome: A manager's dilemma. *Cognition, Technology & Work* 16(4), 541–548. doi: 10.1007/s10111-014-0284-0

Dekker, S. (2016). *Drift into Failure: From Hunting Broken Components to Understanding Complex Systems*. Boca Raton, FL: CRC Press.

Dekker, S., & Pruchnicki, S. (2014). Drifting into failure: Theorising the dynamics of disaster incubation. *Theoretical Issues in Ergonomics Science* 15(6), 534–544.

Dekker, S. W. A. (2005). *Ten Questions about Human Error*. Hillsdale, NJ: Lawrence Erlbaum Associates.

Dekker, S. W. A. (2020). Safety after neoliberalism. *Safety Science* 125, 104630.

Dekker, S. W. A., Cilliers, P., & Hofmeyr, J. H. (2011). The complexity of failure. Implications of complexity theory for safety investigations. *Safety Science* 49(6), 939–945.

Dekker, S. W. A., & Nyce, J. M. (2014). There is safety in power, or power in safety. *Safety Science* 67, 44–49.

Demchak, C. C. (1991). *Military Organizations, Complex Machines: Modernization in the US Armed Services*. Ithaca, NY: Cornell University Press.

Denny, R., & Sunderland, P. (eds.). (2014). *Handbook of Anthropology in Business*. Walnut Creek, CA: Routledge.

Department of Health. (2000). *An Organisation with a Memory.* London: Stationery Office.

Department of Health. (2006). *Best Research for Best Health: A New National Research Strategy.* London: Department of Health.

Dexter, L. (1970). *Elite and Specialized Interviewing.* Evanston, IL: Northwestern University Press.

Dezalay, Y., & Garth, B. G. (1996). *Dealing in Virtue: International Commercial Arbitration and the Construction of a Transnational Legal Order.* Chicago, IL: University of Chicago Press.

Dixon-Woods, M. (2003). What can ethnography do for quality and safety in health care? *Quality and Safety in Health Care* 12(5), 326–327. doi: 10.1136/qhc.12.5.326

Dixon-Woods, M., Bosk, C. L., Aveling, E.-L., Goeschel, C. A., et al. (2011). Explaining Michigan: Developing an ex post theory of a patient safety program. *Milbank Quarterly* 89(2), 167–205.

Dixon-Woods, M., Leslie, M., Tarrant, C., & Bion, J. (2013). Explaining matching Michigan: An ethnographic study of a patient safety program. *Implementation Science* 8, 70.

Dixon-Woods, M., & Martin, G. P. (2016). Does quality improvement improve quality? *Future Hospital Journal* 3(3), 191–194.

Dixon-Woods, M., & Shojania, K. G. (2016). Ethnography as a methodological descriptor: The editors' reply. *BMJ Quality and Safety* 25(7), 555–556. doi: 10.1136/bmjqs-2015-005117

Domeneghetti, B., Benamrane, Y., & Wybo, J.-L. (2018). Analyzing nuclear expertise support to population protection decision making process during nuclear emergencies. *Safety Science* 101, 155–163.

Dourish, P., & Anderson, K. (2006). Collective information practice: Exploring privacy and security as social and cultural phenomena. *Human–Computer Interaction* 21(3), 319–342.

Dourish, P., Grinter, E., Delgado de la Flor, J., & Joseph, M. (2004). Security in the wild: User strategies for managing security as an everyday, practical problem. *Personal and Ubiquitous Computing* 8(6), 391–401.

Downer, J. (2011). "737-Cabriolet": The limits of knowledge and the sociology of inevitable failure 1. *American Journal of Sociology* 117(3), 725–762.

Doyle, L., Brady, A.-M., & Byrne, G. (2009). An overview of mixed methods research. *Journal of Research in Nursing* 14(2), 175–185.

Durkheim, É. (1951). *Suicide: A Study in Sociology.* New York: Free Press.

Eckstein, H. (1975). Case study and theory in political science. In: I. Greenstein & N. W. Polsby (eds.), *Handbook of Political Science.* Reading, MA: Addison-Wesley, pp. 79–133.

Emery, F. E. (1959). *Characteristics of Socio-Technical Systems.* London: Tavistock Documents, no. 527.

Empson, L. (2018). Elite interviewing in professional organizations. *Journal of Professions and Organization* 5(1), 58–69. doi: 10.1093/jpo/jox010

Fairclough, N. (2003). *Analysing Discourse: Textual Analysis for Social Research.* London: Routledge.

Feldman, M. S., Bell, J., & Berger, M. T. (eds.). (2003). *Gaining Access.* Walnut Creek, CA: Altamira Press.

Fleming, M. (2000). *Safety Culture Maturity Model.* HSE Offshore Technology Report. Sudbury, UK.

Flin, R., O'Connor, P., & Mearns, K. (2002). Crew resource management: Improving team work in high reliability industries. *Team Performance Management. Anais: An International Journal* 8, 68–78.

Flynn, L. C., McCulloch, P. G., Morgan, L. J., Robertson, E. R., et al. (2016). The safer delivery of surgical services program (S3): Explaining its differential effectiveness and exploring implications for improving quality in complex systems. *Annals of Surgery* 264(6), 997–1003.

Forsythe, D. (2001). *Studying Those Who Study Us: An Anthropologist in the World of Artificial Intelligence.* Stanford, CA: Stanford University Press.

Fox, N. J. (2006). Postmodern field relations in health research. In: D. Hobbs & R. Wright (eds.), *The Sage Handbook of Fieldwork.* London: SAGE.

Francis, R. (2013). *Report of the Mid Staffordshire NHS Foundation Trust Public Inquiry Executive Summary.* London: Stationery Office.

Francis, R. (2015). *Freedom to Speak Up.* London: TSO. http://webarchive.nationalarchives. gov.uk/20150218150343/https://freedomtospeakup.http://org.uk/wp-content/uploads /2014/07/F2SU_Executive-summary.pdf

Freeman, D. (1983). *Margaret Mead and Samoa: The Making and Unmaking of an Anthropological Myth.* Cambridge: Harvard University Press.

Fruhen, L. S., & Flin, R. (2016). "Chronic unease" for safety in senior managers: An interview study of its components, behaviours and consequences. *Journal of Risk Research* 19(5), 645–663.

Funtowicz, S. O., & Ravetz, J. R. (1992). Three types of risk assessment and the emergence of post-normal science. In: S. Krimsky & D. Golding (eds.), *Social Theories of Risk.* Westport, CT: Praeger, pp. 251–274.

Fylkesmannen i Sør-Trøndelag. (2017). *Styrket involvering av pasienter, tjenestemottakere og pårørende i tilsyn.* Rapport. Trondheim: Fylkesmannen i Sør-Trøndelag.

Gagliardi, P. (ed.). (1990). *Symbols and Artifacts: Views of the Corporate Landscape.* Berlin: Walter de Gruyter.

Garfinkel, H. (1996). Ethnomethodology's program. *Social Psychology Quarterly* 59(1), 5–21.

Ge, J., Xu, K., Wu, C., Xu, Q., Yao, X., Li, L., Xu, X., Sun, E., Li, J., & Li, X. (2019). What is the object of safety science? *Safety Science* 118, 907–914. doi: 10.1016/j.ssci.2019.06.029

Geertz, C. (1973). *The Interpretation of Cultures.* New York: Basic. See especially chapter 1, Thick description: Towards an interpretive theory of culture.

Geertz, C. (2017). *The Interpretation of Cultures* (1st ed.). Perseus.

Gephart, R., Steier, L., & Lawrence, T. (1990). Cultural rationalities in crisis sensemaking: A study of a public inquiry into a major industrial accident. *Industrial Crisis Quarterly* 4, 27–48.

Gerbec, M., & Kontic, B. (2017). Safety related key performance indicators for securing long-term business development – A case study. *Safety Science* 98, 77–88.

Gerring, J. (1999). What makes a concept good. A criterial framework for understanding concept development in the social sciences. *Polity* 31(3) (Spring), 357–393.

Gerring, J., & Baressi, P. (2003). Putting ordinary language to work: A mini-max strategy for concept formation in the social sciences. *Journal of Theoretical Politics* 15(2), 201–232.

Gherardi, S. (1998). A cultural approach to disasters. *Journal of Contingencies and Crisis Management* 6(2), 80–83.

Gherardi, S. (2004). Translating knowledge while mending organizational safety culture. *Risk Management: An International Journal* 6(2), 61–80.

Gherardi, S. (2006). *Organizational Knowledge: The Texture of Workplace Learning.* Oxford: Blackwell.

Gherardi, S. (2019). *How to Conduct a Practice-Based Study: Problems and Methods* (2nd ed.). Cheltenham: Edward Elgar.

Gherardi, S., Jensen, K., & Nerland, M. (2017). Shadow organizing: A metaphor to explore organizing as intra-relating. *Qualitative Research in Organizations and Management: An International Journal* 12(1), 2–17. doi: 10.1108/QROM-06-2016-1385

Gherardi, S., & Nicolini, D. (2000). The organizational learning of safety in communities of practice. *Journal of Management Inquiry* 9(1), 7–18.

Gherardi, S., Nicolini, D., & Odella, F. (1998). What do you mean by safety? Conflicting perspectives on accident causation and safety management in a construction firm. *Journal of Contingencies and Crisis Management* 6(4), 202–213.

Gherardi, S., & Turner, B. A. (1987). *Real Men Don't Collect Soft Data. Publisher.* Università di Trento, Dipartimento di Politica Sociale. http://eprints.biblio.unitn.it/4319/

Giddens, A. (1990). *Modernity and Self-Identity.* Cambridge: CUP.

Giddens, A. (1991). *Modernity and Self-Identity: Self and Society in the Late Modern Age.* Cambridge: Polity Press.

Glaser, B. (1978). *Theoretical Sensitivity Advances in the Methodology of Grounded Theory.* Mill Valley, CA: Sociology Press.

Glaser, B. G., & Strauss, A. L. (1967). *The Discovery of Grounded Theory: Strategies for Qualitative Research.* Chicago, IL: Aldine Publishing Company.

Glaser, B. G., & Strauss, A. L. (1999). *Discovery of Grounded Theory. Strategies for Qualitative Research.* New York: Routledge.

Gluckman, M. (1963). Gossip and scandal. *Current Anthropology* 4(3), 307–316.

Gobo, G. (2008). *Doing Ethnography.* London: SAGE.

Goffman, E. (1959). *The Presentation of Self in Everyday Life.* Garden City, NJ: Doubleday.

Goffman, E. (1968). *Asylums.* Harmondsworth: Penguin.

Goldman, M. (2006). *Imperial Nature: The World Bank and Struggles for Social Justice in the Age of Globalization.* New Haven, CT: Yale University Press.

Goodwin, C. (1994). Professional vision. *American Anthropologist* 96(3), 606–633.

Gould, K. P. (2020). Organizational risk: "Muddling through" 40 years of research. *Risk Analysis: An Official Publication of the Society for Risk Analysis.* doi: 10.1111/risa.13460

Gould, K. P., & Fjaeran, L. (2020). Drift and the social attenuation of risk. In: Le Coze, J. C. (ed.), *Safety Science Research: Evolution, Challenges and New Directions.* Boca Raton, FL: CRC Press, pp. 119–132.

Gouldner, A. W. (1954). *Patterns of Industrial Bureaucracy.* Free Press.

Greenhalgh, T., Howick, J., & Maskrey, N. (2014). Evidence based medicine: A movement in crisis? *BMJ* 348, g3725.

Greenwood, D. J., & Levin, M. (1998). *Introduction to Action Research: Social Research for Social Change* (2nd ed.). Thousand Oaks, CA: SAGE.

Greenwood, D. J., Whyte, W. F., & Harkavy, I. (1993). Participatory action research as a process and as a goal. *Human Relations* 46(2), 175–192. doi: 10.1177/001872679304600203

Grey, C., Anderson-Gough, F., & Robson, K. (1998). *Making Up Accountants: The Organizational and Professional Socialization. of Trainee Chartered Accountants.* London: Ashgate.

Grey, C., & Costas, J. (2016). *Secrecy at Work: The Hidden Architecture of Organizational Life.* Stanford, CA: Stanford University Press.

Guba, E. G., & Lincoln, Y. S. (1994). Competing paradigms in qualitative research. In: *Handbook of Qualitative Research.* Sage Publications, Inc., pp. 105–117.

Haavik, T. K. (2011). On components and relations in sociotechnical systems. *Journal of Contingencies and Crisis Management* 19(2), 99–109.

Haavik, T. K. (2013). *New Tools, Old Tasks: Safety Implications of New Technologies and Work Processes for Integrated Operations in the Petroleum Industry.* Farnham: Ashgate.

Haavik, T. K. (2014). On the ontology of safety. *Safety Science* 67, 37–43.

Haavik, T. K. (2014). Sensework. *Computer Supported Cooperative Work (CSCW)* 23(3), 269–298. doi: 10.1007/s10606-014-9199-9

Haavik, T. K. (2016). Keep your coats on: Augmented reality and sensework in surgery and surgical telemedicine. *Cognition, Technology and Work* 18(1), 175–191.

Haavik, T. K. (2017). Remoteness and sensework in harsh environments. *Safety Science* 95, 150–158.

Haavik, T. K. (2020). Societal resilience – Clarifying the concept and upscaling the scope. *Safety Science* 132. doi: 10.1016/j.ssci.2020.104964

Haavik, T. K., Antonsen, S., Rosness, R., & Hale, A. (2019). HRO and RE: A pragmatic perspective. *Safety Science* 117, 479–489.

Haavik, T. K., Kongsvik, T., Bye, R. J., Røyrvik, J. O. D., and Almklov, P. G. (2017). Johnny was here: From airmanship to airlineship." *Applied Ergonomics* 59(A), 191–202.

Habenstein, R. W. (ed.). (1970). *Pathways to Data – Craft and Methods for Studying Social Organizations*. New York: Routledge.

Hadi, D. N., & Marcus, G. E. (2011). In the green room: An experiment in ethnographic method at the WTO. *PoLAR: Political and Legal Anthropology Review* 34(1), 51–76. doi: 10.1111/j.1555-2934.2011.01138.x

Hale, A. (2000). Culture's confusions. *Safety Science* 34, 1–14.

Hammond, P. E. (ed.). (1964). *Sociologists at Work: Essays Om the Craft of Social Research*. New York: Basic Books, Inc.

Harvey, W. S. (2011). Strategies for conducting elite interviews. *Qualitative Research* 11(4), 431–441. doi: 10.1177/1468794111404329

Haukelid, K. (2008). Theories of (safety) culture revisited – An anthropological approach. *Safety Science* 46, 413–426.

Havinga, J., Dekker, S. W. A., & Rae, A. J. (2017). Everyday work investigations for safety. *Theoretical Issues in Ergonomics Science* 19(2), 1–16. doi: 10.1080/1463922X.2017.1356394

Hayes, J. (2012). Operator competence and capacity – Lessons from the Montara blowout. *Safety Science* 50(3), 563–574.

Hayes, J. (2013). *Operational Decision-Making in High-Hazard Organizations: Drawing a Line in the Sand*. Farnham: Ashgate.

Hayes, J. (2015a). Investigating design office dynamics that support safe design. *Safety Science* 78, 25–34.

Hayes, J. (2015b). Taking responsibility for public safety: How engineers seek to minimise disaster incubation in design of hazardous facilities. *Safety Science* 77, 48–56.

Hayes, J., & Hopkins, A. (2014). *Nightmare Pipeline Failures: Fantasy Planning, Black Swans and Integrity Management*. Sydney: CCH.

Hayes, J., & Maslen, S. (2015). Knowing stories that matter: Learning for effective safety decision-making. *Journal of Risk Research* 18(6), 714–726.

Hayes, J., & Maslen, S. Putting senior management in focus: A critical reflection on the direction of safety science research. In: K. Pettersen Gould & C. Macrae (eds.), *Inside Hazardous Technological Systems: Methodological Foundation, Challenges and Future Directions*. London: CRC Press.

Haynes, A. B., Weiser, T. G., Berry, W. R., Lipsitz, S. R., et al. (2009). A surgical safety checklist to reduce morbidity and mortality in a global population. *New England Journal of Medicine* 360(5), 491–499.

Health Foundation. (2011). *Learning Report: Safer Patients Initiative*. London: Health Foundation.

Helsetilsynet. (2011). System failure at Vestre Viken Hospital. https://www.helsetilsynet.no/h istorisk-arkiv/nyheter-og-pressemeldinger/nyheter-2004-2012/Systemsvikt-ved-Vestre -Viken-Ringerike-sykehus/

Helsetilsynet. (2015a). 4.6.2015. Investigation of adverse event – draft report sent to public hearing. https://www.helsetilsynet.no/no/Tilsyn/Tilsynssaker/Utkast-til-rapport-i-ti lsynssak-dodsfall-etter-postoperative-komplikasjoner-etter-tonsillektomi/

Helsetilsynet. (2015b). 18.11.2015. Final investigation report. https://www.helsetilsynet.no/u pload/tilsyn/varsel_enhet/Danielsaken-endelig-rapport-nov-2015.pdf

Helsetilsynet. (2015). Overview page of the Daniel case. https://www.helsetilsynet.no/tilsyn/ tilsynssaker/daniel-saken-samleside/

Helsetilsynet. (2019). Saman om betre tilsyn. Tilrådingar om brukarinvolvering i tilsyn. Rapport fra Helsetilsynet 2/2019. (Improving supervision together – Recommendation for user involvemen in supervision). https://www.helsetilsynet.no/publikasjoner/ra pport-fra-helsetilsynet/2019/saman-om-betre-tilsyn-tilradingar-om-brukerinvolvering -i-tilsyn/

Henwood, K. L., & Pidgeon, N. F. (1992). Qualitative research and psychological theorising. *British Journal of Psychology* 83, 97–111.

Henwood, K. L., & Pidgeon, N. F. (2014). *Risk and Identity Futures.* Foresight Future Identities Programme. London: Government Office of Science.

Hilgartner, S. (1992). The social construction of risk objects: Or, how to pry open networks of risk. In: J. F. Short & L. B. Clarke (eds.), *Organizations, Uncertainties, and Risk.* Boulder, CO: Westview Press, pp. 39–53.

Hine, C. (2000). *Virtual Ethnography.* SAGE Publications Ltd. doi: 10.4135/9780857020277

Ho, K. (2009). *Liquidated: An Ethnography of Wall Street.* Durham, NC: Duke University Press.

Hoffman, R. R., & Militello, L. G. (2008). *Perspectives on Cognitive Task Analysis: Historical Origins and Modern Communities of Practice.* Psychology Press.

Hollnagel, E. (2006). The myth of human error in risk analysis and safety management. *Keynote Speech at RISIT Programme Research Conference.* Lillehammer.

Hollnagel, E. (2009). *The ETTO Principle: Efficiency-Thoroughness Trade-Off – Why Things that Go Right Sometimes Go Wrong.* Aldershot: Ashgate.

Hollnagel, E. (2012). *FRAM: The Functional Resonance Analysis Method: Modelling Complex Socio-Technical Systems.* Farnham: Ashgate.

Hollnagel, E. (2013). A tale of two safeties. *Nuclear Safety and Simulation* 4(1), 1–9.

Hollnagel, E. (2014). *Safety-I and Safety-II* (New ed.). Ashgate.

Hollnagel, E. (2017, March 2). Why is work-as-imagined different from work-as-done? *Resilient Health Care* 2. doi: 10.1201/9781315605739-24

Hollnagel, E. (2018). *Safety-I and Safety-II: The Past and Future of Safety Management.* Boca Raton, FL: CRC Press.

Hollnagel, E. (2018). *Safety-II in Practice: Developing the Resilience Potentials.* Taylor & Francis. doi: 10.4324/9781315201023

Hollnagel, E., Braithwaite, J., & Wears, R. L. (2013). *Resilient Health Care.* London: Ashgate.

Hollnagel, E., Nemeth, C. P., & Dekker, S. (2008). *Resilience Engineering Perspectives: Remaining Sensitive to the Possibility of Failure.* Aldershot: Ashgate.

Hollnagel, E., Pariès, J., Woods, D. D., & Wreathall, J. (2011). *Resilience Engineering in Practice: A Guidebook.* Farnham: Ashgate.

Hollnagel, E., Wears, R. L., & Braithwaite, J. (2015). *From Safety-I to Safety-II: A White Paper.* University of Southern Denmark, University of Florida, USA, and Macquarie University, Australia.

Hollnagel, E., Woods, D. D., & Leveson, N. (eds.). (2006). *Resilience Engineering: Concepts and Precepts.* Aldershot: Ashgate.

Hollnagel, E., Woods, D. D., & Leveson, N. (2007). *Resilience Engineering: Concepts and Precepts*. Farnham: Ashgate Publishing, Ltd.

Holstein, J. A., & Gubrium, J. F. (2013). *Handbook of Constructionist Research*. New York: Guilford Press.

Hood, C., Jones, D., Pidgeon, N. F., Turner, B., & Gibson, R. (1992). Risk management. Ch 6 of F. Warner (ed.), *Risk Analysis, Perception and Management: Report of a Royal Society Study Group*. London: Royal Society, pp. 135–192.

Hopkins, A. (1999). *Managing Major Hazards: The Lessons of the Moura Mine Disaster*. Sydney, NSW: Allen & Unwin.

Hopkins, A. (1999). The limits of normal accident theory. *Safety Science* 32(2), 93–102.

Hopkins, A. (2000). *Lessons from Longford: The ESSO Gas Plant Explosion*. Sydney, NSW: CCH.

Hopkins, A. (2001). Was three mile island a 'normal accident'? *Journal of Contingencies and Crisis Management* 9(2), 65–72.

Hopkins, A. (2006). Studying organisational cultures and their effects on safety. *Safety Science* 44(10), 875–899.

Hopkins, A. (2007). *The Problem of Defining High Reliability Organisations*. National Research Center for Occupational Safety and Health Regulation. January.

Hopkins, A. (2008). *Failure to Learn: The BP Texas City Refinery Disaster*. Sydney, NSW: CCH.

Hopkins, A. (2012). *Disastrous Decisions: The Human and Organisational Causes of the Gulf of Mexico Blowout*. Sydney, NSW: CCH.

Hopkins, A. (2014). Issues in safety science. *Safety Science* 67, 6–14. doi: 10.1016/j. ssci.2013.01.007

Hopkins, A. (2016). *Quiet Outrage. The Way of a Sociologist*. Sydney, NSW: CCH Press.

Hopkins, A. (2019). *Organising for Safety: How Structure Creates Culture*. Sydney, NSW: CCH.

Hopkins, A., & Maslen, S. (2015). *Risky Rewards: How Company Bonuses Affect Safety*. Surrey: Ashgate.

Horlick-Jones, T. (1996). Is safety a by-product of quality management? In D. Jones & C. Hood (eds.), *Accident and Design*. London: University College London Press.

Horlick-Jones, T., & Rosenhead, J. (2002). Investigating risk, organisations and decision support through action research. *Risk Management* 4(4), 45–63.

Howard-Payne, L. (2015). Glaser or Strauss? Considerations for selecting a grounded theory study. *South African Journal of Psychology*. doi: 10.1177/0081246315593071

Howick, J. H. (2011). *The Philosophy of Evidence-Based Medicine*. Oxford: Wiley.

Hudson, P. (2007). Implementing a safety culture in a major multi-national. *Safety Science* 45, 697–722.

Hughes, T. P. (1994). Technological momentum. In: Marx, L., & Roe Smith, M. (eds.), *Does Technology Drive History? The Dilemma of Technological Determinism*. Cambridge, MA: MIT Press.

Hulme, A, Stanton, N. A., Walker, G., Waterson, P., & Salmon, P. (2019). What do applications of systems thinking accident analysis methods tell us about accident causation? A systematic review of applications between 1990 and 2018. *Safety Science* 117(August 2019), 164–183.

INPO. (2012). *Traits of a Healthy Nuclear Safety Culture*. https://www.nrc.gov/docs/ML130 3/ML13031A707.pdf

Institute of Medicine. (1999). *To Err is Human: Building a Safer Health System*. Washington, DC: National Academy Press.

Isık, I. N., & Atasoylu, E. (2017). Occupational safety and health in North Cyprus: Evaluation of risk assessment. *Safety Science* 94, 17–25.

Jackall, R. (1988). *Moral Mazes: The World of Corporate Managers*. New York: Oxford University Press.

Janis, I. (1982). *Groupthink: Psychological Studies of Policy Decisions and Fiascos*. Boston, MA: Houghton.

Jeffcutt, P. (1999). From the industrial to the post-industrial subculture. *Organization Studies* 20(4), VII.

Jensen, R. S., Bråten, M., Jordfald, B., Leiren, M. D., Nævestad, T.-O., Skollerud, K. H., et al. (2014). *Arbeidsforhold i Gods og Turbil*.

Jones, M. M., Castle-Clarke, S., Manville, C., Gunashekar, S., & Grant, J. (2013). *Assessing Research Impact: An International Review of the Excellence in Innovation for Australia Trial*. Retrieved from Santa Monica, CA: https://www.rand.org/pubs/research_reports /RR278.html

Kalleberg, A. L., & Dunn, M. (2016). Good jobs, bad jobs in the gig economy. *Perspectives on work. 20*, (1–2), 10–74.

Kaspers, S., et al. (2017). Measuring safety in aviation: Empirical results about the relation between safety outcomes and safety management system processes, operational activities and demographic data. *Presario Conference*. https://pdfs.semanticscholar.org/a6 72/aed3cee7a253cd5b1fe3e266de9b384f5ded.pdf?_ga=2.176329765.1417911500.1561 933416-1398805658.1561933416

Kelle, U. (2005). Emergence" vs. "forcing" of empirical data: A crucial problem of "grounded theory" reconsidered. *Forum: Qualitative Social Research* 6(2). http://nbn-resolving.de /urn:nbn:de:0114-fqs0502275.

Kemmis, S., Nixon, R., & McTaggart, R. (2014). *The Action Research Planner. Doing Critical Participatory Action Research*. London: Springer.

Kendra, J. M., & Wachtendorf, T. (2003). Elements of resilience after the world trade center disaster. Reconstituting New York City's emergency operations center. *Disasters* 27(1), 37–53.

Kennedy, I. (2001). *Learning from Bristol: The Report of the Public Inquiry into Children's Heart Surgery at the Bristol Royal Infirmary 1984–1995*. London: Stationary Office, Crown Copyright.

King, N., & Brooks, J. M. (2015). *Template Analysis for Business and Management Students*. London: SAGE Publications.

Kitchin, R. (2014). Big Data, new epistemologies and paradigm shifts. *Big Data & Society* 1(1), 2053951714528481.

Klein, G. (2008). Naturalistic decision making. *Human Factors: The Journal of the Human Factors and Ergonomics Society* 50(3), 456–460.

Knaapen, L. (2013). Being "evidence-based" in the absence of evidence: The management of non-evidence in guideline development. *Social Studies of Science* 43(5), 681–706.

Knowles, W., Prince, D., Hutchison, D., Disso, J. F. P., & Jones, K. (2015). A survey of cyber security management in industrial control systems. *International Journal of Critical Infrastructure Protection* 9(C), 52–80.

Kok, J., Leistikov, I., & Bal, R. (2018). Patient and family engagement in incident investigations: Exploring hospital manager and incident investigators' experiences and challenges. *Journal of Health Services Research & Policy* 23(4), 252–261. doi: 10.1177/1355819618788586

Kongsvik, T. (2003). *Hvilke Barrierer? Ansattes Vurdering av sider ved Sikkerhetskulturen – Snorre A (Which Barriers? Employees' Perception of Aspects of Safety Culture – Snorre Alpha)*. Trondheim: Studio Apertura.

Kongsvik, T., & Bye, R. (2004). Alienation as an explanatory factor for increased risk on service vessels in the North Sea. In: Spitzer, C., Schmocker, U., & Dang, V. N. (eds.), *Probabilistic Safety Assessment and Management*. London: Springer.

Kongsvik, T., Fenstad, J., & Wendelborg, C. (2012). Between a rock and a hard place: Accident and near miss reporting on offshore vessels. *Safety Science* 50, 1839–1846.

Kongsvik, T., Haavik, T. K., Bye, R. J., & Almklov, P. G. (2020). Re-boxing seamanship: From individual to systemic professional competence. *Safety Science* 130, 104871.

Kriegler, R. (1980). *Working for the Company*. Melbourne, VIC: Oxford University Press.

Krimsky, S., & Golding, D. (eds.). (1992). *Social Theories of Risk*. Westport, CT: Praeger Publishers.

La Porte, T. R. (1982). On the design and management of nearly error free organizational control systems. In: Sills, D. L., Wolf, C. P., & Shelanski, V. B. (eds.), *The Accident and Three Mile Island: The Human Dimensions*. Boulder, CO: Westview Press, pp. 185–200.

La Porte, T. R. (1996). High reliability organizations: Unlikely, demanding and at risk. *Journal of Crisis and Contingency Management* 4(2), 60–71.

La Porte, T. R., & Consolini, S. M. (1991). Working in practice but not in theory: Theoretical challenges of "High Reliability Organizations". *Journal of Public Administration Research and Theory* 1(1), 19–47.

La Porte, T. R., & Consolini, P. (2008). Working in practice but not in theory: Theoretical challenges of "high-reliability organizations." In *Crisis Management* (Vol. 2). Sage Publications. http://politicsir.cass.anu.edu.au/staff/hart/pubs/40%20Rosenthal,%20t%20Hart%20&%20Kouzmin.pdf#page=63

La Porte, T. R., Roberts, K., & Rochlin, G. S. (1987). *High Reliability Organizations: The Research Challenge*. Mimeo, working paper.

La Porte, T. R., & Rochlin, G. (1994). A rejoinder to Perrow. *Journal of Contingencies and Crisis Management* 2(4), 221–227.

Lagadec, P. (1982). *Major Technological Risk*. Oxford: Pergamon Press.

Langner, R. (2011). Stuxnet: Dissecting a cyberwarfare weapon. *IEEE Security & Privacy* 9(3), 49–51.

Latour, B. (1986). Visualization and cognition: Thinking with eyes and hands. *Knowledge and Society: Studies in the Sociology of Culture Past and Present* 6, 1–40.

Latour, B. (1987). *Science in Action: How to Follow Scientists and Engineers through Society*. Milton Keynes: Open University Press.

Latour, B. (1992). Where are the missing masses? The sociology of a few mundane artifacts. In: W. E. Bijker & J. Law (eds.), *Shaping Technology/Building Society: Studies in Sociotechnical Change*. Cambridge, MA: MIT Press, pp. 225–264.

Latour, B. (1999). For David Bloor and beyond: A reply to David Bloor's 'anti-Latour'. *Studies in History and Philosophy of Science Part A* 30(1), 113–129.

Latour, B. (1999). *Pandora's Hope: Essays on the Reality of Science Studies*. Cambridge, MA: Harvard University Press.

Latour, B. (2004). *Politics of Nature: How to Bring the Sciences into Democracy*. Cambridge, MA: Harvard University Press.

Latour, B. (2005). *Reassembling the Social: An Introduction to Actor-Network-Theory, Clarendon Lectures in Management Studies*. Oxford: Oxford University Press.

Latour, B., & S. Woolgar. (1979). *Laboratory Life: The Social Construction of Scientific Facts*. Beverly Hills, CA: SAGE.

Latour, B., & S. Woolgar. (1986). *Laboratory Life: The Construction of Scientific Facts* (New ed.). Princeton, NJ: Princeton University Press.

Laurila, J. (1997). Promoting research access and informant rapport in corporate settings: Notes from research on a crisis company. *Scandinavian Journal of Management* 13(4), 407–418. doi: 10.1016/S0956-5221(97)00026-2

Law, J. (1992). Notes on the theory of the actor-network: Ordering, strategy, and heterogeneity. *Systems Practice* 5(4), 379–393.

Law, J. (2002). *Aircraft Stories: Decentering the Object in Technoscience*. Durham, NC: Duke University Press.

Law, J. (2012). Reality failures. In: J.-H. Passoth, B. Peuker, & M. Schillmeier (eds.), *Agency without Actors? New Approaches to Collective Action* (no. 58; issue 58). Oxfordshire: Routledge, pp. 146–160.

Law, J., & Lien, M. E. (2013). Slippery: Field notes in empirical ontology. *Social Studies of Science* 43(3), 363–78.

Lawrence, T., & Suddaby, R. (2006). Institutions and institutional work. In: *Handbook of Organization Studies* (vol. 2), pp. 215–254. doi: 10.4135/9781848608030.n7

Lawton, R., O'Hara, J. K., Sheard, L., Armitage, G., Cocks, K., Buckley, H., et al. (2017). Can patient involvement improve patient safety? A cluster randomised control trial of the patient reporting and action for a safe environment (PRASE) intervention. *BMJ Quality and Safety* 26(8), 622–631. doi: 10.1136/bmjqs-2016-005570

Le Coze, J. C. (2013). New models for new times. An anti-dualist move. *Safety Science* 59, 200–218.

Le Coze, J. C. (2013). Reflecting on Jens Rasmussen's legacy. A strong program for a hard problem. *Safety Science* 71, 123–141.

Le Coze, J. C. (2015). 1984–2014. Normal accident. Was Charles Perrow right for the wrong reasons ? *Journal of Contingencies and Crisis Management* 23(4), 275–286.

Le Coze, J. C. (2015). Reflecting on Jens Rasmussen's legacy (2) behind and beyond," a "constructivist turn". *Applied Ergonomics* 59(Pt B), 558–569.

Le Coze, J. C. (2017). Andrew Hopkins and the sociology of safety. *Journal of Contingencies and Crisis Management* 25(1), 51–53.

Le Coze, J. C. (2017). Globalisation and high risk systems. *Policy and Practice in Health and Safety* 15(1), 57–81.

Le Coze, J. C. (2018). Resilience, reliability and safety: Multilevel research challenges. In: Wiig S., & Fahlbruch B. (eds.), *Exploring Resilience. SpringerBriefs in Applied Sciences and Technology*. Cham: Springer. doi: 10.1007/978-3-030-03189-3_2

Le Coze, J.-C. (ed.). (2019). *Safety Sciences Research: Evolution, Challenges and New Directions*. London: CRC Press.

Le Coze, J. C. (2019). Vive la diversité!. High Reliability Organisation (HRO) and Resilience Engineering (RE). *Safety Science* 117, 469–478.

Le Coze, J. C. (2019). Storytelling or theory building? Hopkins' sociology of safety. *Safety Science* 120, 735–744.

Le Coze, J. C. (2020). *Post Normal Accident. Revisiting Perrow's Classic*. Boca Raton, FL: CRC Press, Taylor and Francis.

Le Coze, J. C., Pettersen, K., & Reiman, T. (2014). The foundations of safety science. *Safety Science* 67, 1–5.

Leape, L. L., Berwick, D. M., & Bates, D. W. (2002). What practices will most improve safety? Evidence-based medicine meets patient safety. *JAMA* 288(4), 501–507.

Lee, C., & Short, J. F. (1993). Social organization and risk: Some current controversies. *Annual Review of Sociology* 19(1), 375–399.

Leonardi, P. M. (2012). Materiality, sociomateriality, and socio-technical systems: What do these terms mean? How are they related? Do we need them? In: P. M. Leonardi, B. A. Nardi, & J. Kallinokos (eds.), *Materiality and Organizing: Social Interaction in a Technological World*. Oxford: Oxford University Press, pp. 25–48.

Leonardi, P. M., & Barley, S. R. (2008). Materiality and change: Challenges to building better theory about technology and organizing. *Information and Organization* 18(3), 159–176.

Leonardi, P. M., & Barley, S. R. (2010). What's under construction here? Social action, materiality, and power in constructivist studies of technology and organizing. *Academy of Management Annals* 4(1), 1–51.

Leplat, J., & Hoc, J.-M. (1983). Tache et Activite Dans l'Analyse Psyholoique Des Situations. *Cahiers de Psychologie Cognitive* 3(1), 49–63.

Leslie, M., Paradis, E., Gropper, M. A., Reeves, S., & Kitto, S. (2014). Applying ethnography to the study of context in healthcare quality and safety. *BMJ Quality and Safety* 23(2), 99–105. doi: 10.1136/bmjqs-2013-002335

Leveson, N. (2004). A new accident model for engineering safer systems. *Safety Science* 42(4), 237–270. doi: 10.1016/S0925-7535(03)00047-X

Leveson, N. (2011). *Engineering a Safer World Systems Thinking Applied to Safety.* Cambridge, MA: MIT Press.

Leveson, N., Dulac, N., Marais, K., & Carroll, J. (2009). Moving beyond normal accidents and high reliability organizations: A systems approach to safety in complex systems. *Organization Studies* 30(2–3), 227–249.

Liang, G., Weller, S. R., Zhao, J., Luo, F., & Dong, Z. Y. (2017). The 2015 Ukraine blackout: Implications for false data injection attacks. *IEEE Transactions on Power Systems* 32(4), 3317–3318.

Lin, C., Wu, S., & Lee, M. (2017). Cyber attack and defense on industry control systems. In: *2017 IEEE Conference on Dependable and Secure Computing*, Taipei, Taiwan, pp. 524–526.

Lincoln, Y., & Guba, E. G. (1985). *Naturalistic Inquiry.* Beverly Hills, CA: SAGE.

Luff, P., Hindmarsh, J., & Heath, C. (2000). *Workplace Studies: Recovering Work Practice and Informing System Design.* Cambridge: Cambridge University Press.

Lundberg, J., Rollenhagen, C., and Hollnagel, E. (2009). What-you-look-for-is-what-you-find-the consequences of underlying accident models in eight accident investigation manuals. *Safety Science* 47(10), 1297–1311.

Maarten, S. (1996). *Selective Ethnographic Analysis - Qualitative Modeling for Work Place Ethnography.* White Plains, NY: NYNEX Science & Technology Collaboration and Practice.

Macrae, C. (2009). Making risks visible: Identifying and interpreting threats to airline flight safety. *Journal of Occupational and Organizational Psychology* 82(2), 273–293.

Macrae, C. (2014). *Close Calls: Managing Risk and Resilience in Airline Flight Safety.* London: Palgrave Macmillan.

Macrae, C. (2014). Early warnings, weak signals, and learning from healthcare disasters. *BMJ Quality and Safety* 23, 440–445. doi: 10.1136/bmjqs-2013-002685

Macrae, C. (2019). Investigating for improvement? Five strategies to ensure national patient safety investigations improve patient safety. *Journal of the Royal Society of Medicine* 22, 1–5. doi: 10.1177/0141076819848114

Macrae, C. (2019). Moments of resilience: Time, space and the organisation of safety in complex sociotechnical systems. In: Wiig, S., & Fahlbruck, B. (eds.), *Exploring Resilience: A Scientific Journey from Practice to Theory.* London: Springer, pp. 15–24.

Macrae, C., & Vincent, C. (2014). Learning from failure: The need for independent safety investigations in healthcare. *Journal of the Royal Society of Medicine* 107(11), 439–443. doi: 10.1177/0141076814555939

Macrae, C., & Vincent, C. (2017). A new national safety investigator for healthcare: The road ahead. *The Royal Society for Medicine* 110(3), 90–92. doi: 10.1177/0141076817694577

Madsen, P., Desai, V., Roberts, K., & Wong, D. (2006). Mitigating hazards through continuing design: The birth and evolution of a pediatric intensive care unit. *Organization Science* 17(2), 239–248.

Makary, M. A., & Daniel, M. (2016). Medical error – The third leading cause of death in the US. *BMJ* 353, i2139.

Malinowski, B. (1922). *Argonauts of the Western Pacific.* London: Routledge.

Malinowski, B. (2013). *Argonauts of the Western Pacific: An Account of Native Enterprise and Adventure in the Archipelagoes of Melanesian New Guinea [1922/1994]*. London: Routledge.

Mannion, R., & Davies, H. T. O. (2015). Cultures of silence and cultures of voice: The role of whistleblowing in healthcare organizations. *International Journal of Health Policy Management* 4(8), 503–505. doi: 10.15171/IJHPM.2015.120

Marchal, B., van Belle, S., van Olmen, J., Hoeree, T., et al. (2012). Is realist evaluation keeping its promise? A review of published empirical studies in the field of health systems research. *Evaluation* 18(2), 192–212.

Marcus, G. E. (1995). Ethnography in/of the world system: The emergence of multi-sited ethnography. *Annual Review of Anthropology* 24, 95–117.

Marcus, G. E. (2013). Experimental forms for the expression of norms in the ethnography of the contemporary. *HAU: Journal of Ethnographic Theory* 3(2), 197–217. doi: 10.14318/hau3.2.011

Marcus, G. E., & Fischer, M. M. J. (1996). *Anthropology as Cultural Critique. An Experimental Moment in the Human Sciences*. Chicago, IL: University of Chicago Press.

Martin, G., Currie, G., & Lockett, A. (2011). Prospects for knowledge exchange in health policy and management: Institutional and epistemic boundaries. *Journal of Health Services Research & Policy* 16(4), 211–217.

Martin, J. (2002). *Organizational Culture: Mapping the Terrain*. doi: 10.4135/9781483328478

Martin, P. Y., & Turner, B. (1986). Grounded theory and organizational research. *Journal of Applied Behavioral Science* 22(2), 141–157. doi: 10.1177/002188638602200207

Martin, R. (2015). Why climate models aren't better. *Technology Review* (November). https://www.technologyreview.com/s/543546/why-climate-models-arent-better/

Martin, R. A. (2001). Managing vulnerabilities in networked systems. *Computer* 34(11), 32–38.

Maslen, S. (2014). Learning to prevent disaster: An investigation into methods for building safety knowledge among new engineers to the Australian gas pipeline industry. *Safety Science* 64, 82–89.

Maslen, S. (2015). Organisational factors for learning in the Australian gas pipeline industry. *Journal of Risk Research* 18(7), 896–909.

Maslen, S., & Hayes, J. (2014). Experts under the microscope: The Wivenhoe Dam case. *Environment Systems and Decisions* 34(2), 183–193.

Maslen, S., & Hayes, J. (2016). Preventing black swans: Incident reporting systems as collective knowledge management. *Journal of Risk Research* 19(10), 1246–1260. doi: 10.1080/13669877.2015.1057204

Maslen, S., & Hopkins, A. (2014). Do incentives work? A qualitative study of managers' motivations in hazardous industries. *Safety Science* 70, 419–428.

Maslen, S., & Ransan-Cooper, H. (2017). Safety framing and compliance in relation to standards: Experience from the Australian gas pipeline industry. *Safety Science* 94, 52–60.

Mayo, E. (1933). The Hawthorne experiment: Western Electric Company. In: *The Human Problems of an Industrial Civilization*. Macmillan Co., pp. 55–76.

McCulloch, P., Morgan, L., Flynn, L., Rivero-Arias, O., et al. (2016). *Safer Delivery of Surgical Services: A Programme of Controlled Before-and-After Intervention Studies with Pre-planned Pooled Data Analysis*. Programme Grants for Applied Research. Southampton: NIHR Journals Library.

McDermott, V., & Hayes, J. (2018). Risk shifting and disorganization in multi-tier contracting chains: The implications for public safety. *Safety Science* 106, 263–272. doi: 10.1016/j.ssci.2016.11.018

McDermott, V., Henne, K., & Hayes, J. (2018). Shifting risk to the frontline: Case studies in different contract working environments. *Journal of Risk Research* 21 (12), 1502–1516.

McDowell, L. (1998). Elites in the city of London: Some methodological considerations. *Environment and Planning A: Economy and Space* 30(12), 2133–2146. doi: 10.1068/a302133

McGlynn, E. A., Asch, S. M., Adams, J., Keesey, J., et al. (2003). The quality of health care delivered to adults in the United States. *New England Journal of Medicine* 348(26), 2635–2645.

McGoey, L. (2019). *The Unknowers: How Strategic Ignorance Rules the World*. London: ZED Books.

Medvedev, Z. (1991). *The Truth about Chernobyl*. New York: Basic Books.

Meisingset, K. (2016). Rusforskning: Big-data og genetikk. [Intoxination research: Big data and genetics – in Norwegian only]. *Rus & Samfunn* 9, Article 9.

Meld. St. 10 (2012–2013). God kvalitet – trygge tjenester. Kvalitet og pasientsikkerhet i helse og omsorgstjenesten. *Det kongelige helse- og omsorgsdepartement*, Oslo.

Meld.St. 11 (2014–2015). Kvalitet og pasientsikkerhet. *Det kongelige helse- og omsorgsdepartement*, Oslo.

Meld.St. 12 (2015–2016). Kvalitet og pasientsikkerhet. *Det kongelige helse- og omsorgsdepartement*, Oslo.

Meld.St. 13 (2016–2017). Kvalitet og pasientsikkerhet. *Det kongelige helse- og omsorgsdepartement*, Oslo.

Melnyk, B. M., & Fineout-Overholt, E. (2011). *Evidence-Based Practice in Nursing and Healthcare: A Guide to Best Practice*. Philadelphia, PA: Wolters-Kluwer.

Mikecz, R. (2012). Interviewing elites: Addressing methodological issues. *Qualitative Inquiry* 18(6), 482–493. doi: 10.1177/1077800412442818

Miles, M. B., & Hubermann, A. M. (1984). *Qualitative Data Analysis: A Methods Sourcebook* (2nd ed.). London: SAGE.

Mills, C. W. (1956). *White Collar: The American Middle Classes*. New York: Oxford University Press.

Mills, C. W. (1957). *The Power Elite*. New York: Oxford University Press.

Mills, C. W. (1959). *The Sociological Imagination*. New York: Oxford University Press.

Mills, C. W. (1962). *Power, Politics, and People: The Collected Essays of C. Wright Mills*. New York: Ballantine Books.

Mitchell, B., Cristancho, S., Nyhof, B. B., & Lingard, L. A. (2017). Mobilising or standing still? A narrative review of surgical safety checklist knowledge as developed in 25 highly cited papers from 2009 to 2016. *BMJ Quality & Safety* 26(10), 837–844.

Mol, A. (2002). *The Body Multiple: Ontology in Medical Practice*. Durham, NC: Duke University Press.

Moller, N., et al. (eds.). (2018). *Handbook of Safety Principles*. Hoboken, NJ: John Wiley & Sons.

Monahan, T., & Fisher, J. A. (2015). Strategies for obtaining access to secretive or guarded organizations. *Journal of Contemporary Ethnography* 44(6), 709–736.

Morgan, G. (1997). *Images of Organization by Gareth Morgan (26 February 1997) Paperback*. Thousand Oaks: SAGE Publications, Inc.

Morse, J. M., Stern, P. N., Corbin, J., Bowers, B., Charmaz, K., & Clarke, A. E. (2009). *Developing Grounded Theory: The Second Generation*. New York: Left Coast Press.

Nader, L. (1972). Up the anthropologist: Perspectives gained from studying up. In: D. Hymes (ed.), *Reinventing Anthropology*. New York: Pantheon Books, pp. 284–311.

Nader, R. (1965). *Unsafe at Any Speed. The Designed-In Dangers of the American Automobile*. New York: Grossman Publishers.

Nadworny, M. J. (1957). Frederick Taylor and Frank Gilbreth: Competition in scientific management. *Business History Review* 31(1), 23–34. doi: 10.2307/3111727

National Academies of Sciences, Engineering, and Medicine. (2018). *Crossing the Global Quality Chasm: Improving Health Care Worldwide.* Washington, DC: National Academies Press. doi: 10.17226/25152

Nichols, T. (1997). *The Sociology of Industrial Injury.* London: Mansell.

Nielsen, K. J., Rasmussen, K., Glasscock, D., & Spangenberg, S. (2008). Changes in safety climate and accidents at two identical manufacturing plants. *Safety Science* 46(3), 440–449.

Nixon, J., & Braithwaite, G. R. (2018). What do aircraft accident investigators do and what makes them good at it? Developing a competency framework for investigators using grounded theory. *Safety Science* 103, 153–161.

NLIA. (2018). Risiko for arbeidsskadedødsfall i det landbaserte arbeidslivet. En sammenligning av norske og utenlandske arbeidstakere. [Risk for work-related fatalities in onshore work. A comparison between Norwegian and foreign workers – in Norwegian only]. *KOMPASS Tema.* Trondheim: Norwegian Labour Inspection Authority.

NOU. (2015:11). *Med åpne kort.* Oslo: DSS. https://www.regjeringen.no/contentassets/daae d86b64c04f79a2790e87d8bb4576/no/pdfs/nou201520150011000dddpdfs.pdf

Nævestad, T. O., Hovi, I. B., Caspersen, E., & Bjørnskau, T. (2014). *Ulykkesrisiko for tunge godsbiler på Norske veger: Sammenlikning av Norske og utenlandske aktører* (1327/2014).

O'Reilly, K. (2015). Ethnography: Telling practice stories. In: *Emerging Trends in the Social and Behavioral Sciences.* American Cancer Society, pp. 1–14. doi: 10.1002/9781118900772.etrds0120

Oakley, A. (2000). *Experiments in Knowing: Gender and Method in the Social Sciences.* Cambridge: Polity Press.

Odendahl, T., & Shaw, A. (2002). Interviewing elites. In: J. Gubrium & J. Holstein (eds.), *Handbook of Interview Research: Context and Method* (pp. 299–331). Thousand Oaks, CA: SAGE.

Orlikowski, W. J., & Scott, S. V. (2008). sociomateriality: Challenging the separation of technology, work and organization. *The Academy of Management Annals* 2(1), 433–474.

Orr, J. E. (1996). *Talking about Machines: An Ethnography of a Modern Job.* Ithaca, NY: Cornell University Press.

Ostrander, S. A. (1993). "SURELY YOU'RE NOT IN THIS JUST TO BE HELPFUL": Access, Rapport, and Interviews in Three Studies of Elites. *Journal of Contemporary Ethnography* 22(1), 7–27. doi: 10.1177/089124193022001002

Oswald, D., Sherratt, F., & Smith, S. (2018). Problems with safety observation reporting: A construction industry case study. *Safety Science* 107, 35–45.

Parker, D., Lawrie, M., & Hudson, P. (2006). A framework for understanding the development of organisational safety culture. *Safety Science* 44(6), 551–562.

Pate-Cornell, M. E., & Murphy, D. M. (1996). Human and management factors in probabilistic risk analysis: The SAM approach and observations from recent applications. *Reliability Engineering and System Safety* 53(2), 115–126.

Patton, M. Q. (2002). *Qualitative Research and Evaluation Methods* (3rd ed.). Thousand Oaks, CA: SAGE.

Pawson, R., & Tilley, N. (1997). *Realistic Evaluation.* London: SAGE.

PCORI. (2018). *PCORI Methodology Standards.* Washington, DC: Patient-Centered Outcomes Research Institute.

Peden, C. J., Stephens, T., Martin, G., Kahan, B. C., et al. (2019). Effectiveness of a national quality improvement programme to improve survival after emergency abdominal surgery (EPOCH): A stepped-wedge cluster-randomised trial. *The Lancet* 393, 2213–2221.

Perin, C. (2005). *Shouldering Risks: The Culture of Control in the Nuclear Power Industry.* Princeton, NJ: Princeton University Press.

Perrow, C. (1983). The organizational context of human factors engineering. *Administrative Science Quarterly* 28(4) 521–541.

Perrow, C. (1984). *Normal Accidents Living with High Risk Technologies.* Princeton, NJ: Princeton University Press.

Perrow, C. (1984). *Normal Accidents: Living with High Risk Systems.* New York: Basic Books.

Perrow, C. (1991). A society of organizations. *Theory and Society* 20(6), 725–762.

Perrow, C. (1994). The limits of safety: The enhancement of a theory of accidents. *Journal of Contingencies and Crisis Management* 4(2), 212–220.

Perrow, C. (1999). *Normal Accidents: Living with High-Risk Technologies.* Princeton, NJ: Princeton University Press.

Perrow, G. (2011). *Normal Accidents: Living with High Risk Technologies* (Updated edition with a New afterword and a new postscript by the author edizione). Princeton University Press.

Peters, T. J., & Waterman, R. H. (1982). *In Search of Excellence Lessons from America's Best-Run Companies.* New York: Harper & Row.

Pettersen, G. K. (2008). *The Social Production of Safety, Theorising the Human Role in Aircraft Line Maintenance* (Doctoral dissertation, Dissertation, University of Stavanger).

Pettersen, G. K. (2020). Organizational risk: "Muddling through" 40 years of research. *Risk Analysis* 1–10. doi: 10.1111/risa.13460

Pettersen, K. A. (2016). Understanding uncertainty: Thinking through in relation to high-risk technologies. In: *Routledge Handbook of Risk Studies.* Routledge; Milton Park, Abingdon, Oxon OX14 4RN, pp. 39–48. ISBN 978-1-138-02286-7.

Pettersen, K. A., & Schulman, P. R. (2019). Drift, adaptation, resilience and reliability: Toward an empirical clarification (special issue on HRO and RE). *Safety Science* 117, 460–468.

Pettersen-Gould, K., & Fjæran, L. (2019). Drift and the social attenuation of risk. In: J.-C. Le Coze (ed.), *Safety Sciences Research: Evolution, Challenges and New Directions.* London: CRC Press, pp. 119–132.

Pidgeon, N. (1995). Safety culture and risk management in organizations, *Journal of Cross-Cultural Psychology* 22(1), 129–140.

Pidgeon, N. (1997). The limits to safety? Culture, politics, learning and man–made disasters. *Journal of Contingencies and Crisis Management* 5(1), 1–14.

Pidgeon, N., & O'Leary, M. (2000). Man-made disasters: Why technology and organizations (sometimes) fail. *Safety Science* 34(1–3), 15–30. doi: 10.1016/S0925-7535(00)00004-7

Pidgeon, N. F. (1998). Safety culture: Key theoretical issues. *Work and Stress* 12(3), 202–216.

Pidgeon, N. F. (2010). Systems thinking, cultures of reliability and safety. *Civil Engineering and Environmental Systems* 27(3), 211–217.

Pidgeon, N. F. (2019). Observing the English weather – A personal journey from safety I to IV. In: J.-C. Le Coze (ed.), *Safety Sciences Research: Evolution, Challenges and New Directions.* London: CRC Press, pp. 269–289.

Pidgeon, N. F., Blockley, D. I., & Turner, B. A. (1986). Design practice and snow loading: Lessons from a roof collapse. *Structural Engineers* 64(A), 67–71.

Pidgeon, N. F., Blockley, D. I., & Turner, B. A. (1987). Reply to discussion of 'Design practice and snow loading'. *The Structural Engineer* 65(A), 239–240.

Pidgeon, N. F., Blockley, D. I., & Turner, B. A. (1988). Site investigations: Lessons from a late discovery of hazardous waste. *Structural Engineers* 66(19), 311–315.

Pidgeon, N. F., & Henwood, K. L. (2004). Grounded theory. In: M. Hardy & A. Bryman (eds.), *Handbook of Data Analysis.* London: SAGE, pp. 625–648.

Pidgeon, N. F., Stone, J., Blockley, D. I., & Turner, B. A. (1990). Management of safety through lessons from case histories. In: M. H. Walter & R. F. Cox (eds.), *Safety and Reliability in the 90s*. London: Elsevier Applied Science, pp. 201–216.

Pidgeon, N. F., Turner, B. A., & Blockley, D. I. (1991). The use of grounded theory for conceptual analysis in knowledge elicitation. *International Journal of Man-Machine Studies* 35(2), 151–173.

Pilbeam, C., Doherty, N., Davidson, R., & Denyer, D. (2016). Safety leadership practices for organizational safety compliance: Developing a research agenda from a review of the literature. *Safety Science* 86, 110–121.

Polanyi, M. (1958). *Personal Knowledge*. London: Routledge.

Pope, C., & Mays, N. (1993). Opening the black box: An encounter in the corridors of health services research. *BMJ* 306, 315–318.

Pope, C., Ziebland, S., & Mays, N. (2006). Analysing qualitative data. In: Pope, C., & Mays, N. (eds.), *Qualitative Research in Health Care*. Oxford: Blackwell Publishing, pp. 63–81.

Power, M. (2004). *The Risk Management of Everything: Rethinking the Politics of Uncertainty*. London: Demos.

Pronovost, P., Needham, D., Berenholtz, S., Sinopoli, D., et al. (2006). An intervention to decrease catheter-related bloodstream infections in the ICU. *New England Journal of Medicine* 355(26), 2725–2732.

Pronovost, P. J., Berenholtz, S. M., & Morlock, L. L. (2011). Is quality of care improving in the UK? *BMJ* 342, c6646.

Prop.68 L (2016–2017). Proposisjon til Stortinget (forslag til lov). *Lov om Statens undersøkelseskommisjon for helse og omsorgstjenesten*. Det kongelige helse og omsorgsdepartementet. https://www.regjeringen.no/no/dokumenter/prop.-68-l-20162017/id25 44823/sec2

Provan, D. J., Rae, A. J., & Dekker, S. W. (2019). An ethnography of the safety professional's dilemma: Safety work or the safety of work? *Safety Science* 117, 276–289.

Quick, O. (2017). *Regulating Patient Safety: The End of Professional Dominance?* Cambridge: Cambridge University Press.

Quinlan, M. (2014). *Ten Pathways to Death and Disaster: Learning from Fatal Incidents in Mines and Other High Hazard Workplaces*. Sydney, NSW: Federation Press.

Rae, A., & Provan, D. (2019). Safety work versus the safety of work. *Safety Science* 111, 119–127.

Rae, A., Provan, D., Aboelssaad, H., & Alexander, R. (2020). A manifesto for reality-based safety science. *Safety Science* 126, 104654. doi: 10.1016/j.ssci.2020.104654

Ragin, C. C., & Becker, H. S. (1992). *What is a Case? Exploring the Foundations of Social Inquiry*. Cambridge: Cambridge University Press.

Randell, R., Greenhalgh, J., Hindmarsh, J., Dowding, D., et al. (2014). Integration of robotic surgery into routine practice and impacts on communication, collaboration, and decision making: A realist process evaluation protocol. *Implementation Science* 9(1), 52.

Rapport, F., Hogden, A., Faris, M., Bierbaum, M., Clay-Williams, R., Long, J., Shih, P., & Braithwaite, J. (2019). *Qualitative Research in Healthcare, Modern Methods, Clear Translation – A White Paper*. Sydney, NSW: Australian Institute of Health Innovation, Macquarie University.

Rasmussen, J. (1997). Risk management in a dynamic society: A modelling problem. *Safety Science* 27(2/3), 183–213.

Rasmussen, J., & Svedung, I. (2000). *Proactive Risk Management in a Dynamic Society*. Karlstad: Swedish Rescue Services Agency.

Ravetz, J. R. (1971). *Scientific Knowledge and Its Social Problems*. Oxford: Clarendon Press.

Reason, J. (1987). The Chernobyl errors. *Bulletin of the British Psychological Society* 40(206), 1–20.

Reason, J. (1990). *Human Error*. Cambridge: Cambridge University Press.

Reason, J. (1990). *Human Error* (1st ed.). Cambridge: Cambridge University Press.

Reason, J. (1997). *Managing the Risks of Organizational Accidents*. Burlington: Ashgate Publishing Limited.

Reason, J. (2013). *A Life in Error: From Little Slips to Big Disasters*. Farnham: Ashgate.

Reason, J. (2016). *Organizational Accidents Revisited*. Boca Raton, FL: CRC Press.

Rees, J. V. (1996). *Hostages of Each Other: The Transformation of Nuclear Safety Since Three Mile Island*. University of Chicago Press.

Reichertz, J. (2010). Abduction: The logic of discovery of grounded theory. *Forum: Qualitative Social Research* 11(1), Article 13.

Reiman, T., & Oedewald, P. (2006). Assessing the maintenance unit of a nuclear power plant – Identifying the cultural conceptions concerning the maintenance work and the maintenance organization. *Safety Science* 44(9), 821–850.

Renn, O. (1998). Three decades of risk research: Accomplishments and new challenges. *Journal of Risk Research* 1(1), 49–71.

Renn, O., Lucas, K., Haas, A., & Jaeger, C. (2019). Things are different today: The challenge of global systemic risks. *Journal of Risk Research* 22(4), 401–415.

Richards, D. (1996). Elite interviewing: Approaches and pitfalls. *Politics* 16(3), 199–204. doi: 10.1111/j.1467-9256.1996.tb00039.x

Rijpma, J. A. (1997). Complexity, tight-coupling and reliability: Connecting normal accidents theory and high reliability theory. *Journal of Contingencies and Crisis Management* 5(1), 15–23.

Rijpma, J. A. (2003). From deadlock to dead end: The normal accidents-high reliability debate revisited. *Journal of Contingencies and Crisis Management* 11(1), 37–45.

Roberts, K. (ed.). (1993). *New Challenges to Understanding Organizations*. New York: Macmillan.

Robson, L. S., Amick III, B. C., Moser, C., Pagell, M., Mansfield, E., Shannon, H. S., South, H. (2016). Important factors in common among organizations making large improvement in OHS performance: Results of an exploratory multiple case study. *Safety Science* 86, 211–227.

Rochlin, G. (1993). Defining high-reliability organizations in practice: A taxonomic Prologomena. In: K. H. Roberts (ed.), *New Challenges to Understanding Organizations*. New York: Macmillan, pp. 11–32.

Rochlin, G. I. (1998). Essential friction: Error-control in organizational behavior. In: Nordal Åkerman (ed.), *The Necessity of Friction*. Boulder, CO: Westview Press, pp. 196–232.

Rochlin, G. I. (1999). Safe operation as a social construct. *Ergonomics* 42(11), 1549–1560.

Rochlin, G. I. (2011). How to hunt a very reliable organization. *Journal of Contingencies and Crisis Management* 19(1), 14–20.

Rochlin, G. I., La Porte, T. R., & Roberts, K. H. (1999). The self-designing high-reliability organization: Aircraft carrier flight operations at sea. *Naval War College Review* 40(4), 76–92.

Rochlin, G. I., & Meier, A. V. (1994). Nuclear power operations: A cross-cultural perspective. *Annual Review of Energy and the Environment* 19(1), 153–187.

Roe, E., & Schulman, P. (2008). *High Reliability Management: Operating on the Edge*. Stanford, CA: Stanford University Press.

Roe, E., & Schulman, P. (2016). *Reliability and Risk: The Challenge of Managing Interconnected Critical Infrastructures*. Stanford, CA: Stanford University Press.

Rosendahl, T., & Hepsø, V. (2013). *Integrated Operations in the Oil and Gas Industry: Sustainability and Capability Development*. Hershey, PA: IGI Global.

Rosness, R., Evjemo, T. E., Haavik, T., & Wærø, I. (2016). Prospective sensemaking in the operating theatre. *Cognition, Technology and Work* 18(1), 53–69.

Rosness, R., Grøtan, T. O., Guttormsen, G., Herrera, I., Steiro, T., Størseth, F., Tinmannsvik, R. K., & Wærø, I. (2010). *Organizational Accidents and Resilient Organizations. Six Perspectives. Revision 2.* Trondheim: Sintef Technology and Society.

Sackett, D. L., Rosenberg, W. M. C., Gray, J. A. M., Haynes, R. B., et al. (1996). Evidence based medicine: What it is and what it isn't. *BMJ* 312, 71–72.

Safety Management International Collaboration Group. (2013). *Safety Management System Evaluation Tool.* https://www.skybrary.aero/bookshelf/books/1774.pdf

Sagan, S. (1993). *The Limits of Safety. Organizations, Accidents, and Nuclear Weapons.* Princeton, NJ: Princeton University Press.

Sagan, S. (1995). *The Limits of Safety: Organisations, Accidents and Nuclear Weapons.* Princeton, NJ: Princeton University Press.

Saldana, J. (2015). *The Coding Manual for Qualitative Researchers* (3rd ed.). SAGE Publications Ltd.

Salzman, P. C. (1996). Methodology. In: A. Barnard & J. Spencer (eds.), *Encyclopedia of Social and Cultural Anthropology.* London: Routledge.

Sanjek, R. (1996). A vocabulary for Fieldnotes. In: R. Sanjek (ed.), *Fieldnotes the Makings of Anthropology.* London: Cornell University Press.

Sassen, S. (2007). *A Sociology of Globalization.* New York: W. W. Norton, & Company.

Schaefer, C., & Wiig, S. (2017). Strategy and practise of external inspection in healthcare services – a Norwegian comparative case study. *BMC Safety in Health* 3, 3. doi: 10.1186/s40886-017-0054-9

Schakel, J.-K., Fenema, P. C., & Faraj, S. (2016). Shots fired! Switching between practices in police work. *Organization Science* 27(2), 391–410.

Schein, E. (1992). *Organizational Culture and Leadership* (2nd ed.). San Francisco, CA: Jossey-Bass.

Schiefloe, P. M., & K. M. Vikland. (2009). Close to catastrophe. Lessons from the Snorre A gas blow-out. In: *25th European Group for Organizational Studies (EGOS).* Barcelona, Spain, 3–4 July 2010.

Schiefloe, P. M., K. M. Vikland, E. B. Ytredal, A. Torsteinsbø, I. O. Moldskred, S. Heggen, D. H. Sleire, S. A. Førsund, and J. E. Syversen. (2005). *Årsaksanalyse etter Snorre A-hendelsen 28.11.2004.* Stavanger, Norway: Statoil.

Schmidt, K. (2014). The concept of "practice": What's the point? In: *Proceeding of the 11th International Conference on the Design-COOP2014* (eds.), C. Rossitto, L. Ciolfi, D. Martin, & B. Conein. Nice, France, 2014 (11th Conference on the design of cooperative systems conference COOP14).

Schoenberger, E. (1991). The corporate interview as a research method in economic geography. *The Professional Geographer* 43(2), 180–189. doi: 10.1111/j.0033-0124.1991.00180.x

Schulman, P. R. (1993). The negotiated order of organizational reliability. *Administration and Society* 25(3), 353–372.

Schulman, P. R., & Roe, E. (2016). *Reliability and Risk: The Challenge of Managing Interconnected Infrastructures.* Stanford, CA: Stanford University Press.

Schuster, M. A., McGlynn, E. A., & Brook, R. H. (2005). How good is the quality of health care in the United States? *Milbank Quarterly* 83(4), 843–895.

Schwartzman, H. B. (1993). *Ethnography in Organizations.* Newbury Park, CA: SAGE.

Seale, C. (2004). *Qualitative Research Practice.* SAGE.

Seale, C., Gobo, G., Gubrium, J. F., & Silverman, D. (eds.). (2004). *Qualitative Research Practice* (1st ed.). Newbury Park, CA: SAGE Publications Ltd.

Selznick, P. (1984). *Leadership in Administration.* Oakland, CA: University of California Press.

Sheard, L., Marsh, C., O'Hara, J., Armitage, G., Wright, J., & Lawton, R. (2017). Exploring how ward staff engage with the implementation of a patient safety intervention: A UK-based qualitative process evaluation. *BMJ Open* 7, e014558.

Shekelle, P. G., Pronovost, P. J., Wachter, R. M., Taylor, S. L., et al. (2011). Advancing the science of patient safety. *Annals of Internal Medicine* 154(10), 693–696.

Shirali, G. A., Salehi, V., Savari, R., & Ahmadiangali, K. (2018). Investigating the effectiveness of safety costs on productivity and quality enhancement by means of a quantitative approach. *Safety Science* 103, 316–322.

Shojania, K. G. (2012). Deaths due to medical error: Jumbo jets or just small propeller planes? *BMJ Quality & Safety* 21(9), 709–712.

Shojania, K. G. (2013). Conventional evaluations of improvement interventions: More trials or just more tribulations? *BMJ Quality & Safety* 22(11), 881–884.

Shojania, K. G., & Dixon-Woods, M. (2017). Estimating deaths due to medical error: The ongoing controversy and why it matters. *BMJ Quality & Safety* 26(5), 423–428.

Shojania, K. G., & Grimshaw, J. M. (2005). Evidence-based quality improvement: The state of the science. *Health Affairs* 24(1), 138–150.

Short, J. F. (1984). The social fabric at risk: Toward the social transformation of risk analysis. *American Sociological Review* 49(6), 711–725.

Short, J. F. (1992). *Organizations, Uncertainties, and Risk*. Boulder, CO: Westview Press.

Short, J. F., & Clarke, L. (1992). Social organization and risk. In: Short, J. F., & Clarke, L. (eds.), *Organizations, Uncertainties, and Risk*. Oxford: Westview Press, pp. 309–321.

Sidney Sidney (2011). Drift into failure. Farnham: Ashgate.

Silverman, D. (2001). *Interpreting Qualitative Data: Methods for Analysing Talk, Text and Interaction*. London: SAGE.

Sismondo, S. (2008). Pharmaceutical company funding and its consequences: A qualitative systematic review. *Contemporary Clinical Trials* 29(2), 109–113.

Sismondo, S. (2010). *An Introduction to Science and Technology Studies*, Vol. 1. Chichester: Wiley-Blackwell.

Sklair, L. (2001). *The Transnational Capitalist Class*. Oxford: Blackwell.

Smith, T. D., Eldridge, F., & DeJoy, D. M. (2016). Safety-specific transformational and passive leadership influences on firefighter safety climate perceptions and safety behavior outcomes. *Safety Science* 86, 92–97.

Smithson, M. (1989). *Ignorance and Uncertainty: Emerging Paradigms*. London: Springer.

Snook, S. A. (2000). *Friendly Fire: The Accidental Shootdown of US Black Hawks over Northern Iraq*. Princeton, NJ: Princeton University Press.

Solem, A., & Kongsvik, T. (2013). Facilitating for cultural change: Lessons learned from a 12-year safety improvement programme. *Safety Science Monitor* 17(1), Article 4.

Squires, A. M. (1986). *The Tender Ship: Governmental Management of Technological Change*. Boston, MA: Springer Science+Business Media, LLC.

Stablein, R. (1999). Data in organization studies. In: Clegg, S. R., & Hardy, C. (eds.), *Studying organizations. Theory & method*. London: SAGE Publications.

Star, S. L., & Strauss, A. (1999). Layers of silence, arenas of voice: The ecology of visible and invisible work. *Computer Supported Cooperative Work* 8(1–2), 9–30.

Stech, F. (1979). *Political and Military Intention Estimation*. Bethesda, MD: MATHTECH, Inc.

Steffy, L. S. (2010). *Drowning in Oil: BP and the Reckless Pursuit of Profit*. London: McGraw-Hill.

Stephens, N. (2007). Collecting data from elites and ultra elites: Telephone and face-to-face interviews with macroeconomists. *Qualitative Research* 7(2), 203–216. doi: 10.1177/1468794107076020

Stephens, T. J., Peden, C. J., Pearse, R. M., Shaw, S. E., et al. (2018). Improving care at scale: Process evaluation of a multi-component quality improvement intervention to reduce mortality after emergency abdominal surgery (EPOCH trial). *Implementation Science* 13(1), 142.

Stevens, S. (1946). On the theory of scales of measurement. *Science* 103, 677–680.

Stewart, A. (1998). *The Ethnographer's Method* (Vol. 46). Thousand Oaks, CA: SAGE.

Stirling, A. (2010). Keep it complex. *Nature* 468, 1029–1031.

Strauss, A. L. (1987). *Qualitative Analysis for Social Scientists*. New York: Cambridge University Press.

Strauss, A. L, & Corbin, J. M. (1998). *Basics of Qualitative Research: Grounded Theory Procedures and Techniques*. Thousand Oaks, CA: SAGE, Inc.

Suchman, L. (1995). Making work visible. *Communications of the ACM* 38(9), 56–64. doi: 10.1145/223248.223263

Sutcliffe, K. M., & K. E. Weick. (2013). Mindful organizing and resilient health care. In: E. Hollnagel, J. Braithwaite, & R. L. Wears (eds.), *Resilient Health Care*. London: Ashgate.

Sørhaug, T. (2004). *Managementaliet og autoirtetens forandring – Ledelse I en kunnskap-søkonomi*. Bergen: Fagbokforlaget.

Tavory, I., & Timmermans, S. (2009). Two cases of ethnography: Grounded theory and the extended case method. *Ethnography* 10(3), 243–263.

Tharaldsen, J.E., Olsen, E., & Rundmo, T. (2008). A longitudinal study of safety climate on the Norwegian continental shelf. *Safety Science*, 46(3), 427–439. https://doi.org/10.1016/j.ssci.2007.05.006.

Thomas, R. (1993). Interviewing important people in big companies. *Journal of Contemporary Ethnography* 22(1), 80–96. doi: 10.1177/089124193022001006

Toft, B., & Reynolds, S. (1994). *Learning from Disasters: A Management Approach*. Oxford: Butterworth-Heinemann.

Toft, B., & Turner, B. A. (1987). The schematic report analysis diagram: A simple aid to learning from large-scale failures. *International CIS Journal* 1(2), 12–23.

Turner, B. (1983). The use of grounded theory for the qualitative analysis of organizational behaviour. *Journal of Management Studies* 20(3), 333–348.

Turner, B. (1988). Connoisseurship in the study of organizational cultures. In: Bryman, A. (ed.), *Doing Research in Organizations*. London: Routledge, pp. 108–122.

Turner, B. (1990). Failed artefacts. In: Gagliardi, P (ed.), *Symbols and Artifacts: Views of the Corporate Landscape*. New York: Walter De Gruyter.

Turner, B. (1992). The sociology of safety. In: Blockley, D. I. (ed.), *Engineering Safety*. Maidenhead: Mc Graw-Hill.

Turner, B. (1994). Causes of disaster: Sloppy management. *British Journal of Management* 5, 215–219. doi: 10.1111/j.1467-8551.1994.tb00172.x

Turner, B. (1995). A personal trajectory through organization studies. *Research in the Sociology of Organizations* 13, 275–301.

Turner, B., & Martin, P. (1986). Grounded theory and organisational research. *The Journal of Applied Behavioral Science* 22(2), 141–157.

Turner, B., & Pidgeon, N. (1997). *Man-Made Disaster. The Failure of Foresight* (2nd ed.). London: Butterworths-Heinmann.

Turner, B. A., Pidgeon, N., Blockley, D., & Toft, B. (1989). *Safety Culture: Its Importance in Future Risk Management. Position Paper for Second World Bank Workshop on Safety Control and Risk Management*. Karlstad, Sweden.

Turner, B. A., Pidgeon, N. F., & Blockley, D. I. (1991). The use of grounded theory for conceptual analysis in knowledge elicitation. *Int. J. Man-Machine Studies* 35, 151–173.

Turner, B. A. (1971). *Exploring the Industrial Subcultures*. London: Macmillan Press.

Turner, B. A. (1976). *The Failure of Foresight.* Ph.D. Thesis, University of Exeter.

Turner, B. A. (1976). The organization and interorganizational development of disasters. *Administrative Science Quarterly* 21(3), 378–397. doi: 10.2307/2391850

Turner, B. A. (1978). *Man-Made Disaster. The Failure of Foresight.* London: Wykeham Science Press.

Turner, B. A. (1981a). Some practical aspects of qualitative data analysis: One way of organizing the cognitive processes associated with the generation of grounded theory. *Quality and Quantity* 15(3), 225–247.

Turner, B. A. (1981b). Some questions about the use of concepts of risk and probability in connection with decisions about hazardous events. Working paper, University of Exeter, September 1981.

Turner, B. A. (1983). Making sense of unseemly behavior in organizations. *International Studies of Management & Organization* 13(3), 164–181.

Turner, B. A. (1983). The use of grounded theory for the qualitative analysis of organizational behavior. *Journal of Management Studies* 20(3), 333–348.

Turner, B. A. (1986). Sociological aspects of organizational symbolism. *Organization Studies* 7(2), 101–115.

Turner, B. A. (1987). Postmodernism and organisational culture. Unpublished manuscript. Exeter University.

Turner, B. A. (1988). Connoisseurship and the study of organizational cultures. In: A. Bryman (ed.), *Doing Research in Organizations.* London: Routledge, pp. 108–122.

Turner, B. A. (1989). How can we design a safe organisation? Paper presented at the *Second International Conference on Industrial Organisation and Crisis Management,* New York University, 3–4, November.

Turner, B. A. (1990). If research methods are commonsense, can they be taught? Paper presented at *International Sociological Association Conference,* Madrid, July.

Turner, B. A. (1990). *Organizational Symbolism.* Berlin: De Gruyter.

Turner, B. A. (1992). The sociology of safety. In: D. Blockley (ed.), *Engineering Safety.* London: McGraw-Hill, pp. 187–205.

Turner, B. A. (1994). Software and contingency: The text and vocabulary of system failure. *Journal of Contingencies and Crisis Management* 2(1), 31–38.

Turner, B. A. (1994). The future for risk research. *Journal of Contingencies and Crisis Management* 2(3), 146–156.

Turner, B. A. (1994). Flexibility and improvisation in emergency response. *Disaster Management* 6(2), 84–90.

Turner, B. A. (1995). A personal trajectory through organization studies. *Research in the Sociology of Organizations* 13, 275–301.

Turner, B. A. (1997). *Man-Made Disasters.[Facsimile].* Boston, MA: Butterworth-Heinemann.

Turner, B. A., & Pidgeon, N. F. (1997). *Man-Made Disasters.* Boston, MA: Butterworth-Heinemann.

Turner, B. A., & Toft, B. (1988). Organizational learning from disasters. In: H. B. F. Gow & R. W. Kay (eds.), *Emergency Planning for Industrial Hazards.* London: Elsevier, pp. 297–313.

Twaalfhoven, S. F. M., & Kortleven, W. J. (2016). The corporate quest for zero accidents: A case study into the response to safety transgressions in the industrial sector. *Safety Science* 86, 57–68.

Underwriters Laboratories. (2013). Using leading and lagging safety indicators to manage workplace health and safety risk. *Underwriters Laboratories.* https://legacy-uploads.ul .com/wp-content/uploads/sites/40/2015/02/UL_WP_Final_Using-Leading-and-Lagg ing-Safety-Indicators-to-Manage-Workplace-Health-and-Safety-Risk_V7-LR1.pdf

Vaughan, D. (1992). Theory elaboration: The heuristic of case analysis. In: C. Ragin, & H. Becker (eds.), *What Is a Case? Exploring the Foundations of Social Enquiry.* Cambridge: Cambridge University Press, pp. 173–192.

Vaughan, D. (1996). *The Challenger Launch Decision: Risky Technology, Culture and Deviance at NASA.* Chicago, IL: NASA, University of Chicago Press.

Vaughan, D. (1997). The Trickle-Down Effect: Policy decisions, risky work and the Challenger tragedy. *California Management Review* 39(2), 80–102.

Vaughan, D. (1999). The dark side of organizations: Mistake, misconduct, and disaster. *Annual Review of Sociology* 25(1), 271–305.

Vaughan, D. (1999). The role of the organization in the production of techno-scientific knowledge. *Social Studies of Science* 29(6), 913–943.

Vaughan, D. (2003). History as cause. Chapter 8. In: *Columbia Accident Investigation Board,* pp. 195–201.

Vaughan, D. (2004). Theorizing disaster: Analogy, historical ethnography, and the challenger accident ethnography. *Ethnography* 5(3), 315–347.

Vaughan, D. (2005). System effects: On slippery slopes, repeating negative patterns, and learning from mistakes? In: H. W. Starbuck, & M. Farjoun (eds.), *Organization at the Limit. Lessons from the Columbia Disaster.* Oxford: Blackwell Publishing, pp. 1997–1999.

Vaughan, D. (2006). NASA revisited: Theory, analogy, and public sociology. *American Journal of Sociology* 112(2), 353–393.

Vaughan, D. (2006). The social shaping of commission reports. *Sociological Forum* 21(2), 291–306.

Vaughan, D. (2007). Beyond macro- and micro-levels of analysis: Organizations and the cultural fix. In: H. N. Pontell, & G. Geis (eds.), *The International Handbook of White-Collar and Corporate Crime.* New York: Springer, pp. 3–24.

Vaughan, D. (2014). Theorizing: Analogy, cases, and comparative social organization. In: R. Swedberg (ed.), *Theorizing in Social Science.* Stanford, CA: Stanford University Press, pp. 61–94.

Vaughan, D. (2016). *The Challenger Launch Decision.* Chicago, IL: University of Chicago Press.

Vidal, R., & Roberts, K. H. (2014). Observing elite firefighting teams: The triad effect. *Journal of Contingencies and Crisis Management* 22(1), 18–28.

Vincent, C., & Amalberti, R. (2016). *Safer Healthcare.* London: Springer Open.

Vincent, C., Carthey, J., Macrae, C., & Amalberti, A. (2017). Safety analysis over time: Seven major changes to adverse event investigation. *Implementation Science* 12, 151. doi: 10.1186/s13012-017-0695-4

Viner, D. (1991). *Accident Analysis and Risk Control.* East Ivanhoe: Viner.

Wachter, R. M. (2010). Patient safety at ten: Unmistakable progress, troubling gaps. *Health Affairs* 29(1), 165–173.

Wackers, G. (2006). Vulnerability and robustness in a complex technological system: Loss of control and recovery in the 2004 Snorre A gas blow-out. *University of Maastricht, Research Report* 42, 2006.

Ward, V., House, A., & Hamer, S. (2009). Knowledge brokering: The missing link in the evidence to action chain? *Evidence & Policy* 5, 267–279.

Waring, A. (2015). Managerial and non-technical factors in the development of human-created disasters: A review and research agenda. *Safety Science* 79, 254–267.

Waring, J., Allen, D., Braithwaite, J., & Sandall, J. (2016). Healthcare quality and safety: A review of policy, practice and research. *Sociology of Health & Illness* 38(2), 198–215.

Waring, J., & Jones, L. (2016). Maintaining the link between methodology and method in ethnographic health research. *BMJ Quality & Safety* 25(7), 556–557.

Waterson, P., Jenkins, D. P., Salmon, P. M., & Underwood, P. (2017). 'Remixing Rasmussen': The evolution of AcciMaps within systemic accident analysis. *Applied Ergonomics* 59, 483–503.

Watson, T. J. (2012). Making organisational ethnography. *Journal of Organizational Ethnography* 1(1), 15–22. doi: 10.1108/20466741211220615

Wears, R. L., & Roberts, K. H. (2019). *Special Issue. Safety Science, High Reliability Organizations and Resilience Engineering.* Elsevier.

Weaver, C. (2008). *Hypocrisy Trap: The World Bank and the Poverty of Reform.* Princeton, NJ: Princeton University Press.

Webb, D. (2011). Foreword. In: *Evidence: Safer Patients Initiative Phase Two.* London: Health Foundation, pp. iv–vii.

Weber, D., Macgregor, S., Provan, D. J., & Rae, A. J. (2018). "We can stop work, but then nothing gets done." Factors that support and hinder a workforce to discontinue work for safety. *Safety Science* 108C, 149–160.

Weick, K. (2011). Organizing for transient reliability: The production of dynamic non-events. *Journal of Contingencies and Crisis Management* 19, 21–27.

Weick, K. E. (1987). Organizational culture as a source of high reliability. *California Management Review* 29(2), 112–128.

Weick, K. E. (1993). The collapse of sensemaking in organizations: The Mann Gulch disaster. *Administrative Science Quarterly* 38(4), 628–652.

Weick, K. E. (1995). *Sensemaking in Organisations.* London: SAGE.

Weick, K. E. (1996). Drop your tools: An allegory for organizational studies. *Administrative Science Quarterly* 41(2), 301–313.

Weick, K. E. (1998). Foresights of failure: An appreciation of Barry Turner. *Journal of Contingencies and Crisis Management* 6(2), 72–75.

Weick, K. E. (2011). Organizing for transient reliability: The production of dynamic non-events. *Journal of Contingencies and Crisis Management* 19(1), 21–27. doi: 10.1111/j.1468-5973.2010.00627.x

Weick, K. E., & Roberts, K. H. (1993). Collective mind in organizations: Heedful interrelating on flight decks. *Administrative Science Quarterly* 38(3), 357–381.

Weick, K. E., & Sutcliffe, K. M. (2001). *Managing the Unexpected: Assuring High Performance in an Age of Complexity.* San Francisco, CA: Jossey-Bass.

Weick, K. E., & Sutcliffe, K. M. (2011). *Managing the Unexpected. Resilient Performance in an Age of Uncertainty.* San Francisco, CA: John Wiley & Sons.

Weiss, C. H. (1995). Nothing as practical as good theory: Exploring theory-based evaluation for comprehensive community initiatives for children and families. In: Connell, J. P., Kubisch, A. C., Schorr, L. B., & Weiss, C. H. (eds.), *New Approaches to Evaluating Community Initiatives: Concepts, Methods, and Contexts.* New York: Aspen Institute, pp. 65–92.

Westrum, R. (1999). *Sidewinder: Creative Missile Design at China Lake.* Annapolis, MD: Naval Institute Press.

Westrum, R. (2004). A typology of organisational cultures. *Quality and Safety in Health Care* 13(Suppl II), ii22–ii27.

Westrum, R. (2014). The study of information flow: A personal journey. *Safety Science* 67(2014), 58–63.

Wiig, S., Aase, K., Johannessen, T., Holen-Rabbersvik, E., Thomsen, L. H., van de Bovenkamp, H., & Ree, E. (2019). How to deal with context? A context-mapping tool for quality and safety in nursing homes and homecare (SAFE-LEAD context). *BMC Research Notes* 12(1), 259.

Wiig, S., Bourrier, M., Aase, K., & Røise, O. (2018). Transparency in health care: Disclosing adverse events to the public. In: Bourrier, M., & Bieder, C. (eds.), *Risk Communication for the Future*. Cham: Springer Open, pp. 111–125.

Wiig, S., & Braut, G. S. (2018). Developments in analysis of severe adverse events in healthcare – policy and practice in Norway. In: Bernatik, A., Locurkova, L., & Jørgensen, K. (eds.), *Prevention of Accidents at Work*. London: CRC Press, pp. 39–45.

Wiig, S., Haraldseid-Driftland, C., Tvete Zachrisen, R., Hannisdal, E., & Schibevaag, L. (2019a). Next of kin involvement in regulatory investigation of adverse events that caused patient death: A process evaluation (part I – The next of kin's perspective). *Journal of Patient Safety*. Published ahead of print. doi: 10.1097/PTS.0000000000000630

Wiig, S., & Lindøe, P. H. (2009). Patient safety in the interface between hospital and risk regulator. *Journal of Risk Research* 12(3–4), 411–27. doi: 10.1080/13669870902952879

Wiig, S., & Macrae, C. (2018). Introducing national healthcare safety investigation bodies. *British Journal of Surgery* 105, 1710–2. doi: 10.1002/bjs.11033

Wiig, S., Schibevaag, L. Tvete Zachrisen, R., Anderson, J. A., Hannisdal, E., & Haraldseid-Driftland, C. (2019b). Next of kin involvement in regulatory investigation of adverse events that caused patient death: A process evaluation (part II – the inspectors' perspective). *Journal of Patient Safety*. Published ahead of print. doi: 10.1097/PTS.0000000000000634

Wikan, U. (1990). *Managing Turbulent Hearts: A Balinese Formula for Living*. Chicago, IL: University of Chicago Press.

Wildavsky, A. B. (1988). *Searching for Safety*. Berkeley, CA: University of California Press.

Willis, J. W., & Edwards, C. L. (2014). The twist and turns of action research history. In: Willig, J. W., & Edwards. C. L. (eds.), *Action Research. Models, Methods and Examples*. Charlotte, NC: Information age publishing, Inc.

Woods, D. (2015). Four concepts for resilience and the implications for resilience engineering. *Reliability Engineering and System Safety* 141, 5–9.

World Health Organization. (2018). *10 Facts on Patient Safety*. https://www.who.int/features/factfiles/patient_safety/en/. Accessed 26 April 2019.

Wright, C. (1994). A fallible safety system: Institutionalized irrationality in the offshore and gas industry, *Sociological Review* 1, 79–103.

Wynne, B. (1992). Risk and social learning: Reification to engagement. In: S. Krimsky, & D. Golding (eds.), *Social Theories of Risk*. Westport, CT: Praeger, pp. 275–300.

Wynne, B. (1992). Uncertainty and environmental learning: Reconceiving science and policy in the preventive paradigm. *Global Environmental Change* 2(2), 111–127.

Ybema, S., Yanow, D., Wels, H., & Kamsteeg, F. H. (eds.). (2009). *Organizational Ethnography: Studying the Complexity of Everyday Life*. London: SAGE.

Yin, R. (2014). *Case Study Research: Design and Methods* (5th ed.). Thousand Oaks, CA: SAGE.

Young, S. A., & Blitvich, J. (2018). Safety in hard times – A qualitative analysis of safety concerns in two industrial plants under financial duress. *Safety Science* 102, 118–124.

Zanutto, A., Rashid, A., Busby, J. S., Frey, S., & Follis, K. (2017). The shadow warriors: In the no man's land between industrial control systems and enterprise IT systems. In: *Thirteenth Symposium on Usable Privacy and Security, SOUPS*, July 12–14, 2017, Santa Clara, CA.

Zwetsloot, G. I. J. M., Kines, P., Ruotsala, R., Drupsteen, L., & Merivirta, M.-L. (2017). The importance of commitment, communication, culture and learning for the implementation of the Zero Accident Vision in 27 companies in Europe. *Safety Science* 96, 22–32.

Index

Printed in the United States
by Baker & Taylor Publisher Services